Soft Matter

Edited by
G. Gompper, M. Schick

Volumes Published in this Series

Volume 1: Polymer Melts and Mixtures

1 Polymer Dynamics in Melts
   Andreas Wischnewski and Dieter Richter

2 Self-Consistent Field Theory and Its Applications
   Mark W. Matsen

3 Comparison of Self-Consistent Field Theory
   and Monte Carlo Simulations
   Marcus Müller

Volume 2: Complex Colloidal Suspensions

1 Phase Behavior of Rod-Like Viruses and Virus–Sphere Mixtures
   Zvonimir Dogic and Seth Fraden

2 Field Theory of Polymer–Colloid Interactions
   Erich Eisenriegler

3 Rod-Like Brownian Particles in Shear Flow
   Jan K. G. Dhont and Wim J. Briels

Soft Matter

Volume 1: Polymer Melts and Mixtures

Edited by
Gerhard Gompper and Michael Schick

WILEY-VCH Verlag GmbH & Co. KGaA

Editors

Prof. Dr. Gerhard Gompper
Institute of Solid State Research
Research Centre Jülich
52525 Jülich
Germany

Prof. Dr. Michael Schick
Department of Physics
University of Washington
Seattle, WA 98195-1560
USA

Cover illustration
The picture illustrates the topological confinement in long-chain polymer melts. Entanglements restrict the lateral motion and allow the red chain to diffuse only along its own contour. In the "reptation" concept, the confinement is represented by a virtual tube which is formed by the adjacent chains. The picture in the background shows the anisotropic scattering intensity from a highlighted deformed chain during relaxation of an uni-axially stretched polymer melt. The scattering intensity allows an unique access to the size of deformed chains as well as their tube diameters. These can be related to mechanical properties of polymer materials. (Original pictures courtesy of W. Pyckhout-Hintzen, A. Wischnewski and D. Richter).

■ All books published by Wiley-VCH are carefully produced. Nevertheless, editors, authors and publisher do not warrant the information contained in these books to be free of errors. Readers are advised to keep in mind that statements, data, illustrations, procedural details or other items may inadvertently be inaccurate.

Library of Congress Card No.: applied for

British Library Cataloguing-in-Publication Data:
A catalogue record for this book is available from the British Library.

Bibliographic information published by Die Deutsche Bibliothek
Die Deutsche Bibliothek lists this publication in the Deutsche Nationalbibliografie; detailed bibliographic data is available in the Internet at http://dnb.ddb.de

© 2006 WILEY-VCH Verlag GmbH & Co KGaA, Weinheim

All rights reserved (including those of translation into other languages). No part of this book may be reproduced in any form – by photocopying, microfilm, or any other means – nor transmitted or translated into a machine language without written permission from the publishers. Registered names, trademarks, etc. used in this book, even when not specifically marked as such, are not to be considered unprotected by law.

Printed in the Federal Republic of Germany
Printed on acid-free and chlorine-free paper

Cover Design: SCHULZ Grafik-Design, Fußgönheim
Composition: Steingraeber Satztechnik GmbH, Ladenburg
Printing: betz-druck GmbH, Darmstadt
Bookbinding: Litges & Dopf Buchbinderei GmbH, Heppenheim

ISBN-13: 978-3-527-30500-1
ISBN-10: 3-527-30500-9

Preface

The distinguished editors, Gerhard Gompper and Michael Schick, have had the inspired idea of doing for the burgeoning field of "Soft Condensed Matter" in the first half of the twenty-first century what the famous Domb–Green–Lebowitz series on "Phase Transitions and Critical Phenomena" did for that subject in the latter half of the twentieth; viz., providing an authoritative account of the field's principles and applications for its current practitioners, and instructing and inspiring a new generation of scientists.

What is "soft condensed matter" and where did the name come from? My own vague memories have now been considerably sharpened by the recollections of P.-G. de Gennes, to whom I am grateful for some of this history. The expression "soft matter" (*matière molle* in French, where it is a "double entendre") very likely originated in the research group of de Gennes, then in Orsay, around 1970. The group included M. Veyssié and others. The expression, as the name of a now hugely important subfield of physics, has been universally accepted.

The status of the field as of 1995 was assessed by de Gennes (1995) in his definitive review, "Soft matter: birth and growth of concepts". The phenomena encompassed by it are characterized by large responses to small perturbations. This is analogous to, but, as de Gennes explains, different from, the large magnetic susceptibility near a Curie point or great compressibility of a fluid near its critical point. "Soft" materials include polymers, liquid crystals, colloids, gels, surfactant phases, and many other such objects of study. One can see "soft matter" in the subtitle, and as the title of Part I, of *Fragile Objects* (de Gennes and Badoz 1996), echoing the title of de Gennes' 1994 Dirac Lecture, *Soft Interfaces* (de Gennes 1997). The program "SOFT COMP", for soft composites, is an EU "Network of Excellence". In the commentary "Viruses and the physics of soft condensed matter" (Zlotnick 2004), commenting on the beautiful article "Origin of icosohedral symmetry in viruses" (Zandi et al. 2004), one reads that: "Physicists studying soft condensed matter provide a framework that will help interpret and rationalize virological experimental observations." As we see, the reach of this subject is vast.

This publishing project is already most promising. The editors, whose own excellent volume *Self-Assembling Amphiphilic Systems* for the Domb–Lebowitz series (Gompper and Schick 1994) would be an appropriate model for these reviews, are to be congratulated for having recruited great experts to write these authoritative articles for this series' inaugural volume. The high quality of this beginning allows me to be confident of the series' success, to which I look forward with much pleasure.

November 2005

B. Widom
Cornell University

References

de Gennes, P.-G., 1995, in *Twentieth Century Physics*, eds. L. M. Brown, A. Pais, and (A.) B. Pippard, Vol. III, pp. 1593–2015. American Institute of Physics, New York.

de Gennes, P.-G., 1997, *Soft Interfaces: The 1994 Dirac Memorial Lecture*. Cambridge University Press, Cambridge.

de Gennes, P.-G. and Badoz, J., 1996, *Fragile Objects: Soft Matter, Hard Science, and the Thrill of Discovery*, transl. A. Reisinger. "Copernicus", Springer-Verlag, Berlin.

Gompper, G. and Schick, M., 1994, *Self-Assembling Amphiphilic Systems*, Vol. 16 of *Phase Transitions and Critical Phenomena*, eds. C. Domb and J. L. Lebowitz. Academic Press, New York.

Zandi, R., Reguera, D., Bruinsma, R. F., Gelbart, W. M., and Rudnick, J., 2004, *Proc. Natl. Acad. Sci. USA* **101**, 15556.

Zlotnick, A., 2004, *Proc. Natl. Acad. Sci. USA* **101**, 15549.

Contents

An Introduction to Soft Matter *1*
Gerhard Gompper and Michael Schick

1   Polymer Dynamics in Melts *17*
Andreas Wischnewski and Dieter Richter
1.1   Introduction *18*
1.2   What is a Polymer? *20*
1.2.1   Synthesis of Polymers *21*
1.3   Experimental Techniques *25*
1.3.1   Neutron Scattering *25*
1.3.2   Rheology *36*
1.4   Static Properties *38*
1.5   Brownian Motion, Viscous and Entropic Forces:
       the Rouse Model *45*
1.5.1   Experimental Studies of the Rouse Model *49*
1.6   Topological Constraints: the Tube Concept *55*
1.6.1   Validating the Tube Concept on a Molecular Scale *59*
1.7   Limiting Mechanisms for Reptation I: CLF *69*
1.8   Limiting Mechanisms for Reptation II: CR *77*
1.9   Summary and Outlook *81*

2   Self-Consistent Field Theory and Its Applications *87*
Mark W. Matsen
2.1   Introduction *87*
2.2   Gaussian Chain *91*
2.3   Gaussian Chain in an External Field *96*
2.4   Strong-Stretching Theory (SST): the Classical Path *102*
2.5   Analogy with Quantum/Classical Mechanics *103*
2.6   Mathematical Techniques and Approximations *105*
2.6.1   Spectral Method *106*

Soft Matter, Vol. 1: Polymer Melts and Mixtures. Edited by G. Gompper and M. Schick
Copyright © 2006 WILEY-VCH Verlag GmbH & Co. KGaA, Weinheim
ISBN: 3-527-30500-9

2.6.2 Ground-State Dominance 107
2.6.3 Fourier Representation 109
2.6.4 Random-Phase Approximation 111
2.7 Polymer Brushes 114
2.7.1 SST for a Brush: the Parabolic Potential 115
2.7.2 Path-Integral Formalism for a Parabolic Potential 117
2.7.3 Diffusion Equation for a Parabolic Potential 120
2.7.4 Self-Consistent Field Theory (SCFT) for a Brush 121
2.7.5 Boundary Conditions 125
2.7.6 Spectral Solution to SCFT 127
2.8 Polymer Blends 133
2.8.1 SCFT for a Polymer Blend 135
2.8.2 Homogeneous Phases and Macrophase Separation 137
2.8.3 Scattering Function for a Homogeneous Blend 140
2.8.4 SCFT for a Homopolymer Interface 143
2.8.5 Interface in a Strongly Segregated Blend 145
2.8.6 Grand-Canonical Ensemble 148
2.9 Block Copolymer Melts 152
2.9.1 SCFT for a Diblock Copolymer Melt 152
2.9.2 Scattering Function for the Disordered Phase 154
2.9.3 Spectral Method for the Ordered Phases 156
2.9.4 SST for the Ordered Phases 163
2.10 Current Track Record and Future Outlook for SCFT 167
2.11 Beyond SCFT: Fluctuation Corrections 169
2.12 Appendix: The Calculus of Functionals 172

3 **Comparison of Self-Consistent Field Theory and Monte Carlo Simulations** 179
Marcus Müller
3.1 Introduction 180
3.2 Polymer Blends 182
3.2.1 Coarse-Grained Polymer Models 182
3.2.2 A Coarse-Grained Model for Binary Polymer Blends 187
3.2.3 Self-Consistent Field (SCF) Theory and Fluctuation Effects 190
3.2.4 Identifying the Effective Flory–Huggins Parameter, $\tilde{\chi}$ 213
3.2.5 Single-Chain Conformations in the Bulk 225
3.2.6 Interfaces 229
3.3 Liquid–Vapor Interfaces 245
3.3.1 Formalism 246
3.3.2 Approximating the Excess Interaction Free-Energy Functional ΔF_{ex} 248
3.3.3 Comparison to Computer Simulations 261

3.4 Outlook *266*
3.5 Appendix *267*
3.5.1 Fluctuations in the Spatially Homogeneous Phase *267*
3.5.2 Alternative Derivation of the SCF Equations for Liquid–Vapor Systems *271*

Index *283*

List of Contributors

Prof. Gerhard Gompper
Institut für Festkörperforschung
Forschungszentrum Jülich
52425 Jülich
Germany

Prof. Mark Matsen
Department of Physics
University of Reading
Reading, RG6 6AF
United Kingdom

Prof. Marcus Müller
Institut für Theoretische Physik
Georg-August-Universität Göttingen
Friedrich-Hund-Platz 1
37077 Göttingen
Germany

Prof. Dieter Richter
Institut für Festkörperforschung
Forschungszentrum Jülich
52425 Jülich
Germany

Prof. Michael Schick
Department of Physics
University of Washington
Seattle, WA 98195-1560
USA

Dr. Andreas Wischnewski
Institut für Festkörperforschung
Forschungszentrum Jülich
52425 Jülich
Germany

An Introduction to Soft Matter

Gerhard Gompper and Michael Schick

What Is Soft Matter?

All matter is made of atoms and molecules, and in most common systems, such as water, silicon, or sodium chloride, the size of these building blocks is on a length scale of ångströms. Such atomic structure cannot be discerned by a microscope with a resolution of tens of nanometers, so that the material looks completely homogeneous. The situation is quite different for a suspension of latex particles, or a mixture of oil, water, and surfactant molecules. Even at a resolution of micrometers, the inhomogeneous structure of these materials is still visible. Figure 1 shows a transmission electron microscopy (TEM) image of a colloidal crystal that clearly shows its structure to be visible on the micrometer scale.

What makes these colloidal systems interesting is that their properties can be quite different from those of liquids or crystals of small molecules. For example, materials that consist of extremely large molecules are usually easily deformable. Hence they are usually referred to as "soft matter". The reason for the softness of such macromolecular materials is readily understood. Let us compare the energies that are needed to deform the surfaces of a common molecular crystal, such as sodium chloride, and of a crystal of spherical colloidal particles with a size of 1 μm, a factor of 10^4 larger than ordinary molecules. The resistance of a crystal with respect to external shear forces is characterized by the shear modulus. For a crystal of linear sizes L_x, L_y, and L_z, to which a force F is applied at the top and bottom surfaces in the x and $-x$ directions, respectively (see Fig. 2), the shear modulus, μ, is defined by

$$\frac{F}{L_x L_y} = \mu \frac{\Delta L_x}{L_z} \qquad (1)$$

Soft Matter, Vol. 1: Polymer Melts and Mixtures. Edited by G. Gompper and M. Schick
Copyright © 2006 WILEY-VCH Verlag GmbH & Co. KGaA, Weinheim
ISBN: 3-527-30500-9

Fig. 1 Transmission electron microscopy (TEM) picture of a colloidal crystal of spherical particles. The figure on the right is an enlargement of the part indicated by the white square in the figure on the left. The ordered arrangement of the colloidal particles is clearly visible.

where ΔL_x is the crystal displacement in the top layer. The shear modulus is an intrinsic property of the material, independent of the system size, and has the dimension energy/(length)^3. We obtain a rough estimate of the shear modulus of a material by inserting the characteristic length and energy scales of the crystal. For a molecular crystal, the energy scale is that of covalent bonds, 10 eV, while for a colloidal crystal the energy scale is *not* that of some microscopic interaction potential, but rather is on the order of the thermal energy $k_\mathrm{B}T$, about (1/40) eV at room temperature. Thus the energy scale is two orders of magnitude smaller in colloidal crystals. The length scale in the colloidal crystal is of the order of its lattice constant, which is four orders of magnitude larger than those of molecular crystals. One therefore estimates that the shear modulus of a colloidal crystal is $10^{-2}/(10^4)^3 = 10^{-14}$ or 14 orders of magnitude smaller than the shear modulus of a molecular crystal. This estimate reproduces quite well experimentally observed values for the shear modulus of colloidal crystals. As the shear modulus is so small, colloidal crystals are extremely soft and can be destroyed by mechanical forces very easily.

The appearance of the thermal energy as the scale of interaction energy is pervasive in soft matter, and reflects the fact that many of the important interactions in these systems are entropic in origin. These effective interactions often arise from microscopic hard-core potentials between the particles themselves, or between the particles and confining walls, or from other constraints. Their net effect is to reduce the allowed configurations of the system

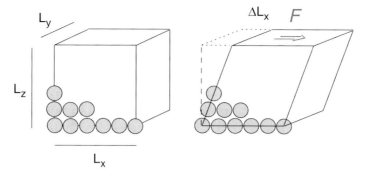

Fig. 2 When a crystal is sheared by a force F applied to its top surface, it is deformed by a distance ΔL_x.

and therefore its entropy, and to increase the system's free energy, an increase that can be assigned to an effective interaction. The presence of a factor of $k_B T$ in the magnitude of such interactions belies their entropic origin (Israelachvili 1992).

In addition to colloids, there are many other kinds of very large "molecules". Aggregates of polymers and of smaller amphiphilic molecules are the most common building blocks of soft matter. These blocks will be considered in more detail in the next section.

For reasons similar to those sketched above for colloidal crystals, one finds that membranes consisting of amphiphilic molecules are also easily deformable. Such biological membranes enclose red blood cells, shown in Fig. 3, whose size is on the order of micrometers. Due to the flexibility of its enclosure, a red blood cell can penetrate through openings that are ten times smaller than itself.

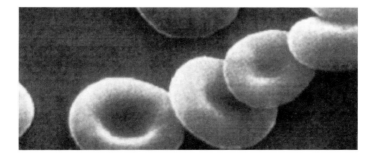

Fig. 3 Red blood cells (erythrocytes) have a characteristic, discocyte shape.

The Classical Systems

The "classical" soft matter systems, which have been studied since the early 19th century, are dispersion colloids, systems of small amphiphiles, and polymers. We briefly introduce these systems here.

Dispersion Colloids

Colloids are rigid particles that have a size in the range of 1 nm to 10 μm. They are usually dispersed in a fluid because, in its absence, the colloidal system behaves as a powder. Such solutions are commonly referred to as *dispersions* or *suspensions*. The colloidal particles are small enough to exhibit thermal motion, and therefore behave as "large molecules". Their thermal motion is commonly referred to as Brownian motion, named after the Scottish botanist Robert Brown, who observed thermal motion of pollen grains through a microscope in 1827.

Since colloidal particles typically have a different dielectric constant than the embedding fluid, they attract each other by van der Waals interactions (Mahanty and Ninham 1976; Israelachvili 1992), with the result that they are unstable to coagulation. Two methods are commonly employed to stabilize such systems. The first is to charge the colloidal particles electrically with charges of the same sign to produce an electrical repulsion. The second is to coat the surfaces with a polymer layer, which leads to a steric repulsion.

The lower limit on the size of a colloidal particle of approximately 1 nm assures that the fluid in which the colloidal particles are embedded behaves for the most part as a structureless continuum. Since the typical size of a solvent molecule is 1 Å, the colloidal particle can usually be regarded as a macroscopic object on the solvent length scale. Moreover, the relaxation times of the degrees of freedom of the solvent molecules are much smaller than those for the colloidal particles. Dynamics and non-equilibrium phenomena can thus be described by equations of motion in which the degrees of freedom of the solvent molecules are integrated out. For both the structure and the dynamics, a colloidal system of identical colloidal particles can therefore be regarded as an effective one-component system. The solvent now appears only via phenomenological parameters, such as the viscosity, and acts as a thermal motor that drives the motion of the colloidal particles.

Colloidal dispersions in some aspects behave much like ordinary condensed matter. Therefore, they provide interesting model systems from which insight can be gained on fundamental processes that apply to molecular systems as well, such as crystallization, critical phenomena, and glass formation. Indeed, as Dogic and Fraden state in their introduction to Chapter 1 of Volume 2, they were originally motivated to study virus suspensions because they

Fig. 4 An *fd* virus has the shape of a long, semi-flexible rod. It has a diameter of about 6 nm, a length of 880 nm, and persistence length of about 2.2 μm.

thought that they were model systems of simple fluids. They acknowledge in the very same sentence, however, that their colloidal system, like so many others, has its own specific properties that have no counterpart in molecular systems. There are several reasons why colloids exhibit such specific phenomena. One is that the particles can exist in many different shapes, some of which have been examined extensively: for example, spheres, such as in silica suspensions (Manoharan et al. 2003); rigid rods, such as tobacco mosaic virus and double-stranded DNA fragments; and semi-flexible rods, such as the *fd* virus shown in Fig. 4. The fascinating phase behavior of this bacteriophage is the subject of Chapter 1 of Volume 2 by Dogic and Fraden. As shown there, these colloids *self-assemble* into various ordered arrays, a property that is shared by other soft matter systems, such as small amphiphiles, and polymers. It is also one of great technological interest, as the periodicity of the structure can often be tuned by varying the constituent particle. The search for a suitable three-dimensional photonic band-gap material has naturally focussed on the above materials (Vos et al. 1996; Pan et al. 1997; Dinsmore et al. 1998). This property of self-assembly, particularly into ordered arrays, is another recurrent theme in soft matter.

Another reason for the particular behavior of colloids is their very slow dynamics. The microstructural arrangement of colloidal particles typically relaxes in a time range of seconds. This results, for example, in a nonlinear response of colloidal systems, like those of rigid rods, to a shear field, a subject treated authoritatively in Chapter 3 of Volume 2 by Dhont and Briels. For molecular systems, such nonlinear response would occur at unrealistically high shear rates and frequencies. In general, the subject of the hydrodynamic behavior of colloidal fluids is one of significant technological, as well as intrinsic, interest.

Colloids are treated extensively in many texts, such as Russel (1987), Russel et al. (1989), van de Ven (1989), Hunter (1989), Evans and Wennerström (1994), Dhont (1996), and Dhont et al. (2002).

Self-Assembling Amphiphilic Systems

Amphiphilic molecules usually consist of a hydrophilic head-group and a hydrophobic tail. The latter typically consists of one or two hydrocarbon chains, which can be totally saturated or partially unsaturated with one or more double bonds. The hydrophilic part consists of either a polar group, whose dipole moment interacts strongly with those of water, or an ionizable group, such as COOH, which dissociates, leaving a residual charge that interacts even more strongly with the solvent dipoles. Such an ionic amphiphile with a tail of a single chain, sodium dodecyl sulfonate, is shown in Fig. 5. An example of a non-ionic, two-chain biological lipid, phosphatidyl choline, is shown in Fig. 6. Note that one chain is saturated, while the other has a sole double bond half-way down the chain. This is a common lipid motif. Molecules such as these have several names: "amphiphilic" from the Greek for "loving both", or "amphipathic", a name favored in the biological community, or "surfactant", a contraction of "surface-active", because these molecules absorb preferentially at interfaces of water with oil or with air and reduce the surface tension.

The basis for the hydrophobic interaction between the tails and water is once again entropic. The tails break up the extensive hydrogen-bonding network of the water, thus reducing its entropy, which can again be viewed as an effective repulsion. The head-groups also disturb this network, but they more

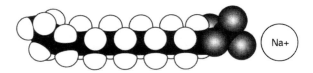

Fig. 5 Sodium dodecyl sulfonate (SDS) is a typical example of an ionic amphiphile. The ionizable head-group (right) is linked to one hydrocarbon chain (left).

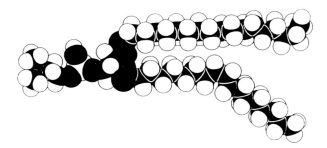

Fig. 6 Phosphatidyl choline (PC) is a typical example of a membrane lipid. The polar head-group (left) is connected to two hydrocarbon tails (right).

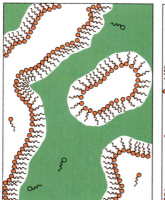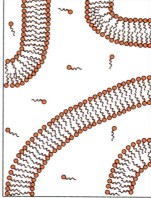

Fig. 7 Amphiphilic molecules self-assemble into complex aggregates due to the hydrophobic effect. In mixtures with oil and water, they form monolayers at the water–oil interface. In mixtures with water, they often form bilayers.

than make up for the entropy reduction by means of their favorable interaction with the water dipoles. Due to these interactions, amphiphiles placed in water will self-assemble and form arrays such that the hydrophilic head-groups are exposed to water while the tails are sequestered from it. The simplest such array is that of a bilayer, shown in Fig. 7, which is a very common structure. Indeed, the basis of all biological membranes is the lipid bilayer. If the amount of amphiphile is increased, the number of bilayers formed is increased, and the system will arrange itself into an ordered lamellar phase. Under various conditions, such as a change in the water content, or a change in the architecture of the amphiphile, several other ordered phases can be formed. Among these are: the cylindrical phase, in which the amphiphiles form *single* sheets that close into cylinders and are packed in a hexagonal array with the water on the outside and the tails hidden on the inside; a body-centered cubic (bcc) phase, in which the sheets form spheres that pack into a bcc array with water on the outside; and, perhaps most bizarre of all, a "gyroid" phase, in which a bilayer membrane divides space into two identical, sample-spanning, volumes. Because of this, the phase is often denoted "bicontinuous". It is a perfectly good cubic phase (space group $Ia\bar{3}d$), and was first identified in small amphiphilic systems by Luzzati and Spegt (1967). Later it was observed in systems of block copolymer (Hadjuk et al. 1994). While not uncommonly exhibited by soft materials, it is not a space group ever found in molecular crystals, probably because the structure would consist of atoms in sites of three-fold coordination creating a rather low-density structure not favored by the strong atomic interactions. A disordered and therefore liquid analog of such a bicontinuous phase is observed in these binary (water and amphiphile) systems, and has been denoted a "sponge" phase (Cates et al. 1988). When water is removed from these systems, *inverted* phases may form in which the reduced amount of water is on the inside of the arrays, and the tails are on the outside (Seddon and Templer 1993).

In addition to the ordered phases at relatively high concentration of amphiphile, there is also a fluid phase at low amphiphile concentration in which these molecules form small, usually spherical, aggregates, called micelles. The head-groups face the water and the tails are again sequestered within.

If a third, oil-like, component is added to these systems, it will swell the regions in which the tails reside, as shown in Fig. 7. Thus one finds exactly the same phases as in the binary system but with at least one new possibility. If the concentration of amphiphile is low in a system in which the solvent is oil, the fluid will contain micelles that are inverted, with the heads in and tails out. As one changes the composition of the solvent from mostly water, containing normal micelles with heads out, to mostly oil, containing inverted micelles, the fluid must pass through a region in which the amphiphiles assemble in some intermediate way. In fact, the system can pass through a composition region in which the oil and water are separated by disordered sheets of amphiphiles, again forming a bicontinuous fluid. Such a phase, which still retains much structure, is denoted a *microemulsion*. In addition to the intrinsic interest of describing a fluid with significant structure, microemulsions have been of technological use, as they can contain in one fluid phase a large

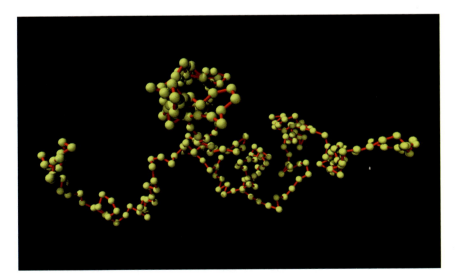

Fig. 8 The typical conformations of long, flexible polymers are random coils. These conformations can be characterized by an end-to-end distance R_e, or, equally well, by a radius of gyration R_g.

amount of two components, here oil and water, which would phase-separate in the absence of the amphiphile.

Self-assembling amphiphilic systems are discussed in the texts by Israelachvili (1992), Gelbart et al. (1994), Gompper and Schick (1994), Safran (1994), Jönsson et al. (1998), and Dhont et al. (2002). Texts devoted to the physics of membranes are Lipowsky and Sackmann (1995), Boal (2002), and Nelson et al. (2004).

Polymers

Polymers are essentially giant molecules, or macromolecules, consisting of an extremely large number of basic units chemically joined together to form one entity. They are ubiquitous. Polyethylene, the practically endless repetition of a simple H–C–H group, forms most of the plastics found in the home, whether in containers, shopping bags, or toys. Polystyrene is the basis for Styrofoam, and also forms the plastic in light switches and the plates surrounding them. Poly(methyl methacrylate) forms the clear plastics Lucite and Plexiglass. Not all polymers, however, are of the technical variety; DNA and RNA are biological polymers consisting of only four distinct monomers, while proteins are also polymers built out of 20 different monomers, the amino acids.

The fact that polymers are ubiquitous should not blind one to the difficulty of accepting the concept of a giant molecule. It is not now, nor was earlier, an obvious one. It was in 1922 that the German organic chemist, Hermann Staudinger, who coined the term "macromolecule", published a theory that rubber was made of a natural polymer, and that it was the entropic properties of such chains that were the source of the elasticity of the material. A long 21 years passed before these insights were recognized by the award of a Nobel Prize in chemistry.

Individual monomers are bound together by covalent bonds, about which rotations are usually possible. Therefore, long polymer chains have a very large number of internal degrees of freedom, which leads to an enormous number of spatial conformations of the chain, and hence a very large entropy. A typical configuration of a flexible polymer is shown in Fig. 8. On distances that are large compared to the size of a monomer, the chemical structure of the building block plays only a minor role, and the properties of the chain are mainly determined by the statistical mechanics of the chain, i.e. essentially by the entropy of a chain. As a consequence, on large distance scales, or for a large number of monomers, N, one expects that polymer properties, rather than depending upon the nature of the monomer, will be *universal*, depending only on the nature of the chain that the monomers form, i.e. whether flexible or semi-flexible, self-intersecting or self-avoiding. The simplest example of such a law is the behavior of the chain's radius of gyration, R_g, with N:

$$R_g \sim N^\nu \qquad (2)$$

where ν equals 1/2 in all dimensions for a self-intersecting, completely flexible chain, and is equal to 1, 3/4, and 0.5886 in one, two (Nienhuis 1982), and three dimensions (Muthukumar and Nickel 1987) respectively for a self-avoiding chain.

The derivation and elucidation of such scaling laws have played an important role in polymer theory (de Gennes 1979; Doi and Edwards 1986; des Cloizeaux and Jannink 1990), and they are an example of the scaling relations emblematic of critical phenomena. Consequently, the theoretical methods characteristic of that area of study, the use of field theory, the renormalization group, of epsilon expansions, etc., have played a major role. These powerful methods are employed in Chapter 2 of Volume 2 by Eisenriegler to capture the behavior of polymers whose entropy is curtailed by walls and other boundaries.

Because different polymers have different desirable properties, it is natural to try to combine them to create a substance with several such features. There are two common ways to do this. The first is the obvious one of making a simple blend of the two polymers. The problem with this approach is that polymers generally phase-separate, which can be seen as follows. The effective interaction between monomers of different polymers is almost invariably repulsive, i.e. monomers of one kind interact more attractively with their own kind than with a different species of monomer. The number of such interactions is proportional to the total number of monomers, nN, where n is the number of chains. However, the entropy is not proportional to the total number of monomers, as it would be were the monomers unlinked; rather, it is only proportional to the number of polymers, n. Thus the entropic tendency to mix is much smaller than the energetic propensity to phase-separate. To counteract this tendency, one plays the same game as in mixtures of oil and water and adds an amphiphile. To solubilize a mixture of A and B homopolymers, one can employ a block copolymer consisting of a block of A monomers chemically joined to a block of B monomers. The diblock will tend to adsorb at the A–B interface and reduce its surface tension. Because the properties of this interface are so important for the blend, it has been the focus of much study. The first technique of choice to be applied to this, as well as most other problems, is that of self-consistent field theory, the subject of Chapter 2 by Matsen, where an entire section devoted to blends can be found. This theory ignores the effects of fluctuations, of course, but their effect can be captured by Monte Carlo simulations, the subject of Chapter 3 by Müller, where a section on blends can also be found. In fact, improvements in both models and theories have led to an exciting synergy of these two methods, as detailed in Chapter 3.

A second approach to produce a material with the desirable properties of more than one homopolymer is to utilize block copolymers consisting of several blocks, of which at least two are those of the desired homopolymers. Like other amphiphiles, these block copolymers self-assemble into a variety of ordered one-, two-, and three-dimensional phases, which are of great potential technological use. To pick at random one application: with the use of block copolymer lithography, ordered hexagonal arrays of holes 20 nm across and 40 nm apart have been fabricated in a silicon-nitride-coated silicon wafer. The template is made from a polystyrene–polybutadiene diblock that spontaneously self-assembles into a hexagonal phase (Park et al. 1997). It is in the accurate description of such ordered phases, over a wide range of temperatures, that self-consistent field theory has made great strides recently. Again, these are ably reviewed in Chapter 2 by Matsen.

The dynamics of such large polymeric molecules depends, of course, on just how large they are. For long chains, topological constraints absent in normal molecular fluids dominate the dynamics. The constraints are that the polymers cannot break into smaller pieces and cannot cross one another. As a consequence, the polymers become entangled. A particular polymer cannot move normal to its own contour, but only slide along the "tube" formed by the other polymers (de Gennes 1981). Given that the polymer within one tube is part of the boundary of yet another tube makes the problem all the more complex. This difficult but fascinating area of polymer dynamics forms the subject of Chapter 1 by Wischnewski and Richter.

Many fine texts are devoted to polymers. Among them are Flory (1969), de Gennes (1979), Ferry (1980), Doi and Edwards (1986), des Cloizeaux and Jannink (1990), Grosberg and Khokhlov (1997), Hamley (1998), and Dhont et al. (2002).

Why Soft Matter?

In the past 10–20 years, the research areas of colloids, membranes, microemulsions, and polymers, which were previously largely independent, have been integrated into the single field of *soft matter*. This is not just a matter of old wine in new bottles; rather, it has been realized that many phenomena in these systems have the same underlying physical mechanisms, so that similar effects do not have to be discovered and understood anew in each subfield. Moreover, combinations of basic elements, such as polymers and colloids, exhibit new properties, which are not found in either system separately. From the brief overview of the classical systems above, we can extract several unifying themes.

Model Systems for Atomic Condensed Matter

Colloidal particles exhibit many of the physical phenomena that are found in ordinary molecular systems. Due to their large size and slow dynamics, however, the experimental study of colloids is usually much simpler than for molecular systems. For example, instead of X-ray or neutron scattering utilized in molecular systems, light scattering can often be employed. The time scales of dynamical phenomena are often in the range of seconds, much longer than the scales of molecular systems.

Tunable Interactions

In a soft matter system, the interaction between macromolecules arises indirectly from the basic Coulomb and quantum-mechanical interaction of their constituents and those of the surrounding solvent. As the solvent has many degrees of freedom that are subject to thermal fluctuations, the resulting *effective interactions* are usually temperature-dependent. They can often be controlled and tuned by additives to the solvent.

A good example is the effective interaction potential between colloidal particles, which can be manipulated to vary from a purely long-range repulsion to a very short-range attraction. The tuning of this potential can be accomplished in several ways:

1. *Grafting polymers and varying solvent quality.* The surface of a colloidal particle can be covered by polymer chains. When the solvent is a good one for the polymers attached to the surface, and the polymer chains are very short in comparison to the size of the core of the particle, the chains do not interpenetrate much, as this would cause a loss of entropy. As a result, the interaction is effectively a hard-core potential: essentially infinite when the inter-particle separation is smaller than the size of the core, and zero otherwise. On the other hand, if the solvent is a poor one for the polymer, in addition to the hard-core repulsion, the interaction has a short-range attractive part because the chains will interdigitate to some extent in order to escape the effects of the solvent. The quality of the solvent, which affects the interaction between the polymers anchored to the colloidal particles, can be varied simply by changing the temperature. One effectively changes the depth of the attractive well between particles.

2. *Employing the depletion interaction.* Another possibility is to add long, free, polymers to the solvent. Because they are excluded from the region between particles when those particles come close to one another, there results an

imbalance in the osmotic pressure, leading to an effective attraction. This attraction, which is completely entropic in origin, is denoted the depletion interaction. The range of the attraction is set by the size of the excluded polymers, while the interaction strength is set by their number concentration, as would be expected from the simple van't Hoff relation for osmotic pressure. This polymer-induced interaction is discussed in detail by Eisenriegler in Chapter 2 of Volume 2.

3. *Charging the colloids.* If the surface of the colloidal particles is charged, one can change the range of the repulsion by adding or removing salt from the solvent, and thereby changing the Debye screening length.

4. *Manipulating the dielectric properties.* The refractive index of the solvent can also be varied, which changes the strength of the van der Waals attractions between the cores of the colloidal particles.

Such mechanisms for tuning effective interactions are not restricted to colloids. In a very similar way, the interactions between membranes can be varied by anchoring polymers to them, or by adding polymers to the solution, or by adding salt if the membranes are charged.

Universality

Many phenomena of soft matter are independent of the particular chemical structure of the molecular building blocks. An example discussed above is the relation between the size of a polymer, as measured by its average end-to-end distance, and the physical length of the polymer, which is determined by the number of its monomers. Similar scaling relations appear in the description of the behavior of membranes and colloids.

In addition to this universality, there is also a universality of phase behavior among the different classes of materials. We have already discussed the fact that colloids, amphiphiles, and block copolymers all self-assemble into ordered phases. This is a consequence of effective interactions that are repulsive at short distances and attractive at long distances. Consequently, there is an intermediate scale that is favored by the system, one which sets the scale for the ordered phases. Although the sequence of ordered phases is most commonly manifest in a multitude of soft matter systems, they are also displayed in the small quantum systems of two-dimensional electron gases where the mobile charges form ordered states (Fogler et al. 1996) right up to the astronomical systems of neutron stars in which protons and electrons order (Pethick and Potekhin 1998)! (Obviously they attract one another at large distances. They also repel one another at short distances if they do not have sufficient energy to form a neutron at the top of the Fermi sea.)

Another example of different systems with the same behavior is that of rod-like micelles, which can be so long that they exceed their persistence length many times over and then behave as "living polymers". What is meant is that in many respects they behave as polymers, so that many of their properties can be predicted readily. However, their dynamics are not the same because long micelles can fission into shorter ones and vice versa; i.e. they do not have the same topological constraints as ordinary polymers.

Soft Matter Composites

More and more, systems are being investigated that contain several components of the classical materials, such as mixtures of colloids and polymers, polymers attached to membranes, colloids in microemulsions, and emulsion droplets in a nematic solvent. In order to understand the behavior of these systems, knowledge must be drawn from the various subfields of soft matter. A very nice example of the complexity that can arise in soft matter systems is the mixture of rod-like colloids and polymers (Dogic and Fraden 2001) discussed by Dogic and Fraden in Chapter 1 of Volume 2. In the regime of very small colloid concentrations, the colloid particles are found to form a thermally fluctuating membrane. The interaction that keeps the colloids together is the depletion interaction, discussed briefly above. The colloidal rods in this experiment are the *fd* viruses shown in Fig. 4. Thus, these are not stiff rods, but semi-flexible objects, and therefore display polymer-like behavior. Note that the thickness of this membrane is now given by the length of the *fd* virus, i.e. about 1 μm; this is about three orders of magnitude larger than the thickness of usual lipid bilayer membranes. Finally, the *fd* virus itself is a biological polymer, which is enclosed by a self-assembled layer of amphiphilic proteins. Thus, this system behaves as a colloidal, an amphiphilic, and a polymeric system on several different scales.

Loomings

It is clear that there are at least two very strong currents in contemporary soft matter research. The first is that of nanotechnology and new materials; the second is that of biology. Both of these exploit the self-assembling nature of systems, and the malleability and variability of composites. In this, the first of a series of volumes devoted to pedagogical reviews of new and exciting topics in soft matter, we have tried to assemble some of the basics required for our journey of exploration. We can only surmise where this newly launched vessel will carry us, but we can surely guarantee that the voyage itself will be an exciting one!

References

Boal, D. H., 2002, *Mechanics of the Cell*. Cambridge University Press, Cambridge.
Cates, M. E., Roux, D., Andelman, D., Milner, S. T., and Safran, S. A., 1988, *Europhys. Lett.* **5**, 733.
de Gennes, P.-G., 1979, *Scaling Concepts in Polymer Physics*. Cornell University Press, Ithaca, NJ.
de Gennes, P.-G., 1981, *Macromolecules* **14**, 1637.
des Cloizeaux, J. and Jannink, G., 1990, *Polymers in Solution*. Clarendon Press, Oxford.
Dhont, J. K. G., 1996, *An Introduction to Dynamics of Colloids*. Elsevier, Amsterdam.
Dhont, J. K. G., Gompper, G., and Richter, D. (eds.), 2002, *Soft Matter – Complex Materials on Mesoscopic Scales*, Vol. 10 of *Matter and Materials*. Forschungszentrum Jülich, Jülich.
Dinsmore, A. D., Crocker, J. C., and Yodh, A. G., 1998, *Curr. Opin. Coll. Interface Sci.* **3**, 5.
Dogic, Z. and Fraden, S., 2001, *Phil. Trans. R. Soc. Lond. A* **359**, 997.
Doi, M. and Edwards, S. F., 1986, *The Theory of Polymer Dynamics*. Clarendon Press, Oxford.
Evans, D. F. and Wennerström, H., 1994, *The Colloidal Domain, Where Physics, Chemistry, Biology and Technology Meet*. VCH Publishers, New York.
Ferry, J. D., 1980, *Viscoelastic Properties of Polymers*. John Wiley & Sons, New York.
Flory, P. J., 1969, *Statistical Mechanics of Chain Molecules*. Interscience, New York.
Fogler, M. M., Koulakov, A. A., and Shklovskii, B. I., 1996, *Phys. Rev. B* **54**, 1853.
Gelbart, W. M., Ben-Shaul, A., and Roux, D. (eds.), 1994, *Micelles, Membranes, Microemulsions, and Monolayers*. Springer-Verlag, Berlin.
Gompper, G. and Schick, M., 1994, in *Phase Transitions and Critical Phenomena*, eds. C. Domb and J. Lebowitz, Vol. 16, pp. 1–176. Academic Press, London.
Grosberg, A. Y. and Khokhlov, A. R., 1997, *Statistical Physics of Macromolecules*. Springer, New York.
Hadjuk, D. A., Harper, P. E., Gruner, S. M., Honeker, C. C., Kim, G., Thomas, E. L., and Fetters, L. J., 1994, *Macromolecules* **27**, 4063.
Hamley, I. W., 1998, *The Physics of Block Copolymers*. Oxford University Press, Oxford.
Hunter, R. J., 1989, *Foundations of Colloid Science*. Clarendon Press, Oxford.

Israelachvili, J. N., 1992, *Intermolecular and Surface Forces*. Academic Press, London.

Jönsson, B., Lindman, B., Holmberg, K., and Kronberg, B., 1998, *Surfactants and Polymers in Aqueous Solution*. John Wiley, Chichester.

Lipowsky, R. and Sackmann, E. (eds.), 1995, *Structure and Dynamics of Membranes – From Cells to Vesicles*, Vol. 1 of *Handbook of Biological Physics*. Elsevier, Amsterdam.

Luzzati, V. and Spegt, P. A., 1967, *Nature* **215**, 701.

Mahanty, J. and Ninham, B. W., 1976, *Dispersion Forces*. Academic Press, London.

Manoharan, V. N., Elsesser, M. T., and Pine, D. J., 2003, *Science* **301**, 483.

Muthukumar, M. and Nickel, B. G., 1987, *J. Chem. Phys.* **86**, 460.

Nelson, D., Piran, T., and Weinberg, S. (eds.), 2004, *Statistical Mechanics of Membranes and Surfaces*, 2nd edn. World Scientific, Singapore.

Nienhuis, B., 1982, *Phys. Rev. Lett.* **49**, 1062.

Pan, G., Kesavamoorthy, R., and Asher, S. A., 1997, *Phys. Rev. Lett.* **78**, 3860.

Park, M., Harrison, C., Chaikin, P. M., Register, R. A., and Adamson, D. H., 1997, *Science* **276**, 1401.

Pethick, C. J. and Potekhin, A. Y., 1998, *Phys. Lett. B* **427**, 7.

Russel, W. B., 1987, *The Dynamics of Colloidal Systems*. University of Wisconsin Press, London.

Russel, W. B., Saville, D. A., and Schowalter, W. R., 1989, *Colloidal Dispersions*. Cambridge University Press, Cambridge.

Safran, S. A., 1994, *Statistical Thermodynamics of Surfaces, Interfaces, and Membranes*. Addison-Wesley, Reading, MA.

Seddon, J. M. and Templer, R. H., 1993, *Phil. Trans. R. Soc. Lond. A* **334**, 377.

van de Ven, T. G. M., 1989, *Colloidal Hydrodynamics*. Academic Press, London.

Vos, W. L., Sprik, R., van Blaaderen, A., Imhof, A., Lagendijk, A., and Wegdam, G. H., 1996, *Phys. Rev. B* **53**, 16231.

1
Polymer Dynamics in Melts

Andreas Wischnewski and Dieter Richter

Abstract

The dynamics of linear polymer chains in the melt depends strongly on the chain length: for short, unentangled chains, the dynamics is determined by a balance of viscous and entropic forces; for long chains, topological constraints are dominant. In this chapter, the experimental exploration of chain dynamics is introduced and discussed in detail. The focus is on neutron spin-echo (NSE) spectroscopy, which is one of the most powerful tools to explore the different dynamic regimes in polymer melts on a microscopic scale. It allows direct observation of the transition from a regime of free relaxation at short times, which can be described in terms of the Rouse model, to constrained motion at longer times. The constrained motion is caused by the entanglements that emerge in long-chain polymer systems. The tube concept models these topological confinements by assuming the chain to be confined in a virtual tube formed by adjacent chains. This concept of chains reptating in a tube is strongly supported by experiments in the limit of long chains.

However, there is also strong experimental evidence that the tube model starts to fail if the polymer chains become shorter. In this regime of intermediate chain length, neither the Rouse model nor the pure reptation concept are applicable. A close comparison of linear rheology data with the predictions of the reptation model indicates the existence of additional degrees of freedom that release the topological confinement. Fluctuating chain ends that destroy the tube confinement starting from both ends were proposed as one candidate. This process, called contour length fluctuations (CLF), indeed accounts for the observed behavior of the mechanical relaxation function. In this chapter, we present a systematic study of this mechanism on a microscopic scale.

A second relaxation process that appears to limit the topological confinement in polymer melts is the relaxation of the tube itself (constraint release). Since the tube is formed by adjacent chains, which themselves accomplish all motions that are permitted for an observed test chain, the topological constraints that are represented by the tube are not fixed in time. In this chapter, the loosening of confinement by the constraint release effect, which can be observed on a molecular scale, will be investigated.

1.1
Introduction

In this chapter, the dynamics of flexible polymers in the melt – one of the most fascinating topics in the field of polymer science – will be discussed. The structure of polymer chains has been the subject of intensive scientific work at both theoretical and experimental levels, so it is a logical consequence of the activity in that field to focus on the problem of how these large molecules move. This question becomes all the more challenging because the static properties alone are surprising. Drawing the structural formula of a polymer on paper, one could be tempted to conclude that the chains in reality are prolate objects. The fundamental investigation of Flory (1953) has shown that this is not the case. His prediction that the spatial extent of a large chain molecule should be coil-like rather than rod-like (stretched) was confirmed by neutron scattering experiments (Kirste et al. 1973). This was at the same time one of the first applications of neutron scattering to polymer science. The coil conformation is a consequence of the large number of internal degrees of freedom, a common property of soft matter systems. Further "soft matter properties" of polymers are the weak interaction between the structural units (chains) and the significant role of entropy.

The mechanical properties of polymers are extraordinarily diverse. Addressing the response to strain, condensed matter is generally divided into solids and liquids: for solids, the stress is proportional to the strain at least for small deformations (elastic behavior); for liquids, the stress is proportional to the change in strain, i.e. viscous behavior is observed. For a polymer, the mechanical response can be solid-like, rubber-like or viscous depending on the temperature or load time. For long-chain polymer melts at intermediate frequencies, a plateau in the relaxation function reflects elastic behavior. In the low-frequency region, i.e. at longer times, the same material behaves in a viscous manner. This "viscoelastic" behavior is a good reason for comparing a long-chain polymer melt with a network, where chains are chemically cross-linked. In a melt, there are no chemical cross-links; however, if the chains are long enough, they build entanglements, which are not stable but may act temporarily as network points. In this picture, the plateau in the relaxation function is caused by a temporary network and, in analogy to the mesh size in a real network, one can define a distance between the entanglements in the melt: typical distances are in the order of some nanometers. Compared with the segment length (around 0.5 nm) and the radius of the coil of the entire chain (around 100 nm), this distance defines an intermediate length scale.

We have seen that the dynamics of polymers strongly depend on the time scales and temperature under consideration. One may also focus on different length scales. If a polymer is subjected to a mechanical strain, complex molecular rearrangements are provoked. This relates not only to molecular dimensions but also to individual bonds. Since local bond dynamics is governed by local potentials, the rearrangements are comparably fast to a normal solid (picosecond range). For distances between the bond length and the entanglement distance, entropic forces are dominant. For even larger distances, the motion of the chain is restricted due to the entanglements, and the chain is localized. All these dynamic processes, starting from lengths of about 0.1 nm and times in the picosecond range up to the size of the polymer chain (about 100 nm) and up to macroscopic times, determine the viscoelastic properties of a polymer system.

Thus, the dynamics of polymers is manifold: the relevant length scales vary from atomic distances to the length of the polymer macromolecule. This leads to time scales that are comparable to the characteristic time scales of atomic vibrations up to very long "macroscopic" times. The wide range of relevant time and length scales for polymers leads to their various mechanical properties. The huge variety of applications in conjunction with their simple production and processing makes polymers particular and interesting. Polymers are an integral part of our daily lives and they are of utmost importance for industry.

It is evident that a discussion of polymer dynamics over the entire range of time scales is far beyond the scope of this chapter. Here, we will focus on polymers at high temperatures, far beyond the glass temperature T_g, where a polymer is in the liquid-like "melt state". Furthermore, we will not discuss the local dynamics or glass aspects, but concentrate on the mesoscopic dynamics, i.e. on intermediate to large length scales that are in the order of the entire polymer chain, or some units of it. The manifold dynamic behavior on mesoscopic length scales is completely unknown from conventional solid-state physics. Consequently, specific experimental techniques and theoretical approaches are required. For instance, if a macromolecule that "prefers" the coil conformation is stretched a little bit – say by Brownian motion of neighboring molecules – the entropy will try to bend it back. On the other hand, the molecule has to overcome friction if it wants to move. How can this behavior be described? Another aspect to address is the question of how these large molecules diffuse and on what time scale. As mentioned above, long polymer chains will build entanglements, or packing constraints will strongly affect the freedom to move. What kind of chain motion will this condition allow?

One fascinating property of polymers on large length scales is universality. The principles we are going to discuss in this chapter do not depend on local, chemical details. This leads to the fact that we can focus our studies on one or two polymers to elucidate the main features of dynamical behavior.

One ultimate goal of polymer science is an understanding of macroscopic dynamic and mechanical properties on a molecular basis. The hope is that revealing all dynamic processes in polymer systems not only gives an essential contribution to fundamental research but also allows the design of materials with specific macroscopic properties by selectively manipulating microscopic parameters like architecture or composition. This "molecular design" presumes an understanding of all relevant relaxation processes in polymers, and we hope that the following pages serve as a first step in this direction.

In Section 1.2 the basic properties of polymer chains are discussed. Here, a brief introduction to the synthesis of polymer chains and their characterization is also given. In Section 1.3, experimental techniques that allow one to access the mechanical and dynamic properties will be introduced. Sections 1.4 to 1.6 describe theoretical approaches that are available and can explain the main features of polymer melt systems, starting from the Brownian dynamics of a single segment of a polymer chain to the large-scale diffusion of an entire macromolecule. These concepts will be scrutinized by experiments. The results demand the consideration of secondary relaxation processes, which will be discussed and again tested by experiments in Sections 1.7 and 1.8. Finally, in Section 1.9 the chapter will be summarized and an outlook will be given.

1.2
What is a Polymer?

Polymers are large molecules that are built from a repeat unit, the monomer. If all monomers are of the same species, the polymer is called a homopolymer. If a chain consists of several parts (blocks) that are chemically different, they are denoted as diblock or multiblock copolymers. The segments may be linked one after another, which leads to linear chains, or, by introducing branching points, versatile complex architectures can be produced. In polymer science, star polymers with different numbers of "arms" (one branching point), structures with one backbone chain and two (H-shaped) or more (pom-poms) arms at each end (two branching points), and even more complex molecules with a

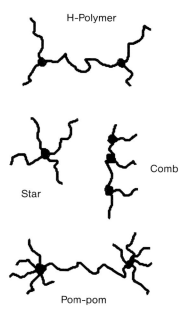

Fig. 1.1 Schematic representation of different polymer architectures.

comb-like structure have been considered and intensively investigated in the past decades (see Fig. 1.1).

In addition to the variation of the chemical structure within one chain or the architecture of the molecules, interesting features are found in mixed systems, e.g. in polymer blends that consist of different homopolymers, chains with different length (polydisperse systems) or architecture. Also, polymer solutions should be mentioned here.

Converging to an understanding of the basic dynamic properties in polymers presumes an understanding of the simplest systems. Therefore, we will focus in this chapter on homopolymers with a linear architecture in the melt state. They are the ideal probe to investigate the motions on the scale of segments and chains, respectively. In the following sections, we will see that the main features of polymer dynamics strongly depend on the length of the chains or the number of monomers N, respectively.

1.2.1
Synthesis of Polymers

Modern polymer chemistry generally distinguishes between two basic polymerization mechanisms (Odian 1991). These are either step growth or chain growth polymerizations. Step growth polymerization proceeds by the stepwise reaction between the functional groups of reactants, which can be either monomers, oligomers or polymers. In such polymerizations, the size of the polymer increases at a relatively slow rate and reaches a high molecular weight only if high conversion is reached. Typical examples for step growth polymerizations are the formation of polyamides by the reaction of diamines with dicarboxylic acids or the formation of polyurethanes by reaction of diisocyanates with diols. In contrast to step growth, chain growth polymerizations require an initiator. The initiator produces a reactive center, which may be either a free radical, cation, or anion. The monomer reacts exclusively with the reactive center and not with other monomers, oligomers or polymers. Chain growth polymerization can be subdivided into four elementary steps, i.e. initiation, propagation, chain transfer, and termination.

- During the initiation step, a reactive center, I^*, is formed, which initiates chain growth by reaction with the monomer M:

$$I \longrightarrow I^*$$

$$I^* + M \longrightarrow I\text{-}M^*$$

- During propagation, the reactive species repeatedly reacts with the monomer:

$$I\text{-}M^* + nM \longrightarrow I\text{-}(M)_n\text{-}M^*$$

where n denotes the number of monomers incorporated into the growing chain and $n + 1 = N$ is the degree of polymerization (number of monomers of the terminated chain).

- Termination is the step where the reactive center is destroyed by reaction with a compound S. At this point, the polymer chain, P, is inactive and ceases to grow:

$$I\text{-}(M)_n\text{-}M^* + S \longrightarrow P$$

- An additional reaction step that has to be considered is chain transfer. This step involves reaction with a compound S–X leading to a terminated polymer chain P–X, but at the same time to a new reactive species S^*, which itself acts as an initiator for chain growth:

$$I\text{-}(M)_n\text{-}M^* + S\text{-}X \longrightarrow P\text{-}X + S^*$$

A special case of chain growth polymerization is obtained when termination and chain transfer can be suppressed. This process is called *living polymerization*, which in particular is essential for the preparation of model polymers as outlined in the following paragraph. A detailed description of the definitions of, criteria for and consequences of living polymerizations can be found in the literature (Hsieh and Quirk 1996; Matyjaszewski and Sawamoto 1996; Kamigaito et al. 2001).

Under living polymerization conditions it is possible to synthesize model polymers with well-defined composition and structure, e.g. block copolymers by consecutive polymerization of two or more monomers, or star polymers by reaction with an appropriate coupling agent. The molecular weight of the polymers can be anticipated simply by the ratio of the mass of monomer to the moles of initiator provided that 100% conversion is reached. Polymers with narrow molecular-weight distribution can be prepared if the initiation rate is fast or comparable to the propagation rate and a rapid mixing of the reactants is ensured. Hence, polymerizations accomplished under such conditions yield model polymers suitable in particular for fundamental research in polymer physics. In this respect anionic polymerization of styrene and dienes with alkyllithium initiators has turned out to be the most powerful method (Young et al. 1984), even though living cationic and living radical polymerization have become increasingly important in recent years. In the following we will focus on the preparation and properties of some polyolefins, which are of particular importance for this chapter.

Polyolefins are saturated hydrocarbon polymers that exhibit superior stability toward thermal, oxidative, and radiation-induced degradation. They are therefore suitable for fundamental studies under extreme conditions, e.g. in the high-T limit. Polyolefins can be prepared by Ziegler–Natta and metallocene catalysts or by free-radical polymerization. Unfortunately, the polymers prepared by these methods usually possess a relatively broad polydispersity and the molecular architecture is difficult to control. In order to produce model polyolefins with narrow molecular-weight distribution, predefined molecular weights, and controlled architecture, the synthetic strategy involves the preparation of polydienes by living anionic polymerization, which are subsequently saturated with hydrogen to the corresponding polyolefins, e.g. polyethylene (PE) and poly(ethylene-*alt*-propylene) (PEP) can be made from polybutadiene and polyisoprene, respectively (see Fig. 1.2). Such polymerized polyolefins differ from pure linear chains by the appearance of a certain amount of side chains. This is due to the microstructure of the parent polydienes, which consists of four modes: *cis*-1,4-, and *trans*-1,4-, 1,2- and 3,4-. Polymerization of 1,3-butadiene in hydrocarbon solvents (benzene, cyclohexane) with alkyllithium initiator concentrations below 10^{-2} mol l^{-1} leads to a microstructure of about 51% *trans*-1,4-, 42% *cis*-1,4- and 7% 1,2- addition. The 1,2- and 3,4- isomers are identical in polybutadiene.

24 | 1 Polymer Dynamics in Melts

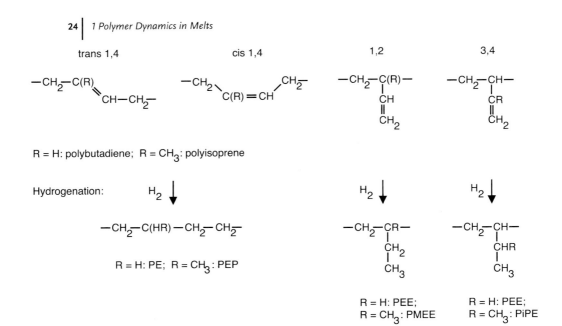

Fig. 1.2 Schematic illustration of possible microstructures of polyisoprene and polybutadiene and their transformation to the respective polyolefins by hydrogenation.

The microstructure of polyisoprene prepared under similar conditions consists of about 70% *cis*-1,4-, 23% *trans*-1,4- and 7% 3,4- units. The 1,2- addition does not appear in hydrocarbons. The composition of the microstructure changes drastically when polymerizations are performed in polar solvents like tetrahydrofuran or diethyl ether. The *cis*-1,4- and *trans*-1,4- units are clearly reduced, while beside an increased amount of 3,4- also 1,2- units are obtained. In the case of polybutadiene, the presence of 1,2- addition can be driven to almost 100% with dipiperidinoethane as polar cosolvent in a molar excess of at least 4 over the initiator. The polydienes can be saturated in a post-polymerization reaction by addition of hydrogen to the corresponding polyolefins. Several homogeneous and heterogeneous catalytic methods have been applied to get polyolefins with the same molecular characteristics as the parent polydienes. It turns out that heterogeneous catalytic hydrogenation by means of a palladium/barium sulfate catalyst is the most effective for this requirement. In particular, for the hydrogenation of polybutadienes and polyisoprenes, no chain scission and complete saturation are observed. There is also no metal contamination due to the ease of catalyst extraction.

The amount and type of side chains in the polyolefins prepared via this route obviously depend on the microstructure of the polydiene precursors. The hydrogenation product from *cis*-1,4- and *trans*-1,4-polyisoprene is an alternating poly(ethylene–propylene) (PEP) copolymer, the 1,2- structure transforms into poly(1-methyl-1-ethylethylene) (PMEE) and the 3,4- structure into poly(isopropylethylene) (PiPE). Hence, the hydrogenation product from polyisoprene prepared in a hydrocarbon solvent is in fact an alternating copolymer of ethylene and propylene with a random distribution of 7% isopropylethylene units. However, the abbreviation PEP has been accepted for this material and will be also used in this chapter.

A similar but less complicated scenario is obtained for polybutadiene. The *cis*-1,4- and *trans*-1,4- structures yield linear polyethylene (PE), and the 1,2- structure poly(ethylethylene) (PEE). PEE is formally the polymerization product of 1-butene. Therefore, the hydrogenation product of polybutadiene is a random copolymer of ethylene and 1-butene. The widely accepted abbreviation for this polymer is PEB-x [poly(ethylene-*co*-butene)-x], where the integer x denotes the number of ethyl side branches per 100 backbone carbons. According to this nomenclature, the hydrogenation product of a polybutadiene with 93% 1,4- and 7% 1,2- units is PEB-2 and that of 100% 1,2- polybutadiene is PEB-50.

Due to the presence of side chains, the average molar mass of a repeat unit M_0 is calculated as typically shown for PEB-2 in the following: If we cut from a long PEB-2 polymer chain a subchain with 100 carbon atoms ($M = 1200\,\text{g}\,\text{mol}^{-1}$), it has $200 - 2 = 198$ hydrogen atoms ($M = 198\,\text{g}\,\text{mol}^{-1}$) and two ethyl branches C_2H_5 ($M = 58$ g mol^{-1}), which yields a molecular weight of 1456 g mol^{-1}, i.e. $M_0 \approx 14.6$ g mol^{-1}. In analogy to PEP and for the sake of clarity, we use PE for PEB-2 and PEE for PEB-50 throughout this chapter.

1.3
Experimental Techniques

1.3.1
Neutron Scattering

Neutron scattering (Squires 1978) is a powerful tool for the investigation of the structure and dynamics in condensed matter samples at atomic and intermediate scales. Neutrons may be produced by a nuclear chain reaction in a reactor. They are moderated by, for example, D_2O before they react with the next nucleus. To produce slow ("cold") neutrons, a cold source is inserted into the reactor, which moderates the thermal neutrons to low temperatures (e.g. by hydrogen with a temperature of 20 K). An alternative method for the production of neutrons is spallation, where a high-energy beam of hydrogen ions hits a heavy-metal target so that neutrons evaporate from the cores. For a pulsed operation of the ion accelerator, a broad band of wavelengths may be used for the scattering experiments.

Thermal and cold neutrons have de Broglie wavelengths from $\lambda = 0.1$ nm up to 2 nm, corresponding to typical distances in condensed matter systems. The kinetic energy of the neutrons compares with the excitation energies of atomic or molecular motions like vibrations or phonons. Therefore, motions of the scatterers in condensed matter samples are detectable by a velocity change of the neutron. The spatial character of the motion may be inferred from the angular distribution of the scattered neutrons. The fact that neutrons deliver information about the structure and the dynamics at the same time makes them one of the most important tools in condensed matter research.

The energy E and momentum p of a neutron can be defined by the velocity v (here, the neutron is seen as a particle) or the wavenumber $k = 2\pi/\lambda$ (here, the neutron is seen as a wave with wavelength λ):

$$\boldsymbol{p} = m_\mathrm{n}\boldsymbol{v} = \hbar\boldsymbol{k} \tag{1.1}$$

$$E = \frac{m_\mathrm{n}}{2}v^2 = \frac{(\hbar k)^2}{2m_\mathrm{n}} \tag{1.2}$$

where m_n is the mass of the neutron. Fig. 1.3 shows the principle of a scattering experiment. A neutron beam of intensity I_0 with neutrons of energy E_i and a wavevector $\boldsymbol{k}_\mathrm{i}$ is scattered at the sample. The interaction of the neutron with the sample can be divided into two types. The first type is the magnetic interaction, where the magnetic moment of the neutron interacts with the

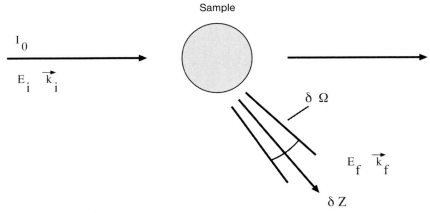

Fig. 1.3 Principle of a scattering experiment.

magnetic moment of the electrons or the cores. This type of interaction has no relevance for the topic discussed here, so we will focus on the second type, the scattering of the neutron at the core potential by the strong interaction. The strength of this interaction is represented by the scattering length b_j, a complex number that varies with the mass number of the core in an unsystematic way. The real part of b_j represents the scattering, and the imaginary part the absorption of the neutrons.

In the solid angle $\delta\Omega$ steradians, the scattered neutrons δZ can be characterized by energy E_f and wavevector $\mathbf{k}_f$. The energy and momentum that were transferred by the scattering process is given by

$$\Delta E = \hbar\omega = E_i - E_f \tag{1.3}$$

$$\Delta \mathbf{p} = \hbar \mathbf{Q} = \hbar(\mathbf{k}_i - \mathbf{k}_f) \tag{1.4}$$

where $\mathbf{Q}$ is the scattering vector. If $E_i = E_f$, the scattering process is called elastic. Following Eq. (1.2), this also means that $k_i = |\mathbf{k}_i| = k_f = |\mathbf{k}_f| = 2\pi/\lambda$.

We have to distinguish between coherent and incoherent scattering. The nucleus as well as the isotope distribution determine how much of these fractions contribute to the scattering. For elements that have a nucleus with spin zero and only one isotope, the scattering is purely coherent, i.e. the scattering of different atoms interferes. If the spin is not zero, the scattering amplitude may depend on the relative orientation of the nuclear and neutron spins, respectively. Then we are dealing with a mean scattering amplitude, which contains a coherent part and fluctuations of the scattering amplitude that contribute to the intensity without interference of waves emitted from different atoms. The latter represents the behavior of single atoms and is called incoherent. For vanadium or hydrogen, for example, incoherent scattering is significantly stronger than coherent scattering.

Let us describe coherent scattering of an atom by the mean scattering length squared $\bar{b}^2$ and incoherent scattering by the scattering cross-section $\sigma_{\text{inc}} = 4\pi(\overline{b^2} - \bar{b}^2)$. The coherent part of the double differential cross-section of N equal atoms is given by (Marshall and Lovesey 1971)

$$\left(\frac{\mathrm{d}^2\sigma}{\mathrm{d}\omega\,\mathrm{d}\Omega}\right)_{\text{coh}} = \frac{k_f}{k_i} \frac{1}{2\pi} \int_{-\infty}^{\infty} \mathrm{d}t\, e^{-i\omega t} \bar{b}^2 \sum_{i,j=1}^{N} \left\langle e^{-i\mathbf{Q}\mathbf{R}_i(0)} e^{i\mathbf{Q}\mathbf{R}_j(t)} \right\rangle \tag{1.5}$$

The angle brackets denote an ensemble average. The incoherent part of the double differential cross-section is given by

$$\left(\frac{\mathrm{d}^2\sigma}{\mathrm{d}\omega\,\mathrm{d}\Omega}\right)_{\text{inc}} = \frac{k_f}{k_i} \frac{1}{2\pi} \int_{-\infty}^{\infty} \mathrm{d}t\, e^{-i\omega t} \frac{\sigma_{\text{inc}}}{4\pi} \sum_{i=1}^{N} \left\langle e^{-i\mathbf{Q}\mathbf{R}_i(0)} e^{i\mathbf{Q}\mathbf{R}_i(t)} \right\rangle \tag{1.6}$$

Now we can express the cross-sections by a normalized dynamic structure factor $S(\boldsymbol{Q},\omega)$:

$$S_{\text{coh}}(\boldsymbol{Q},\omega) = \frac{k_i}{k_f} \frac{1}{N\overline{b}^2} \left(\frac{d^2\sigma}{d\omega\, d\Omega}\right)_{\text{coh}} \tag{1.7}$$

The polymer systems that will be discussed in this chapter are isotropic without any preferential direction. Thus, we may use

$$S_{\text{coh}}(Q,\omega) = S_{\text{coh}}(\boldsymbol{Q},\omega) \tag{1.8}$$

For incoherent scattering $S_{\text{inc}}(Q,\omega)$ is

$$S_{\text{inc}}(Q,\omega) = \frac{k_i}{k_f} \frac{4\pi}{N\sigma_{\text{inc}}} \left(\frac{d^2\sigma}{d\omega\, d\Omega}\right)_{\text{inc}} \tag{1.9}$$

It can be shown that the coherent dynamic structure factor is the Fourier transform (with respect to space and time) of the van Hove pair correlation function $G_{\text{pair}}(R,t)$ (van Hove 1954). In a classical interpretation this function stands for the probability of finding an atom j at position R_j at time t, if any atom i has been at position $R_i = 0$ at time $t = 0$. In analogy to the pair correlation function, the self-correlation function $G_{\text{self}}(R,t)$ is obtained by a Fourier transform of the incoherent scattering factor. Then $G_{\text{self}}(R,t)$ represents the probability of finding the atom i at position R_i at time t if the same atom has been at $R_i = 0$ at time $t = 0$.

Finally, before discussing two important neutron scattering techniques, which are vital for the understanding of the dynamics in polymer systems, we define the elastic scattering function as $S(Q,0) \equiv S(Q,\omega{=}0)$ and the static scattering function $S(Q)$, which is directly measured by small-angle neutron scattering (SANS):

$$S(Q) \equiv S(Q, t{=}0) = \int_{-\infty}^{\infty} S(Q,\omega)\, d\omega \tag{1.10}$$

As we will see below, in the neutron spin-echo technique, the Fourier transform of $S(Q,\omega)$ can be measured directly. The coherent scattering function in the time domain, also called the intermediate scattering function, can be calculated by the Fourier transform of Eq. (1.7) and using Eq. (1.8):

$$S_{\text{coh}}(Q,t) = \frac{1}{N} \sum_{i,j=1}^{N} \left\langle e^{-i\boldsymbol{Q}\boldsymbol{R}_i(0)} e^{i\boldsymbol{Q}\boldsymbol{R}_j(t)} \right\rangle \tag{1.11}$$

The respective incoherent scattering function is given by the Fourier transform of Eq. (1.9):

$$S_{\text{inc}}(Q,t) = \frac{1}{N} \sum_{i=1}^{N} \left\langle e^{-i\bm{Q}\bm{R}_i(0)} e^{i\bm{Q}\bm{R}_i(t)} \right\rangle \tag{1.12}$$

Small-Angle Neutron Scattering (SANS)

At a given wavelength of the probing radiation, large objects scatter into small angles. Long polymer chains are much larger than single atoms – their size is mesoscopic. To obtain information on the chain conformation, small scattering angles and therefore small Q-values have to be resolved. This can be realized by small-angle neutron scattering (SANS). In this technique, no energy analysis of the scattered neutrons is performed; the integral over all energies is detected. The measured scattering function $S(Q)$ therefore contains information about the structure of the sample but not about the dynamics.

The schematic picture of the experimental setup is shown in Fig. 1.4. A polychromatic neutron beam from a cold moderator is monochromated by a neutron velocity selector. This is a rotating cylinder with tilted absorbing lamellas. Only neutrons with a defined wavelength λ ($\Delta\lambda/\lambda \approx 0.1$) can pass. Neutron guides bring the neutrons to the collimation aperture. Then, the neutrons propagate freely to the sample aperture, which defines the divergence of the beam. The neutrons hit the sample and some are scattered. The transmitted (non-scattered) neutrons hit the beam stop on the detector and are used to measure the transmission of the sample. The scattered neutrons are detected on a position-sensitive detector, which is used to measure the cross-section of the sample. The collimation and detector distances are varied to achieve lower or higher resolution.

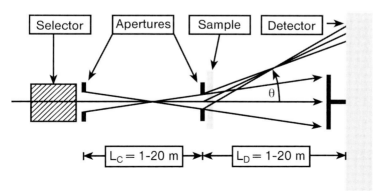

Fig. 1.4 Schematic picture of a SANS experiment.

One assumes that the size of the sample is much smaller than the distances from the source to the sample and from the sample to the detector. The wave fields of the incident and scattered beams are described by the wavevectors k_i and k_f respectively. The scattering vector is defined as the difference between the incident and the scattered wavevectors. The absolute value of the scattering vector for an elastic scattering process is given by

$$Q = |k_i - k_f| = 2k \sin \frac{\theta}{2} = \frac{4\pi}{\lambda} \sin \frac{\theta}{2} \qquad (1.13)$$

where λ is the wavelength of the neutrons adjusted by the velocity selector. The scattering angle θ is depicted in Fig. 1.4.

The elastic scattering function $S(Q,0)$ and the static scattering function $S(Q)$ have been defined above. SANS does not analyze the energy of the scattered neutrons and therefore cannot distinguish between elastic and inelastic scattering processes, i.e. it measures the integral over all energies $S(Q)$. Why can we then use Eq. (1.13), which assumes elastic scattering $k_i = k_f$? SANS accesses large objects or large volumes in real space. As an example, the fast atomic vibrations of a single atom in a polymer chain are not detected by SANS, while the slow motion of the entire macromolecule is. The energy transfer that is detectable in this Q-value regime is in the range $\Delta E < \mu eV$, while the neutrons have energies in the meV range. Therefore, it is justified to assume the scattering to be elastic, although the measured function is $S(Q)$ and not $S(Q,0)$. Equation (1.13) is also applicable for the neutron spin-echo (NSE) technique to be discussed in Section 1.3.1, because $k_i \approx k_f$ still holds. However, as will be explained in detail later, the NSE technique is able to detect the very tiny energy transfers in this Q-value regime by exploiting a property of the neutrons that has not been considered so far.[1]

Since for SANS the detected volumes in the relevant Q-value regime are significantly larger than the volumes of single atoms, the description of the scattering in terms of atomic scattering lengths is not appropriate. As shown below, the coherent scattering is determined by different scattering length densities of volumes which are – with respect to their size – relevant for the Q-values under consideration.

As mentioned above, the Q dependence of the scattering gives information about the structure within the sample. Imagine, for instance, the Bragg equation (first order):

$$\lambda = 2d \sin \frac{\theta}{2} \qquad (1.14)$$

---

[1] Note, however, that even at small angles inelastic scattering processes originating from higher momentum transfers may contribute to $S(Q)$ if the primary forward scattering is not very strong.

Here, d is the distance between lattice planes. Combining Eq. (1.13) with the Bragg equation (1.14) yields $Q = 2\pi/d$, illustrating the relation between the scattering vector and a "typical" distance in the sample.

The resulting scattering intensity should be treated by corresponding reduction procedures to obtain the scattering cross-section, which is independent of the experimental setup and the background. The measured intensity is connected to the cross-section $d\Sigma/d\Omega$ by

$$I = I_i D_e \Delta\Omega A T d \left(\frac{d\Sigma}{d\Omega}\right) \tag{1.15}$$

where I_i is the incident beam intensity, D_e is the detector efficiency, A is the irradiated sample area, d is the sample thickness, T is the sample transmission, and $\Delta\Omega$ is the angle of one detector element. For the absolute calibration, a reference material with a flat cross-section in the measured Q range can be used (e.g. Lupolene).

The structure factor is given by

$$S(Q) = \frac{d\Sigma/d\Omega}{K} \tag{1.16}$$

where K is the contrast factor, which describes the interaction of the neutrons with the sample. Let us, for instance, consider a two-component system like a polymer in a solution. Then K is given by $\Delta\rho^2$, where $\Delta\rho$ is the difference in the coherent scattering length densities of the polymer and the solution.

Neutron Spin-Echo (NSE) Spectroscopy

How do these large objects move? As we have seen above, we need to detect the intensity scattered at small angles. However, if we would like to follow the motion of such an object, what is the time scale involved? One can imagine that – in contrast to atomic vibrations, for example – the velocity of these large objects is very low. Imagine a drop of honey that flows from a spoon – it may have a velocity of a few millimeters per second. In contrast to that, neutrons have a velocity of a few hundred meters per second, which in turn means that we would like to detect a velocity change of neutrons in the order of 10^{-5}, at small angles.

In this section we will introduce neutron spin-echo (NSE) spectroscopy, which combines small-angle scattering with the desired energy resolution. To get that high energy resolution, a trick is needed where the neutron spin plays an important role. Instead of using a trick, we could also try to cut from an incoming neutron beam – with a Maxwellian distribution of velocities – a band that is narrow enough to allow for a detection of velocity changes in the order of 10^{-5} by removing all neutrons with unwanted velocity. But then the remaining intensity would be so poor that a scattering experiment would not be feasible in a realistic time frame.

The essence of the NSE technique is a method to decouple the detectability of tiny velocity changes caused by the scattering process from the width of the incoming velocity distribution. This allows one to run NSE instruments with 10–20% width of the velocity distribution yielding about 10^4 times more neutrons in the primary beam than the direct 10^{-5} filtering. The neutron spin is the key element to realize this decoupling. The basic NSE instrument invented by F. Mezei (Mezei 1980) works as follows (see also Fig. 1.5). A beam of longitudinally polarized neutrons, i.e. neutrons with spins pointing into the beam direction, enters the instrument and traverses a $\pi/2$ flipper located in a low longitudinal magnetic field. During the passage of this flipper, the neutron spins are rotated by $\pi/2$ and are then perpendicular to the beam, e.g. the spins are all pointing upward. Immediately after leaving the flipper, they start to precess around the longitudinal field generated by the primary precession solenoid. As they proceed into the precession coil, the Larmor frequency Ω_L, which is proportional to the field, increases up to several MHz (e.g. about one turn per 0.1 mm length of path) in the middle of the solenoid. The field and Ω_L decrease to low values again on the way to the sample. Upon arriving at the sample, the neutron may have performed many thousand precessions. Keeping in mind that different neutrons with velocities different from the incoming 10–20% distribution have total precession angles that differ proportionally, the ensemble of neutron spins at the sample contains any spin direction perpendicular to the longitudinal field with virtually equal probability.

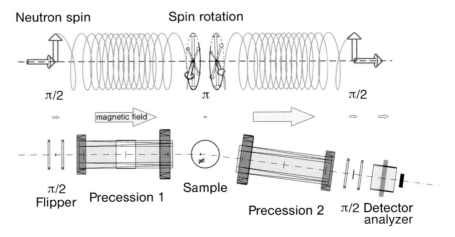

Fig. 1.5 Schematic picture of the geometry of a NSE spectrometer (Monkenbusch et al. 1997).

Nevertheless, each single neutron is tagged with respect to its velocity by the individual precession angle (modulo 2π) of its spin. The spin therefore can be viewed as the neutron's own stopwatch. Now near to the sample position there is a π flipper which, during passage of the neutron, turns its spin by π around an upward-pointing axis. Thereby the sign of the accumulated precession angle is reversed, i.e. the "spin stopwatch" times are set to their negative values. Then the neutrons enter the secondary part of the spectrometer, which is symmetric to the primary part. During the passage of the second main solenoid, provided the sample did not change the neutrons' velocity, each spin undergoes exactly the same number of precessions as in the primary part. Due to the reversing action of the π flipper, this leads to the result that all neutron spins arrive at the same precession angle, pointing upwards, at the second $\pi/2$ flipper, irrespective of their individual velocity. This effect is called the *spin-echo* in analogy to similar phenomena in conventional nuclear resonance experiments. The $\pi/2$ flipper turns these spins by $90°$ into the longitudinal direction. This switches off any effect of the precession on the longitudinal spin polarization. Further down the neutrons enter an analyzer that transmits only neutrons with axially parallel (or antiparallel) spins to the detector. If now the neutrons undergo a velocity change Δv_s in the course of scattering by the sample, the final spin direction is rotated by an amount proportional to Δv_s with respect to the upward direction of the echo. The second $\pi/2$ flipper rotates the upward (echo) direction into the longitudinal direction. Thereby the precession angle with the echo direction is preserved (as the angle of the precession cone after the $\pi/2$ flipper) and the time measurement of the "spin stopwatch" is effectively stopped. The subsequent analyzer has a transmission that depends on the longitudinal component of the spin.

Since only the cosine of the pointer angle counts, the analyzer works as a cosine modulating filter. The filter period Δv_c is controlled by the average number of precessions, e.g. by the average neutron velocity and the magnetic field inside the main precession solenoids. Due to the cosine modulating filter function, the NSE instrument measures the cosine transform of $S(Q,\omega)$, the so-called intermediate scattering function $S(Q,t)$:

$$S(Q,t) \propto \frac{1}{2}\left[S(Q,0) \pm \int \cos\left(\underbrace{J\lambda^3 \gamma \frac{m_n^2}{2\pi h^2} \omega}_{t}\right) S(Q,\omega)\,\mathrm{d}\omega\right] \quad (1.17)$$

where $J = \int_{\mathrm{path}} |\boldsymbol{B}|\,\mathrm{d}l$ is the integral of the magnetic induction along the flight path of the neutron from the $\pi/2$ flipper to the sample, and $\gamma = 1.830\,33 \times 10^8$ radian s^{-1} T^{-1} is the gyromagnetic ratio. The sign of the integral depends on the type of analyzer and on the choice of the sign of the flipping angle of the secondary $\pi/2$ flipper. The time parameter,

$t = J\lambda^3 \gamma m_\mathrm{n}^2/(2\pi h^2)$, may be easily scanned by varying the main solenoid current I_0 to which J is (approximately) proportional.

Note that the maximum achievable time t, e.g. the resolution, depends linearly on the (maximum) field integral and on the cube of the neutron wavelength λ!

The fact that the instrument measures the intermediate scattering function directly makes it especially useful for relaxation-type scattering because the relaxation function is measured directly as a function of time. Furthermore, the correction for instrumental resolution is easier. For instruments that measure $S(Q,\omega)$, the correction for the instrumental resolution function $R'(Q,\omega)$ has to be realized by a tedious deconvolution. For NSE, a simple point-by-point division by the result of a measurement of an elastically scattering reference sample is sufficient (Mezei 1980; Monkenbusch et al. 2004). The resolution leads to a decreasing $R(Q,t)$ with increasing Fourier time for purely elastic scattering, where it would be unity for ideal instrument resolution. This is mainly caused by the fact that the neutrons travel slightly different paths leading to different field integrals and therefore to a dephasing of the echo signal. This decay is accounted for by dividing the sample signal $S(Q,t)$ by $R(Q,t)$.

What is the meaning of the scattering function measured by NSE? This depends on the sample. For a deuterated polymer matrix that contains some (about 10–15%) protonated chains, $S(Q,t)$ is dominated by the *coherent* single-chain dynamic structure factor in the SANS regime. The scattering is determined by the difference in scattering length density between the protonated test chains and the environment (the deuterated matrix) as explained in the section about the SANS technique. For low Q-values the coherent scattering is dominant (see Fig. 1.6). The higher the Q-values, the more important becomes the incoherent background from the protonated chains as well as from the smaller incoherent scattering of deuterium. At the intersection of the dashed and thick solid lines in Fig. 1.6, the two contributions have equal magnitude and, since the incoherent signal is phase-shifted by π (see below), the echo signal may even vanish in a certain region of Q and t.

For a sample that contains only fully protonated chains, the *incoherent* scattering function is measured, which reflects the self-correlation of the protons – or in the Q-value range discussed here – of the chain segments. We may call this the segmental self-correlation function. It can be derived easily from Eq. (1.12) in Section 1.3.1 for a Gaussian chain. The derivation and the definition of a Gaussian chain will be discussed in detail below. The result for the incoherent scattering function may be anticipated at this point:

$$S_{\mathrm{inc}}(Q,t) = \exp[-\tfrac{1}{6}Q^2 \langle r^2(t) \rangle] \qquad (1.18)$$

From this scattering function, the segmental mean square displacement $\langle r^2(t) \rangle$ can easily be extracted.

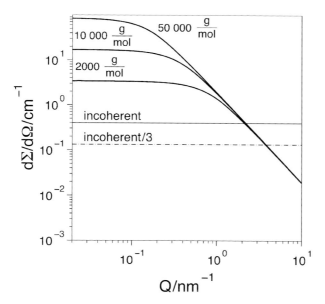

Fig. 1.6 Distribution of coherent and incoherent scattering intensity computed for polyethylene (PE) melts of $M_w = 50\,000$, 10000 and 2000 g mol^{-1} containing 15% (volume fraction) of H-labeled chains and the levels of incoherent background. At the intersection of the polymer structure factors (thick solid lines) with the dashed line, representing one-third of the incoherent scattering (spin flip, see text), the two contributions have equal magnitude. Increasing the amount of labeled component shifts the limit to higher Q-values. For a fully (100%) protonated sample, the incoherent level is higher by a factor of 100/15. On the other hand, to have a reasonable transmission, the sample volume has to be lower by a factor of 10. Therefore, the lines give a realistic estimate for the ratio of the coherently (with 15% protonated chains in a deuterated matrix) and the incoherently scattered intensity.

However, incoherent scattering has some difficulties. Two-thirds of the spin incoherent scattering events flip the neutron spin (Squires 1978); the one-third spin-flipped neutrons compensate the one-third that are not flipped. In the end the resulting intensity consists of one-third spin-flipped neutrons (i.e. the echo is phase-shifted by π compared to a coherent signal) on a background of two-thirds depolarized neutrons:

$$1\uparrow \xrightarrow{\text{incoherent scattering}} \tfrac{2}{3}\downarrow + \tfrac{1}{3}\uparrow = \tfrac{1}{3}\downarrow + \text{depolarized background} \quad (1.19)$$

What is even more important is the fact that the incoherent intensity lacks any amplifying interference effects and, as a sum from N independent point-like centers, it is scattered into the full solid angle, in contrast to the coherent scattering, which focusses most of the intensity in the low-Q region according to the coherent structure factor. Fig. 1.6 shows calculated scattering intensities for a polymer with three different molecular weights. The shape of the scattering function is discussed in Section 1.4; here we focus on the intensities. As described above, the dashed line in Fig. 1.6 refers to the incoherent scattering of protonated chains (15%) in a deuterated matrix (taking into account the factor 1/3 due to the spin flip). The incoherent scattering of a fully protonated sample will be higher by a factor 100/15. On the other hand, the thickness of typical incoherent samples must be lower by a factor of about 10 to get a reasonable transmission. This leads to the fact that the dashed line in Fig. 1.6 gives a good estimate for the incoherent scattering intensity of a fully protonated sample. Comparing the two intensities, we realize a difference of up to two orders in magnitude (depending on the molecular weight of the labeled chains). Due to the low incoherent intensity, the sample and the background have to be measured carefully and for long times. This requires an extremely stable instrument configuration.

1.3.2
Rheology

Now we will make an excursion from molecules to the macroscopic properties of polymers. The term "rheology" is derived from the Greek word *rheos* (flow); it measures and describes the relation between elongation or shear displacement and stress or torque for different materials.

Looking at a polymer at low temperature, we mainly observe elastic properties. Heating up the material, however, reveals liquid-like characteristics. Instead of varying the temperature, we may focus on different time scales. A very short, shock-like impact on a water surface illustrates that a liquid can be very stiff at short time scales. Elasticity and viscosity – viscoelasticity – are the keywords related to macroscopic properties of polymer systems which can be explored by observing the response of the material to mechanical forces over a broad frequency range. The elasticity leads to deformation, while the viscosity causes the flow of the material. Elastic deformation is a change to the molecular configuration associated with the storage of energy (like in a spring) while flow is an irreversible dissipative process.

A typical mechanical perturbation that can be applied to the system is a small-amplitude oscillatory shearing. This is realized in a rheometer with, for example, parallel disks, where one of the disks rotates with frequency ω. The oscillatory deformation (strain) is then

$$\gamma(t) = \gamma_0 \sin(\omega t) \qquad (1.20)$$

By definition and using the Boltzmann linear superposition of the strain history for a viscoelastic body (Ferry 1970), the resulting time-dependent stress is given by

$$\sigma(t) = \gamma_0[G'(\omega)\sin(\omega t) + G''(\omega)\cos(\omega t)] =: \gamma_0 G(t). \qquad (1.21)$$

The frequency-dependent moduli are obtained by a Fourier transformation of the relaxation function $G(t)$:

$$G'(\omega) = \omega \int_0^\infty G(t)\sin(\omega t)\,dt \qquad (1.22)$$

$$G''(\omega) = \omega \int_0^\infty G(t)\cos(\omega t)\,dt \qquad (1.23)$$

Here, $G'(\omega)$ is a measure for the elastic energy stored in the sample, and is therefore called the *storage modulus*. It is in phase with the strain. $G''(\omega)$ is a measure for the energy that is dissipated in the system (viscous dissipation), and is called the *loss modulus*. It is out of phase with the strain by $\pi/2$. We will focus here on small-amplitude deformations, i.e. on the linear viscoelastic regime.

Let us investigate the response in the two extreme cases of a Newtonian fluid and an elastic solid. For the Newtonian fluid, the shear stress is proportional to the shear rate, with the viscosity η as the constant of proportionality. This means that the modulus G', which reflects the elastic response, vanishes, while $G'' = \eta\omega$ is a linear function of the frequency.

For the second extreme case of an ideal elastic solid, the shear stress is proportional to the strain (Hooke's law for small deformations), with a constant of proportionality G_0. Here, $G'' = 0$ and $G' = G_0$. For a viscoelastic material, we expect a viscous behavior at low frequencies (long times) and an elastic behavior at high frequencies (short times), which should be reflected in a constant G' and for low frequencies in a linear increase of G'' with frequency.

To explore the short-(long-)time behavior, first the frequency window of the rheometer is exploited. If the maximum frequency ω_{max} of the rheometer is reached, one could fix the frequency at ω_{max} and lower the temperature. Then $1/\omega_{\text{max}}$ fixes the time scale and, by decreasing the temperature, faster processes are shifted into this particular time window, i.e. they become "visible". Qualitatively it is obvious that the response now refers to a virtual higher frequency, because the system becomes stiffer with decreasing temperature. The same holds for increasing the temperature at fixed ω_{min}, which relates to "virtually" lower frequencies. Then $1/\omega_{\text{min}}$ determines the longest time scale, and increasing the temperature speeds up slow processes, which thereby shift into the accessible time window. If the relation between time (or frequency) and temperature is known for all relaxation processes in the sample, the variation of temperature allows the accessible frequency range to be expanded to typically eight to ten decades instead of three.

This time–temperature superposition is valid for most polymers above their glass transition temperature. Rheologically simple behavior means that a single horizontal shift factor a_T for all relaxations in these viscoelastic systems yields a master curve, if, additionally, a vertical shift factor b_T, which relates the strength of the relaxations to temperature, is applied (Ferry 1970; Gotro and Graessley 1984). Usually the Williams–Landel–Ferry (WLF) law for a_T applies (Ferry 1970):

$$\log(a_T) = \left[\frac{-c_1(T - T_0)}{c_2 + (T - T_0)}\right] \qquad (1.24)$$

At $T = T_0$, where T_0 is the reference temperature, a_T as well as b_T are unity; c_1 and c_2 are material-specific constants that themselves also depend on T_0. Fig. 1.7 shows how the typical frequency range of an instrument is extended by performing a time–temperature sweep and applying time–temperature superposition.

The lower part of Fig. 1.7 shows the time–temperature superposition for a polyisoprene sample. The storage modulus shows a pronounced plateau at intermediate frequencies. This plateau is a signature of elasticity before the polymer melt starts to flow at lower frequencies (or longer times). It has been demonstrated that the plateau region expands to lower frequencies when the molecular weight is increased while the high-frequency behavior remains constant (Onogi et al. 1970). The plateau, i.e. the elastic behavior at intermediate times, which becomes more dominant the longer the chains are, is one of the most important features in long-chain polymer melts. It will play a fundamental role in this chapter.

1.4
Static Properties

To describe the basic static properties of a polymer chain, let us take polyethylene as an example. There the CH_2 units are linked together as shown in Fig. 1.8. The bond length l_0 between the carbon atoms is fixed. We now consider the model of a freely jointed chain as shown schematically in Fig. 1.9. It assumes N consecutive backbone bonds that have no restrictions on the bond angles, the only condition being that l_0 is fixed. One may then ask for the end-to-end distance vector:

$$\boldsymbol{R}_{ee} = \sum_{i=1}^{N} \boldsymbol{r}_i \qquad (1.25)$$

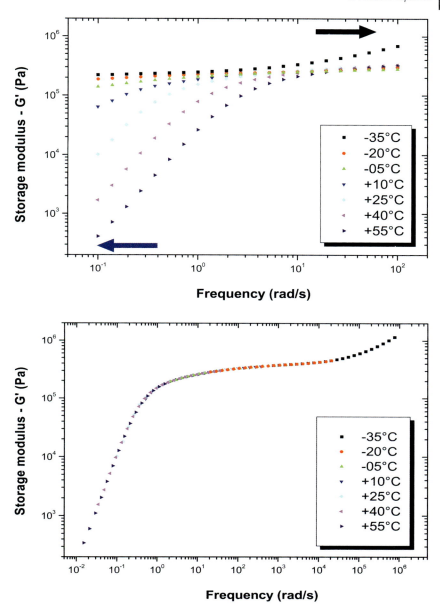

Fig. 1.7 Principle of time–temperature superposition. The frequency range of the instrument is three orders of magnitude (upper figure). By measuring at different temperatures and applying horizontal time–temperature superposition to a reference temperature (here 25 °C) as well as a vertical T-dependent correction, the range is extended to about eight orders of magnitude (lower figure). Data are from a polyisoprene sample with a molecular weight of $M_\mathrm{w} = 250$ kg mol^{-1} (Blanchard 2004).

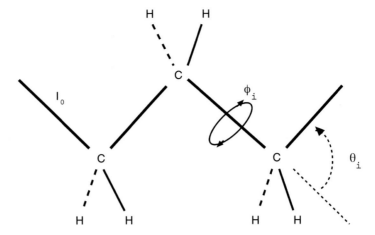

Fig. 1.8 Schematic sketch of a polyethylene chain with definitions of angles: θ_i is the bond angle, and ϕ_i is the torsional angle.

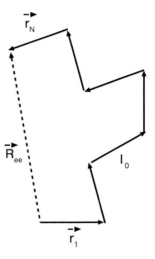

Fig. 1.9 A sketch of a freely jointed chain. The bond vector has a constant length l_0, but the angle is totally free. The end-to-end distance $\boldsymbol{R}_{ee}$ is described by the sum of all bond vectors.

The end-to-end distance vector is defined as the sum of all the bond vectors connecting two consecutive monomers (since $\boldsymbol{r}_{i,j}$ connects the monomers i and j, we define $\boldsymbol{r}_i \equiv \boldsymbol{r}_{i-1,i}$). Since there is no restriction concerning the bond angles – only the bond length is fixed – the chain has no preferential direction. This means that the average of the end-to-end distance – and of every other vector in the system – has to be zero.

The second moment is the average of

$$R_{ee}^2 = \sum_{i,j=1}^{N} \boldsymbol{r}_i \cdot \boldsymbol{r}_j = \sum_{i=1}^{N} r_i^2 + 2 \sum_{1 \leq i < j \leq N} \boldsymbol{r}_i \cdot \boldsymbol{r}_j \qquad (1.26)$$

Since in our simple model we have no correlation of different bonds, the averaging leaves a contribution only from the first part of Eq. (1.26):

$$\langle R_{ee}^2 \rangle = \sum_{i=1}^{N} \langle r_i^2 \rangle + 2 \times 0 = N l_0^2 \qquad (1.27)$$

The result of our simple model, that the end-to-end distance is proportional to the square root of N, holds also for more general models. Assuming a chain with fixed bond length *and* fixed angle between the bonds, which can rotate freely around the bonds, again gives the same scaling of the end-to-end distance. In fact, one can show that in general, as long as a chain is subject to any local restriction, in all models for flexible polymer chains the characteristic size increases with the square root of N.

This dependence on the number of segments shows that the characteristic size of a chain is much smaller than the full extension (a stretched chain). Only for a rigid rod-like chain is the end-to-end distance just the number of segments times their length, i.e. it is proportional to N.

At this point it might be helpful to remember the basic law for the diffusion of a particle, which is given by a continuous random walk. To travel a distance R, the particle must make

$$N = \left(\frac{R}{l_0}\right)^2 \qquad (1.28)$$

steps, where l_0 is the mean free path. That means that the chain conformation is nothing other than a random walk with a "step length" given by the bond length l_0 and the number of steps given by the number of bonds N. The end-to-end distance then represents the distance the particle has traveled.

Next to the end-to-end distance, the radius of gyration is often used to describe the chain size. It measures the size of a molecule in the manner of a moment of inertia. The vectors of all monomers relative to the center of mass are squared (assuming the same weight of all monomers). This sum is normalized by the number of monomers. The radius of gyration therefore measures the average extension of a chain relative to the center of mass. Without derivation we write

$$\langle R_g^2 \rangle = \tfrac{1}{6} \langle R_{ee}^2 \rangle \qquad (1.29)$$

It is obvious that real chains are not random walks where the angle between two steps is randomly distributed in a mathematical sense. The probability for high angles is lower compared to low angles due to the stiffness of the chains. The stiffness of a chain will depend on its chemical structure, side groups, etc. Therefore we will introduce a stiffness parameter C_∞ with

$$l^2 = C_\infty l_0^2 \tag{1.30}$$

and call l the effective bond length.

Before discussing the Gaussian chain, we define the frequently used Kuhn segment length. A model-independent parameter for polymer chains is the contour length L, which is calculated by multiplying the number of effective bonds by their length, $L = Nl$. Now, it is obvious that one can divide the chain into (arbitrary) larger segments l' in such a way that the contour length $L = N'l'$ is constant. One important definition of such an l' is the Kuhn segment length l_K, which is a measure for the distance over which correlations between bond vectors are lost. The idea is quite simple. If the number of bonds per segment l_K is large enough to erase any correlation between different Kuhn segments, the chain can be regarded as a freely jointed chain with N_K segments of length l_K. This means that real polymer chains, as long as they are long enough, can be mapped to a freely jointed chain. The Kuhn segment length is defined by

$$l_K = C_\infty l_0 \tag{1.31}$$

The Gaussian chain is a more general model of a freely jointed chain. The idea is that the number of bonds in a segment of a Gaussian chain is already sufficiently large that the distribution function of a single segment can be approximated by a Gaussian distribution. Then the distribution function of such N_K segments can be written as

$$W(\boldsymbol{r}_1,\ldots,\boldsymbol{r}_{N_K}) = (\tfrac{2}{3}\pi l_K^2)^{-3N_K/2} \exp\left(-\frac{3}{2l_K^2}\sum_i r_i^2\right) \tag{1.32}$$

or, using the difference of monomer vectors instead of the effective bond vectors,

$$W(\boldsymbol{R}_0,\ldots,\boldsymbol{R}_{N_K}) = (\tfrac{2}{3}\pi l_K^2)^{-3N_K/2} \exp\left(-\frac{3}{2l_K^2}\sum_i (\boldsymbol{R}_i - \boldsymbol{R}_{i-1})^2\right) \tag{1.33}$$

This distribution function can be compared with a thermodynamic partition function:

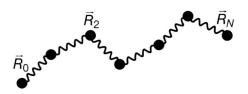

Fig. 1.10 A Gaussian chain represented by a number of effective monomers, which are connected by elastic springs (the spring–bead model).

$$W \propto \exp\left(-\frac{1}{k_\mathrm{B}T} \underbrace{\frac{3k_\mathrm{B}T}{2l_\mathrm{K}^2}\sum_i(\boldsymbol{R}_i - \boldsymbol{R}_{i-1})^2}_{\text{Hamiltonian}}\right) \qquad (1.34)$$

where T is the temperature and k_B the Boltzmann constant. The Hamiltonian in Eq. (1.34) describes N_K Hookean entropic springs with a single temperature-dependent spring constant. The Gaussian chain can therefore be described as beads connected by N_K hypothetical springs, the spring–bead model (see Fig. 1.10).

Let us now discuss how the size of a polymer chain can be measured. As discussed in Section 1.3 the structure factor $S(Q)$ can be measured by SANS. We will see in the following that the size of a polymer in terms of, for example, the radius of gyration R_g can be directly extracted from the measured $S(Q)$. The static structure factor for a single chain is given by a sum over all possible pairs of monomers i,j reflecting all possible interferences of elementary waves originating from monomer pairs. We calculate $S(Q)$ for a Gaussian distribution of the $\boldsymbol{R}_i - \boldsymbol{R}_j$ (Gaussian chain):

$$S(Q) = \frac{1}{N}\sum_{i,j}^N \langle \exp[\mathrm{i}\boldsymbol{Q}(\boldsymbol{R}_i - \boldsymbol{R}_j)]\rangle \qquad (1.35)$$

$$= \frac{1}{N}\sum_{i,j}^N \exp\langle -\tfrac{1}{2}[\boldsymbol{Q}(\boldsymbol{R}_i - \boldsymbol{R}_j)]^2\rangle \qquad (1.36)$$

$$= \frac{1}{N}\sum_{i,j}^N \exp\langle -\tfrac{1}{6}\boldsymbol{Q}^2(\boldsymbol{R}_i - \boldsymbol{R}_j)^2\rangle \qquad (1.37)$$

$$= \frac{1}{N}\sum_{i,j}^N \exp[-\tfrac{1}{6}l^2 Q^2 |i-j|] \qquad (1.38)$$

$$= \frac{1}{N}\int_0^N \mathrm{d}i \int_0^N \mathrm{d}j\, \exp(-\tfrac{1}{6}Q^2|i-j|l^2) \qquad (1.39)$$

We will discuss Eqs. (1.35) to (1.39) step by step. Equation (1.36) is obtained from Eq. (1.35) by Gaussian transformation (only valid for Gaussian chains); the next step is the result of an average projection. The averaging is executed from Eq. (1.37) to Eq. (1.38). Finally, we replace the summation by an integration, which is correct for a Gaussian chain with a large number of monomers N. The result of this double integral reads

$$S(Q) = \frac{2N}{R_g^4 Q^4}[\exp(-R_g^2 Q^2) - 1 + (R_g^2 Q^2)] \qquad (1.40)$$

which is the so-called Debye function. The asymptotic form of $S(Q)$ is given by

$$S(Q) = \begin{cases} N(1 - R_g^2 Q^2/3) & \text{for } QR_g \ll 1 \\ 2N/R_g^2 Q^2 & \text{for } QR_g \gg 1 \end{cases} \qquad (1.41)$$

The Debye function may be approximated by the simpler form:

$$S(Q) = \frac{N}{1 + Q^2 R_g^2/2} \qquad (1.42)$$

Fig. 1.11 shows the Debye function and different approximations.

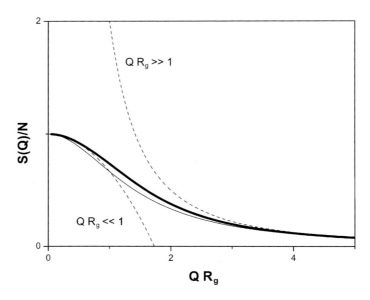

Fig. 1.11 The Debye function normalized by N versus QR_g (thick solid line). The dashed lines represent the asymptotic forms, and the thin solid line is the approximation of Eq. (1.42).

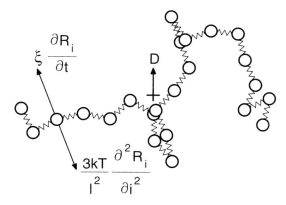

Fig. 1.12 Spring–bead model of a Gaussian chain as assumed in the Rouse model. The two arrows on the left side illustrate the forces as indicated on the left-hand side of Eq. (1.44). D represents the center-of-mass diffusion.

1.5
Brownian Motion, Viscous and Entropic Forces: the Rouse Model

We now consider the dynamics of a Gaussian chain in terms of the so-called Rouse model. We will start with a Gaussian chain, which we now consider as a coarse-grained polymer model where beads represent several monomers connected by hypothetical springs (see Fig. 1.12). It has been shown in Section 1.4 that the spring–bead model is an equivalent description of a Gaussian chain.

The equation of motion contains the entropic springs $k = 3k_\mathrm{B}T/l^2$ as the source of restoring forces, a simple local friction ξ as the sole interaction of the chain with the embedding melt of identical chains, and a random force $\boldsymbol{f}_i(t)$ with

$$\langle f_{i\alpha}(t_1) f_{j\beta}(t_2)\rangle = 2k_\mathrm{B}T\xi \delta_{ij}\delta_{\alpha\beta}\delta(t_1 - t_2)$$

representing the interaction with the heat bath. Here, α and β denote the Cartesian components. The resulting Langevin equation is

$$\xi\frac{\mathrm{d}\boldsymbol{R}_i}{\mathrm{d}t} - \frac{3k_\mathrm{B}T}{l^2}(\boldsymbol{R}_{i+1} - 2\boldsymbol{R}_i + \boldsymbol{R}_{i-1}) = \boldsymbol{f}_i(t) \qquad (1.43)$$

Assuming a continuous index variable i we obtain

$$\xi\frac{\partial \boldsymbol{R}_i}{\partial t} - \frac{3k_\mathrm{B}T}{l^2}\frac{\partial^2 \boldsymbol{R}_i}{\partial i^2} = \boldsymbol{f}_i(t) \qquad (1.44)$$

with the boundary conditions $(\partial \mathbf{R}_i/\partial i)|_{i=0,N} = 0$. The solution is obtained by a transformation to normal coordinates

$$\mathbf{X}_p(t) = \frac{1}{N} \int_0^N di \, \cos\left(\frac{p\pi i}{N}\right) \mathbf{R}_i(t) \tag{1.45}$$

yielding

$$2N\xi \frac{\partial \mathbf{X}_p}{\partial t} + \frac{6k_\mathrm{B}T\pi^2}{Nb^2} p^2 \mathbf{X}_p = \mathbf{f}_p(t) \tag{1.46}$$

Here, $\mathbf{f}_p(t)$ are again random forces that are independent of each other. The transformation of Eq. (1.44), which describes coupled oscillators, to Eq. (1.46), is realized by a set of normal coordinates $\mathbf{X}_p(t)$, which can move independently, and therefore allows a description of the dynamics of the polymer chain by independent motions, which are called "modes" p.

The correlation functions for $p > 0$ are given by

$$\langle X_{p\alpha}(t) X_{q\beta}(t) \rangle = \delta_{\alpha\beta} \delta_{pq} k_\mathrm{B}T \frac{Nl^2}{3k_\mathrm{B}T 2\pi p^2} \exp\left(-t\frac{p^2}{\tau_\mathrm{R}}\right) \tag{1.47}$$

with

$$\tau_\mathrm{R} = \frac{\xi N^2 l^2}{3\pi^2 k_\mathrm{B}T} \tag{1.48}$$

where τ_R is called the Rouse time.

Mode $p = 0$ denotes the center-of-mass diffusion, which is exactly the Einstein expression for the diffusion of a particle with friction coefficient $N\xi$:

$$\langle X_{0\alpha}(t) X_{0\beta}(0) \rangle = \delta_{\alpha\beta} \frac{2k_\mathrm{B}T}{N\xi} t \tag{1.49}$$

The scattering of the polymer chain is obtained by the summation of the segmental scattering amplitudes lumped into the beads with the proper phase factors:

$$S(Q,t) = \frac{1}{N} \left\langle \sum_{i,j=1}^N \exp[i\mathbf{Q} \cdot (\mathbf{R}_i(t) - \mathbf{R}_j(0))] \right\rangle \tag{1.50}$$

This is the scattering function defined in Eq. (1.35) but with time-dependent position vectors. In analogy to the calculation in Section 1.4, we can now, under the assumption of a Gaussian chain, calculate the averages $\langle \; \rangle$:

$$\langle \exp[i\mathbf{Q} \cdot (\mathbf{R}_i(t) - \mathbf{R}_j(0))] \rangle = \exp\left[-\tfrac{1}{6}Q^2 \underbrace{\langle (\mathbf{R}_i(t) - \mathbf{R}_j(0))^2 \rangle}_{\Phi_{ij}(t)}\right] \tag{1.51}$$

Repeating the steps demonstrated in Section 1.4 in Eqs. (1.35) to (1.38) yields the scattering function:

$$S(Q,t) = \frac{1}{N} \sum_{i,j} \exp[-\tfrac{1}{6} Q^2 \Phi_{ij}(t)] \tag{1.52}$$

The mean square displacement $\Phi_{ij}(t)$ may be decomposed into three contributions:

$$\Phi_{ij}(t) = \Phi_D(t) + \Phi_{ij}^0 + \Phi_{ij}^1(t) \tag{1.53}$$

where $\Phi_{ij}^0 = |i-j|\, l^2$ describes the correlation due to the structure of the Gaussian chain. Inserting this expression for $\Phi_{ij}(t)$ in Eq. (1.52) reproduces the static structure factor as derived in Section 1.4 (Eq. 1.38), which finally gives the Debye function (Eq. 1.40.)

The dynamics is taken into account by two contributions. The first, $\Phi_D(t) = 6Dt$, is the center-of-mass diffusion, which is, as for any diffusing object, represented by a common factor:

$$S_{\text{diffusion}}(Q,t) = \exp(-Q^2 Dt) \tag{1.54}$$

The second contribution, $\Phi_{ij}^1(t)$, is obtained by first calculating the inverse transform of the normal coordinates as defined in Eq. (1.45):

$$\boldsymbol{R}_i(t) = \boldsymbol{X}_0 + 2 \sum_{p=1} \boldsymbol{X}_p(t) \cos\left(\frac{p\pi i}{N}\right) \tag{1.55}$$

By inserting Eq. (1.55) in $\Phi_{ij}(t)$ as defined in Eq. (1.51) and omitting the diffusion and structural correlations, we get, after some algebraic transformations,

$$\Phi_{ij}^1 = \frac{4Nl^2}{\pi^2} \sum_{p=1} \frac{1}{p^2} \cos\left(\frac{p\pi j}{N}\right) \cos\left(\frac{p\pi i}{N}\right) \left[1 - \exp\left(-\frac{tp^2}{\tau_R}\right)\right] \tag{1.56}$$

representing the internal dynamics of the chain, which vanishes for $t = 0$: $\Phi_{ij}^1(t{=}0) = 0$. Inserting all contributions of $\Phi_{ij}(t)$ (three addends in Eq. 1.53) into Eq. 1.52, we finally get the Rouse scattering function:

$$S(Q,t) = \frac{1}{N} \sum_{i,j} \exp\Bigg\{ -Q^2 Dt - \tfrac{1}{6} Q^2 |i-j| l^2 \tag{1.57}$$

$$- \frac{2Q^2 N l^2}{3\pi^2} \sum_{p=1} \frac{1}{p^2} \cos\left(\frac{p\pi j}{N}\right) \cos\left(\frac{p\pi i}{N}\right) \left[1 - \exp\left(-\frac{tp^2}{\tau_R}\right)\right] \Bigg\}$$

The Rouse model provides a good description of the dynamics of a Gaussian chain. It is valid for real linear polymer chains on an *intermediate* length scale. The specific (chemical) properties of a polymer enter only in terms of two parameters $Nl^2 = 6R_g^2$ and l^2/ξ, i.e. the dimension of the chain and an effective friction. This of course means that for high Q-values, where neutrons start to detect the chemical structure of single monomers, one expects deviations from pure Rouse behavior. Indeed this has been found in simulation and experiments (Paul et al. 1998; Richter et al. 1999).

The friction parameter ξ/l^2 is often expressed in terms of the so-called *Rouse rate*,

$$Wl^4 = 3k_B T l^2/\xi \tag{1.58}$$

and the center-of-mass diffusion may be expressed as

$$D = \frac{k_B T (l^2/\xi)}{6R_g^2} = \frac{Wl^4}{18R_g^2}. \tag{1.59}$$

In modeling the Rouse expression (at intermediate Q), the parameters N and l are somewhat arbitrary as long as the physical values l^2/ξ and R_g are kept constant.

What is the static structure factor $S(Q, t=0) = S(Q)$ in the framework of the Rouse theory? Starting from Eq. (1.57) we get

$$S(Q) = \frac{1}{N} \sum_{i,j} \exp(-\tfrac{1}{6} Q^2 |i-j| l^2) \tag{1.60}$$

Replacing the summations by integrals, we end up with what was derived in Section 1.4 (see Eq. 1.39), i.e. we get the Debye function.

Before comparing the model to experimental data, let us consider the mean square displacement of a single segment in the framework of the Rouse model for very short times. Neglecting the diffusion for short times, we start from Eq. (1.56) and get, for one segment n,

$$\Phi_n^1 = \frac{4Nl^2}{\pi^2} \sum_{p=1} \frac{1}{p^2} \cos^2\left(\frac{p\pi n}{N}\right) \left[1 - \exp\left(-\frac{tp^2}{\tau_R}\right)\right] \tag{1.61}$$

Since we are interested in short times, the large p-numbers dominate, and we can replace the $\cos^2$ by the average $1/2$. Transforming the sum into an integral finally yields:[2]

---

[2] Note that the approximate prefactor of Eq. (6.104) in Doi and Edwards (1986) differs from the exact result given here by a factor $\sqrt{12/\pi}$.

$$\Phi_n^1 = \frac{4Nl^2}{\pi^2} \int_0^\infty dp \, \frac{1}{p^2} \frac{1}{2} \left[1 - \exp\left(-\frac{tp^2}{\tau_R}\right) \right] \quad (1.62)$$

$$= \left(\frac{12k_B T l^2 t}{\xi \pi}\right)^{1/2} = \left(\frac{4Wl^4 t}{\pi}\right)^{1/2} \quad (1.63)$$

The main result of this approximation is that the mean square displacement of a segment at short times is $\propto t^{1/2}$ in the Rouse model. This represents a sub-Fickian motion due to correlations of displacements along the chain away from the probe monomer n. With Eqs. (1.18) and (1.63) the incoherent scattering function for Rouse dynamics is given by

$$S_{\text{inc}}(Q,t) = \exp\left[-\frac{Q^2}{3}\left(\frac{Wl^4 t}{\pi}\right)^{1/2}\right] \quad (1.64)$$

1.5.1
Experimental Studies of the Rouse Model

We have seen in Section 1.3 that the mean square segment displacement can be extracted directly from the incoherent scattering function $S(Q,t)$ by $-6\ln[S_{\text{inc}}(Q,t)]/Q^2 = \langle r^2(t) \rangle$ (Eq. 1.18).

Incoherent NSE experiments have been performed at the NSE spectrometer at the DIDO research reactor FRJ2 in Jülich, Germany (Monkenbusch et al. 1997) on a fully protonated poly(ethylene–propylene) (PEP) sample of 0.3–0.4 mm thickness and a molecular weight of $M_w = 80$ kg mol^{-1}. The wavelength of the incoming neutrons was $\lambda = 0.8$ nm. The data will be discussed in more detail in the next section. Here, we focus on the mean square displacement at short Fourier times. Fig. 1.13 shows the data. Using $Wl^4(T=492\,\text{K}) = 3.26$ nm^4 ns^{-1} (Richter et al. 1993), they are in excellent agreement with the Rouse prediction of Eq. (1.63). Note that the line in Fig. 1.13 was calculated without any adjustable parameter.

Now we will consider the single-chain dynamic structure factor, measured by coherent NSE. In the high-Q regime, where $S(Q) = 2N/(R_g^2 Q^2)$ (see Eq. 1.41), and neglecting the diffusion, the following form for $S(Q,t)$ can be derived (Doi and Edwards 1986):

$$S(Q,t) = \frac{12}{Q^2 l^2} \int_0^\infty du \, \exp\left[-u - \sqrt{\Gamma(Q)t}\, h\bigl(u/\sqrt{\Gamma(Q)t}\bigr)\right] \quad (1.65)$$

which contains the characteristic relaxation rate

$$\Gamma(Q) = \frac{k_B T}{12\xi} Q^4 l^2 = \frac{Wl^4 Q^4}{36} \quad (1.66)$$

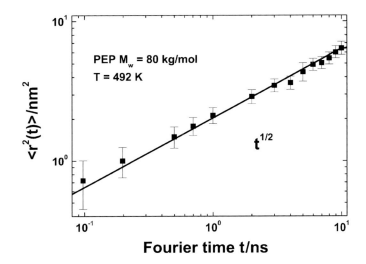

Fig. 1.13 NSE data of protonated PEP (H-PEP) with $M_w = 80$ kg mol^{-1} at 492 K. The scattering vector is $Q = 1$ nm^{-1}. The solid line shows the Rouse prediction.

and

$$h(u) = \frac{2}{\pi} \int_0^\infty dx \, \cos(xu) \frac{(1 - e^{-x^2})}{x^2}$$

$$= \frac{2}{\sqrt{\pi}} e^{-u^2/4} + u[\text{erf}(u/2) - 1] \tag{1.67}$$

For $t = 0$, Eq. (1.65) is not the Debye function but yields its high-Q limiting behavior $\propto Q^{-2}$. It is only valid for $QR_g \gg 1$. In that regime the form of $\Gamma(Q)$ immediately reveals that the local (intra-chain) relaxation increases $\propto Q^4$ in contrast to normal diffusion $\propto Q^2$. The form of $S(Q,t)/S(Q)$ obtained from Eq. (1.65) depends on $x = \sqrt{\Gamma(Q)t}$ only. NSE data obtained for different Q plotted versus x should collapse onto a common master curve if the Rouse model is valid, and they indeed do. Fig. 1.14 shows the single-chain dynamic structure factor of 10% H-PEE in a deuterated matrix (D-PEE) in the Rouse scaling representation $[S(Q,t)/S(Q)$ versus the so-called Rouse variable $Q^2 l^2 (Wt)^{1/2}]$ for several Q-values at $T = 473$ K. All data merge onto one master curve. Fig. 1.15 shows $\Gamma(Q)$ as derived from a Rouse fit in the short-time regime of the same data compared to Eq. (1.66) (dashed line). Taking into account center-of-mass diffusion (solid line), excellent agreement is found (Montes et al. 1999).

Fig. 1.16 shows NSE data of PE with $M_w = 2$ kg mol^{-1}, taken at the MESS NSE spectrometer at the LLB, Saclay, France, in the representation

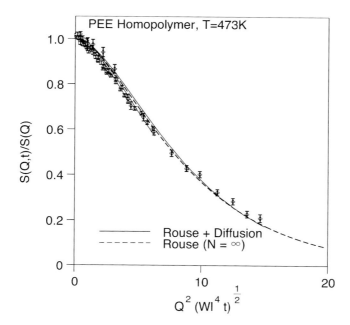

Fig. 1.14 Single-chain dynamic structure factor measured on a melt of 10% protonated polyethylethylene (H-PEE, $M_w = 21.5$ kg mol^{-1}) in a deuterated matrix of polyethylethylene (D-PEE) ($M_w = 24.5$ kg mol^{-1}) in a Rouse scaling plot (Monkenbusch et al. 1997; Montes et al. 1999). Shown are seven different Q-values between $Q = 0.5$ and 2 nm^{-1}.

$S(Q, t)/S(Q)$ versus Fourier time t. The data were taken at $T = 509$ K for five different Q-values with a neutron wavelength of $\lambda = 0.6$ nm. The solid lines represent a fit with the Rouse model (Eq. 1.57). Since the segment length $l = l_0\sqrt{C_\infty} = 4.12$ Å (Boothroyd et al. 1991) for polyethylene is known and the number of segments can easily be calculated by $N = M_w/M_0 = 137$, the only free parameter to fit all Q-values simultaneously is the Rouse rate Wl^4 or the friction coefficient ξ. Fig. 1.16 shows that the data are described perfectly. In conjunction with the agreement demonstrated in Fig. 1.13 for the incoherent data, the dynamic structure factor results demonstrate the validity of the rather simple concept of a chain in a heat bath where the relaxation of thermally activated fluctuations is determined by a balance of viscous forces (velocity, friction) and entropic forces.

For Q-values larger than Q_{max} (interestingly $Q_{max} \approx 1.5$ nm^{-1} seems to be a more or less universal value for many polymers), deviations from the Rouse model are observed. At high Q, effects of local chain stiffness and internal viscosity start to play a role (Richter et al. 1999). However, in the Q-value range shown in Fig. 1.16, the local structure of the polymer chain does not play any role and the Rouse model is valid.

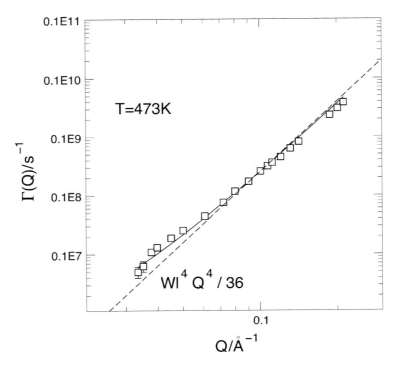

Fig. 1.15 Initial slope extracted from the same PEE data as shown in Fig. 1.14 as a function of Q. The dashed line represents $\Gamma(Q)$ as calculated by Eq. (1.66). The solid line takes translational diffusion into account (Montes et al. 1999).

Fig. 1.17 shows data of PE with $M_w = 12$ kg mol^{-1} measured at the IN15 NSE spectrometer at the ILL, Grenoble, France, in the representation $S(Q,t)/S(Q)$ versus Fourier time t. Due to the availability of a sufficiently high intensity of long-wavelength neutrons [in the case of the data shown in Fig. 1.17, the wavelength is $\lambda = 1.5$ nm (open symbols)], the IN15 holds the world record with respect to resolution (Schleger et al. 1999). Following Eq. (1.17) the maximum Fourier time is proportional to the wavelength cubed. The accessible Fourier time for $\lambda = 1.5$ nm is about 165 ns. However, if the sample is a strong scatterer, one can use neutrons with even longer wavelength (at the cost of intensity), e.g. $\lambda = 2$ nm, which would yield a maximum Fourier time of 390 ns.

The data in Fig. 1.17 were fitted with the Rouse model. Forcing an agreement at very short Fourier times by restricting the fits to 5 or 10 ns, it is evident that the data strongly deviate from the Rouse prediction at longer times. However, the only difference between the samples shown in Figs. 1.16 and 1.17 is the molecular weight, the chains in the latter sample being six times longer

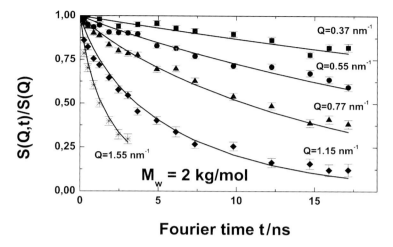

Fig. 1.16 NSE data of PE with $M_\mathrm{w} = 2$ kg mol^{-1} at $T = 509$ K for various Q-values (Richter et al. 1994). The solid lines represent a fit with the Rouse model.

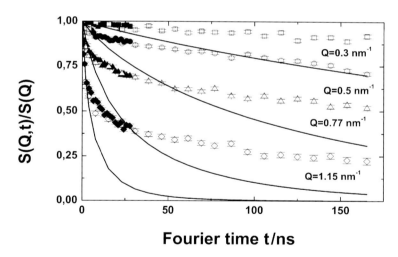

Fig. 1.17 NSE data of PE with $M_\mathrm{w} = 12$ kg mol^{-1} at 509 K for various Q-values. Filled symbols: wavelength of the incoming neutrons $\lambda = 0.82$ nm (maximum Fourier time 28 ns); open symbols: $\lambda = 1.5$ nm (maximum Fourier time 165 ns). The solid lines represent a fit with the Rouse model.

than in the former. For the longer chains the free three-dimensional Rouse-like relaxation is obviously perturbed. After an initial decay of $S(Q,t)$, the relaxation slows down, and we will see later that for even longer chains the relaxation in the NSE time window stops completely.

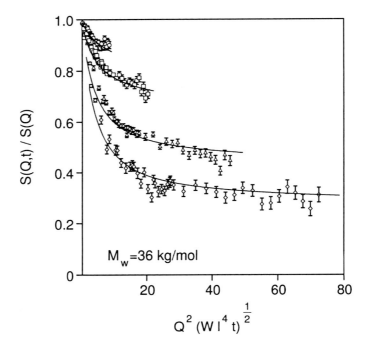

Fig. 1.18 NSE data of PE with $M_w = 36$ kg mol^{-1} at 509 K for various Q-values in a Rouse scaling plot. Lines are guides to the eye.

Fig. 1.18 shows how NSE data from long PE chains look if plotted in the Rouse scaling. In contrast to the data presented in Fig. 1.14 for PEE, here the four shown Q-values of a sample with molecular weight $M_w = 36$ kg mol^{-1} do not at all merge onto one master curve. They show a distinct splitting for different Q-values already at short Fourier times, pointing to a significant deviation from the Rouse picture.

The observation in NSE experiments as illustrated in Figs. 1.17 and 1.18 is in agreement with what has been observed in the storage modulus of long-chain polymer melts. As already briefly discussed in Section 1.3.2, in the range of intermediate frequencies $G'(\omega)$ shows a plateau, which is an indication for elastic behavior. Indeed we will see that the slowing down of relaxation in NSE experiments, on the one hand, and the existence of a plateau in the storage modulus, on the other, have the same origin. They are both indications of what is dominating the dynamics of long-chain polymer systems: topological constraints in terms of entanglements. These prevent the chain from relaxing completely via the Rouse modes by building a kind of temporary network which leads to the observed network-like response in rheology.

In the next section the most successful model to describe these topological constraints is discussed: the tube concept.

1.6
Topological Constraints: the Tube Concept

As we have seen above, in the melt, long-chain polymers interpenetrate each other and restrict their motions. This restriction originates from topological constraints that arise due to the formation of entanglements by the long chains.

In his famous reptation model, P.-G. de Gennes described the effect of these entanglements by a virtual tube along the coarse-grained chain profile, localizing the chain and confining the chain motion (de Gennes 1981; Doi and Edwards 1986) (see Fig. 1.19). The tube follows a random walk and represents the topological confinements. The tube in this concept is not meant as a "chemical" tube, which would have a diameter in the range of the chain–chain distance, i.e. a few tenths of a nanometer. We will see that the diameter d of the virtual tube is much larger.

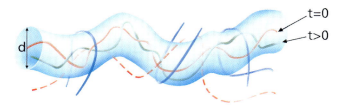

Fig. 1.19 Schematic illustration of the tube concept. The lateral confinement of a polymer chain is represented by a tube with diameter d formed by adjacent chains.

Let us now take a "test chain" confined in such a virtual tube. What kind of motion can this chain execute? First of all, we would expect that our chain does not know that it is confined at very short times. If the mean square displacement of the segments is smaller than half the tube diameter d, no contact of the segments with the virtual walls of the tube will have happened. Therefore, normal Rouse dynamics should be expected in the short-time regime. After this contact has taken place, after a time that we call the "entanglement time" τ_e, the Rouse dynamics can only take place in one dimension along the tube profile, a kind of curvilinear version of the Rouse motion that is called "local reptation" in de Gennes' model.

What happens with the center-of-mass diffusion of the entire chain? Of course, provided the tube exists, diffusion also can take place only along the tube, in one dimension. Since the tube itself is defined as a kind of envelope of the chain, the diffusing molecule follows, and has to follow, its own contour. It is a snake-like motion, which is in the end the reason for calling this the "reptation" model. We will see that this third process is very slow – too slow to be observed in NSE, even if we use the IN15, which provides the highest resolution in time available in the world.

The introduction of the tube concept for long-chain polymer systems is inevitably connected with the introduction of a new parameter besides the segment length: a second length scale that describes the geometry of the tube. The tube is thought to represent entanglements. These should be characterized by yet another parameter: the distance between these entanglements. If we assume N_e segments between two entanglements, the distance d_e between them is given by the end-to-end distance spanned by these N_e segments (provided they can be represented by a Gaussian chain): $d_e = \sqrt{N_e l^2}$, where l is the segment length. The end-to-end distance of the entire chain is given by $\sqrt{Nl^2} = \sqrt{Ll}$, where $L = Nl$ is the contour length. Let us accomplish a coarse-graining such that the new "segment length" is the entanglement distance d_e. The end-to-end distance must remain the same as before:

$$R_{ee}^2 = Ll = L'd_e \tag{1.68}$$

We define a virtual "thickness" of the new contour length L' (the so-called primitive path), which should be again the end-to-end distance of the segments between two entanglements, d_e. It is an intuitive step to define this thickness as the tube diameter:

$$d \equiv d_e = \sqrt{N_e l^2} \tag{1.69}$$

Given that the tube is a virtual object representing the topological constraints, it is not an obvious one. However, we will show that the introduction of *one* additional length scale is sufficient to obtain a consistent description of the dynamics in long-chain polymer melts.

The dynamics of a chain in a tube described above may be summarized as in Fig. 1.20, which shows schematically the segmental mean square displacements versus time in a double logarithmic representation. For times $t < \tau_e$ we expect Rouse behavior, which gives, as already derived above, a power law $\langle r^2(t) \rangle \propto t^{1/2}$ reflecting the sub-Fickian motion due to correlations of displacements along the chain away from the probe monomer [see Eq. (1.63)]. In the previous section an incoherent experiment on H-PEP was presented that corroborates this Rouse prediction at short times.

For times $t > \tau_e$ the Rouse dynamics is constricted to the tube, which represents basically a random walk. The mean square displacement can be calculated by first switching to the coordinates $s_n(t)$ along the primitive path of the tube. Then as in Eqs. (1.62) and (1.63) the mean square displacement along the tube profile for $t < \tau_R$ can be calculated:

$$\langle [s_n(t) - s_n(0)]^2 \rangle = \left(\frac{4k_B T l^2 t}{3\xi\pi} \right)^{1/2} = \left(\frac{4Wl^4 t}{9\pi} \right)^{1/2} \tag{1.70}$$

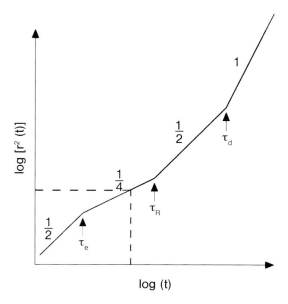

Fig. 1.20 Schematic representation of the segmental mean square displacements versus time in a double logarithmic plot. The white rectangle indicates the time range accessible by NSE spectroscopy.

The factor $1/3$ compared to Eq. (1.63) accounts for the fact that the motion is restricted to one dimension. Since the tube is a random walk with step length d, the mean square displacement in three-dimensional real space is given by $\Phi_n = Zd^2 = Ld$, where Z is the number of steps and $L = |s_n(t) - s_n(0)|$ is the contour length of the primitive path:

$$\Phi_n = d\langle|s_n(t) - s_n(0)|\rangle \qquad (1.71)$$

$$= d(\langle[s_n(t) - s_n(0)]^2\rangle)^{1/2} \qquad (1.72)$$

and with Eq. (1.70) we get[3]

$$\Phi_n = \left(\frac{4k_BTl^2d^4t}{3\pi\xi}\right)^{1/4} = \left(\frac{4Wl^4d^4t}{9\pi}\right)^{1/4} \qquad (1.73)$$

If we force the sub-Fickian motion to take place along a random walk, which alone gives rise to a behavior $\langle r^2(t)\rangle \propto t^{1/2}$, it is not surprising that the two $t^{1/2}$ laws finally yield a power law $\langle r^2(t)\rangle \propto t^{1/4}$ in the regime of local reptation.

---

3) Note that the approximate prefactor of Eq. (6.110) in Doi and Edwards (1986) differs from the exact result given here (Eq. 1.73) by a factor $[4/(3\pi)]^{1/4}$.

For times longer than τ_R the dynamics is dominated by center-of-mass diffusion, which gives $\langle r^2(t)\rangle \propto t$. Of course, we have to apply the arguments as for the Rouse motion in the tube, since the center-of-mass diffusion is again forced to take place along the random-walk profile of the tube. This leads to $\langle r^2(t)\rangle \propto t^{1/2}$ for $\tau_R < t < \tau_d$, where τ_d is the disentanglement time. Note that this power law is the same as for the Rouse regime ($t < \tau_e$), but for a different reason.

Finally, for times $t > \tau_d$, the chain has escaped completely from the tube and three-dimensional Fickian diffusion can take place, i.e. we expect $\langle r^2(t)\rangle \propto t$.

What is the Entanglement Time τ_e?
Since there is quite a bit of confusion in the literature, the definition of τ_e deserves its own small subsection. Note that we will imply here and in the following the validity of Eq. (1.69). There are two classical ways of defining this time scale, in addition to the one we shall use in this chapter.

- The first assumes τ_e to be the Rouse time τ_R of a chain with an end-to-end distance $R_{ee} = \sqrt{N_e}\, l$, where N_e is the number of segments between two entanglements. This τ_e refers via τ_R to the slowest Rouse mode of a chain with an end-to-end distance R_{ee} fitting into the tube:

$$\tau_e' = \frac{\tau_R N_e^2}{N^2} = \frac{N_e^2 l^2 \xi}{3\pi^2 k_B T} = \frac{d^4}{\pi^2 W l^4} \qquad (1.74)$$

- The second way is also quite evident: τ_e is the time when the mean square displacement of a segment equals the tube diameter squared. This definition results by equating Φ with d^2 and adopting τ_e in Eq. (1.63):

$$\tau_e'' = \frac{\pi d^4 \xi}{12 k_B T l^2} = \frac{\pi d^4}{4 W l^4} \qquad (1.75)$$

This is the definition in Doi and Edwards (1986), after applying the correction mentioned with respect to Eq. (1.63).

- The third definition is the one we will use in this chapter. Combining the time dependence of the mean square displacement as defined within the Rouse model (Eq. 1.63) and that of the reptation model for the local reptation regime (Eq. 1.73), τ_e follows by calculating the intersection point of these two straight lines (if plotted as shown schematically in Fig. 1.20). Equating (1.63) and (1.73) finally yields:

$$\tau_e = \frac{\pi d^4}{36 W l^4} \qquad (1.76)$$

This point in time marks the transition from free Rouse behavior to restricted motion, and is therefore the best choice for a definition of τ_e. We will use this definition in the following. However, it deviates by just $\approx 15\%$ from the first definition. Note that inserting τ_e in Eq. (1.63) for the Rouse regime or into Eq. (1.73) for the local reptation regime yields a mean square displacement $\Phi_n = d^2/3$ at τ_e.

1.6.1
Validating the Tube Concept on a Molecular Scale

NSE measurements of the self-correlation function have been performed at the NSE spectrometer at the DIDO research reactor FRJ2 in Jülich (Monkenbusch et al. 1997) on a fully protonated PE sample of 0.3–0.4 mm thickness and a molecular weight $M_w = 190$ kg mol^{-1} (Wischnewski et al. 2003). Fig. 1.21 shows the data for $T = 509$ K at $Q = 1$ and 1.5 nm^{-1}.

The total measuring time for each single (Q, t) pair (including background) was $\simeq 5$ h, requiring a very high temporal stability of the instrument. The neutron wavelength of $\lambda = 0.8$ nm allowed for a Fourier time range 0.1 ns $\leq t \leq 22$ ns in the normal instrument setup, and 0.01 ns ≤ 0.13 ns in a special short-time configuration with small precession coils in the sample region. All data shown stem from an integrated detector area of 615 cm^2. The solid lines

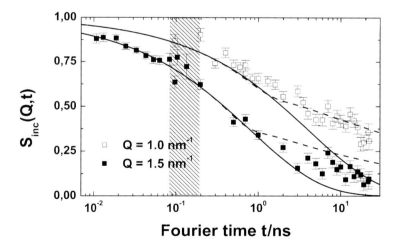

Fig. 1.21 NSE data obtained from the incoherent scattering from a fully protonated PE melt with $M_w = 190$ kg mol^{-1} at $T = 509$ K for $Q = 1.0$ and 1.5 nm^{-1} (Wischnewski et al. 2003). At the range boundaries (shaded gray bar) of the two spectrometer configurations (short or normal, see text), the data quality is worse than the bulk of the data points, as seen by the sizes of the error bars. The solid and dashed lines represent the predictions of Eqs. (1.64) and (1.78), respectively (see text after Eq. 1.78 for more details).

in Fig. 1.21 represent the Rouse prediction. For short times they agree nicely with the experimental data.

For the very long PE chains used here, virtually all of the scattering intensity stems from "inner" segments. These should exhibit the same segmental diffusion behavior, and if the assumption of a Gaussian shape of the diffusive displacement probability distribution holds for all times, the mean square segment displacement can be extracted directly from $S(Q,t)$ as defined in Eq. (1.18):

$$\langle r^2(t) \rangle = -6 \ln[S_{\text{inc}}(Q,t)]/Q^2 \qquad (1.77)$$

Fig. 1.22 displays the H-PE data in this representation.

The mean square displacement in the Rouse regime has been derived in Eq. (1.63). Inserting the previously determined value for the Rouse rate $W(509\,\text{K})l^4 = 7 \pm 0.7$ nm^4 ns^{-1} (Richter et al. 1993) from the analysis of the single-chain structure factor of low-molecular-weight PE melts, Eq. (1.63) is quantitatively corroborated as already seen for the H-PEP data in the last section. Also a transition to a regime $\propto t^{1/4}$ is clearly visible. The crossover time can be extracted from Fig. 1.22 by fitting the data with Eq. (1.63) for the Rouse regime and Eq. (1.73) for the local reptation regime. The only free parameter is then the crossover time, yielding $\tau_e \simeq 1$ ns corresponding to a tube diameter of 3 nm (Eq. 1.76).

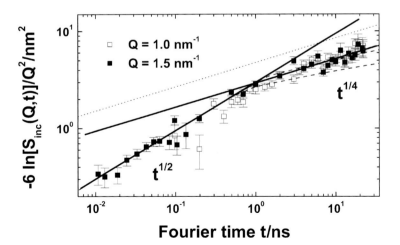

Fig. 1.22 Same data as shown in Fig. 1.21 in a representation of $-6\ln[S_{\text{inc}}(Q,t)]/Q^2$ versus time, i.e. the mean square displacement $\langle r^2(t) \rangle$ as long as the Gaussian approximation holds (Wischnewski et al. 2003). The solid lines describe the asymptotic power laws $\langle r^2(t) \rangle \propto t^{1/2}$ and $t^{1/4}$, respectively; the dotted line results by inserting $d = 4.8$ nm in Eq. (1.73); and the dashed lines represent the prediction of Eq. (1.78) (see text for more details).

On the other hand, this value may be calculated by using Eq. (1.76), where the elementary step length of the Gaussian contorted virtual tube is identified with the tube diameter, which can be extracted from NSE experiments on the pair correlation function. These measurements will be discussed in detail below, but the resulting tube diameter may be anticipated at this point: $d = 4.8$ nm. Equation (1.76) then yields $\tau_e \simeq 7$ ns. The dotted line in Fig. 1.22 results by inserting $d = 4.8$ nm in Eq. (1.73) for the local reptation regime. The intersection with the $\propto t^{1/2}$ line defines $\tau_e = \tau_e = 7$ ns as calculated above.

The apparent strong discrepancy between the crossover time of $\tau_e = 1$ ns with a tube diameter of 3 nm extracted from the incoherent data and the values $\tau_e \simeq 7$ ns and $d = 4.8$ nm from the coherent data will be discussed below.

First, we apply the same evaluation procedure as described above to the H-PEP data, revealing the mean square displacements of H-PEP. They have already been discussed for short times in the last section. Fig. 1.23 displays the results for the entire time range.

Again, a deviation from $\langle r^2(t) \rangle \propto t^{1/2}$ is clearly visible for $t \gtrsim 10$ ns, while the data are in agreement with Eq. (1.63) for $t \lesssim 10$ ns as already shown in the last section. Using $W(492\,\mathrm{K})l^4 = 3.26$ nm^4 ns^{-1} (Richter et al. 1993), the data were fitted with Eqs. (1.63) and (1.73), yielding a crossover time of $\tau_e = 8.4$ ns and a tube diameter of about 4.3 nm.

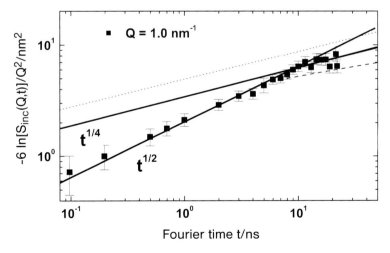

Fig. 1.23 Data of H-PEP in the representation of $-6\ln[S_{\mathrm{inc}}(Q,t)]/Q^2$ versus time for $T = 492$ K (Wischnewski et al. 2003). As before, the solid lines describe the asymptotic power laws; the dotted line results from Eq. (1.73); and the dashed line is from Eq. (1.78) (see text for details).

To compare this result with the tube diameter obtained from the single-chain dynamic structure factor of PEP, again we will anticipate the results of the evaluation discussed in detail below: $d_{\text{PEP}} = 6$ nm, i.e. $[\tau_e] = 40$ ns (see dotted line in Fig. 1.23), again in contradiction to $d = 4.3$ nm and $\tau_e = 8.4$ ns, obtained from an interpretation of the self-correlation data in terms of the Gaussian approximation.

In the derivation of Eq. (1.73), the Gaussian width after a diffusion time t of the single segment distribution along the one-dimensional tube contour, the path coordinate s is taken to be the time-dependent displacement. Projecting this on the Gaussian contorted tube again corresponds to a Gaussian sublinear diffusion in real space (Eq. 1.73). However, the real process has to be modeled by projecting the segment probability distribution due to curvilinear Rouse motion on the linear coordinate s onto the random-walk-like contour path of the contorted tube, leading to a non-Gaussian probability distribution of the segments at times $t > \tau_e$. The necessity to perform the proper averaging was first shown by Fatkullin and Kimmich (1995) in the context of the interpretation of field-gradient NMR diffusometry data (Fischer et al. 1999), which yield results that are analogous to the incoherent neutron scattering functions. However, they are in another time and space regime covering mainly the regime $\tau_R < t < \tau_d$. Their result

$$S_{\text{inc}}(Q, t > \tau_e) = \exp\left[\frac{Q^4 d^2}{72} \frac{\langle r^2(t)\rangle}{3}\right] \text{erfc}\left[\frac{Q^2 d}{6\sqrt{2}} \sqrt{\frac{\langle r^2(t)\rangle}{3}}\right] \quad (1.78)$$

invalidates the Gaussian approximation (Eq. 1.64) for times above τ_e. We note that Eq. (1.78) is strictly valid only for $t \gg \tau_e$ when $\langle r^2(t)\rangle \gg d^2$. The effect on the scattering function is that, if (wrongly) interpreted in terms of the Gaussian approximation, the crossover to local reptation appears to occur at significantly lower values of τ_e.

Figs. 1.21 to 1.23 show a comparison of the scattering function $S_{\text{inc}}(Q, t)$ as predicted by Eqs. (1.64) (Rouse regime) and (1.78) (dashed lines, local reptation regime) with the NSE data. The parameters Wl^4 and d were fixed to the values taken from the single-chain structure factor measurements. For PE and $Q = 1$ nm^{-1}, the free Rouse regime ($t < \tau_e$) as well as the local reptation regime are perfectly reproduced. For $Q = 1.5$ nm^{-1} in the case of PE (lower dashed line in Figs. 1.21 and 1.22) and for $Q = 1$nm^{-1} for PEP, the prediction of Eq. (1.78) lies slightly outside the error band of the data points.

The spatial resolution increases with increasing Q-values. Agreement with theory may only be expected for Q-values less than $2\pi/d$. For the tube diameter of 4.8 nm in PE, the "limiting" wavevector would be 1.3 nm^{-1}, which may explain why deviations become visible at $Q = 1.5$ nm^{-1}. The same holds for PEP, where, with $d = 6$ nm, the typical wavevector would be about $Q = 1$ nm^{-1}.

Up to now we have considered the segmental mean square displacement in a long-chain polymer melt. Now we will investigate the single-chain dynamic structure factor. As already demonstrated in Section 1.5, the Rouse model does not describe the dynamics of long chains in the melt. After an initial decay of $S(Q, t)$ by a free three-dimensional Rouse motion, the relaxation slows down. We have seen above that this slowing down starts when the mean square displacement of the chain segments becomes comparable to the diameter squared of the virtual tube described in the reptation model. The tube concept provides a model for the pair correlation scattering function of a single chain in the melt. It is given by (de Gennes 1981; Doi and Edwards 1986):

$$\frac{S(Q,t)}{S(Q)} = [1 - F(Q)]S_{\text{locrep}}(Q,t) + F(Q)S_{\text{esc}}(Q,t) \tag{1.79}$$

Here, $F(Q)$ is the (cross-sectional) form factor of the tube:

$$F(Q) = \exp[-(Qd/6)^2] \tag{1.80}$$

The two processes that determine the relaxation of a chain in a tube are local reptation and the creeping of the entire chain out of the tube where the chain follows its own profile:

$$S_{\text{locrep}}(Q,t) = \exp(t/\tau_0)\,\text{erfc}\left(\sqrt{t/\tau_0}\right) \tag{1.81}$$

$$S_{\text{esc}}(Q,t) = \sum_{p=1}^{\infty} \frac{2AN\mu}{\alpha_p^2(\mu^2 + \alpha_p^2 + \mu)} \sin^2(\alpha_p) \exp\left(-\frac{4t\alpha_p^2}{\pi^2 \tau_d}\right) \tag{1.82}$$

where $\mu = Q^2 N l^2/12$ and α_p are the solutions of the equation $\alpha_p \tan(\alpha_p) = \mu$; A is a normalization constant, so that $S_{\text{esc}}(Q,0) = 1$. The two time scales in Eqs. (1.81) and (1.82) are given as follows: the first, $\tau_0 = 36/(Wl^4 Q^4)$, is for Rouse-type segment diffusion along the tube; the second is the so-called disentanglement time $\tau_d = 3N^3 l^2/(\pi^2 W d^2)$ for reptation-type escape of the chain from the tube (creep).

In the high-Q limit, i.e. $R_g Q \gg 1$, $\alpha_p \approx (p-\frac{1}{2})\pi$ and $S_{\text{esc}}(Q,t)$ in Eq. (1.82) can be approximated by

$$S_{\text{esc}}(Q,t) = \frac{8}{\pi^2} \sum_{p,\text{odd}} \frac{1}{p^2} \exp\left(-\frac{p^2 t}{\tau_d}\right) =: \mu_{\text{rep}}(t) \tag{1.83}$$

Note that $S_{\text{esc}}(Q,0)$ is again normalized to unity. In this equation, $\mu_{\text{rep}}(t)$ is called the "tube survival probability" and is directly related to the relaxation

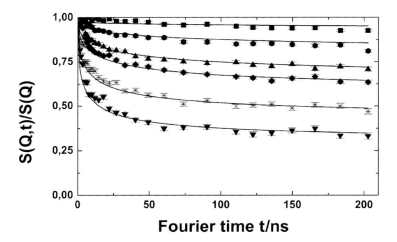

Fig. 1.24 Plot of $S(Q,t)/S(Q)$ from a $M_w = 200$ kg mol^{-1} PEP melt at $T = 492$ K for the scattering wavevectors $Q = 0.3, 0.5, 0.68, 0.77, 0.96,$ and 1.15 nm^{-1}, from above (Wischnewski et al. 2003). The solid lines represent a fit with Eq. (1.79).

function, $G(t)/G_e = \mu_{\text{rep}}(t)$, where G_e is the plateau modulus. By Fourier transformation of $G(t)$ and replacing the sum in Eq. (1.83) by an integral, it can be shown that, for times $t < \tau_d$ or $\omega > \omega_d = 1/\tau_d$ (Doi and Edwards 1986; Milner and McLeish 1998; McLeish 2002),

$$G''_{\text{rep}}(\omega) \propto \omega^{-1/2} \tag{1.84}$$

Now we will describe NSE results from high-molecular-weight PE and PEP samples by the single-chain dynamic structure factor of Eqs. (1.79) to (1.82). Inspecting the formulas we recognize that all the parameters needed can easily be calculated or are already known from the evaluation of NSE data of the respective short-chain systems: the segment length l of PE [$l = 0.41$ nm (Boothroyd et al. 1991)] and PEP [$l = \sqrt{l_0^2 C_\infty n} = 0.74$ nm with $n = 3.86$ the effective bond number per monomer (Richter et al. 1992; Richter et al. 1993)], the Rouse variables [$Wl^4 = 7$ nm^4 ns^{-1} for PE at $T = 509$ K, and $Wl^4 = 3.26$ nm^4 ns^{-1} for PEP at $T = 492$ K (Richter et al. 1993)] and the number of segments [$N = 13\,013$ for the $M_w = 190$ kg mol^{-1} PE sample, and $N = 2857$ for the $M_w = 200$ kg mol^{-1} PEP sample]. There is only one parameter left that is not known: the diameter of the virtual tube d, representing the topological constraints due to entanglements in long-chain polymer systems. In Eq. (1.69) d was defined as the end-to-end distance of a chain with N_e segments, where N_e is the number of segments between two entanglements.

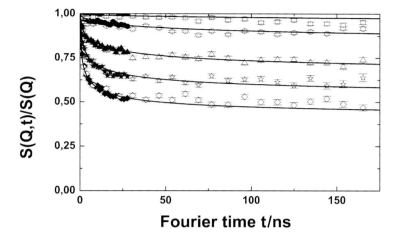

Fig. 1.25 Plot of $S(Q,t)/S(Q)$ from a $M_\mathrm{w} = 190$ kg mol^{-1} PE melt at $T = 509$ K for the scattering wavevectors $Q = 0.3, 0.5, 0.77, 0.96$, and 1.15 nm^{-1}, from above. The data were taken with a wavelength of the incoming neutrons of $\lambda = 0.8$ nm (filled symbols) and $\lambda = 1.5$ nm (open symbols). Solid lines represent a fit with Eq. (1.79).

Fig. 1.24 shows the NSE data for PEP with $M_\mathrm{w} = 200$ kg mol^{-1} at $T = 492$ K (about 10% H-PEP in D-PEP). The data were taken at the IN15 spectrometer (ILL, Grenoble) and corrected for background and resolution. They show a dramatic slowing down of the relaxation after an initial decay. The curves end up in a plateau for all Q-values at high Fourier times, reflecting the topological constraints. The solid lines in Fig. 1.24 represent a fit with Eq. (1.79). As explained above, the only free parameter to fit all Q-values simultaneously is the tube diameter d. The description of the data is excellent over the whole range of Fourier times and Q-values. The resulting diameter is $d = 6$ nm.

Fig. 1.25 shows the results of a comparable measurement for H-PE in D-PE with a molecular weight $M_\mathrm{w} = 190$ kg mol^{-1} at $T = 509$ K. Again, the data show the characteristic plateaus at long times for all Q-values, and again the description by the reptation model is excellent having in mind that only one parameter can be varied. The tube diameter for this PE sample is $d = 4.8$ nm. What do we learn from these two experiments?

- The reptation model (Eqs. 1.79 to 1.82) yields an excellent description of the PE and PEP data at high molecular weights over the entire range of Fourier times and scattering vectors. For the fit, only one parameter is free, the tube diameter d, for which we obtain reasonable values ($d = 6$ nm for PEP and $d = 4.8$ nm for PE).
- Calculating the disentanglement time $\tau_\mathrm{d} = 3N^3 l^2/(\pi^2 W d^2)$, we obtain $\tau_\mathrm{d} = 0.02$ s (with $W = 242$ ns^{-1}) for PE and $\tau_\mathrm{d} = 0.01$ s (with $W = 10.9$ ns^{-1}) for PEP. Since the disentanglement time depends on the chain

length to the third power ($\tau_d \propto N^3$) in the accessible NSE time range (some hundreds of nanoseconds), the tube escape term (with a time constant in the range of 10 ms) is virtually constant for the high-molecular-weight samples. In conclusion, the local reptation process dominates the polymer dynamics in the NSE time range for long-chain polymer systems.

- Assuming that S_{esc} is still time-independent in the time window of NSE and that S_{locrep} has essentially decayed to zero in the plateau region, we find from Eq. (1.79) that $S(Q,t) \approx F(Q)$. Therefore reading the plateau values permits the direct extraction of the tube diameter via $\exp(-Q^2 d^2 / 36)$.

- The initial decay is dominated by Rouse dynamics. We have seen from the incoherent experiments that the segments undergo free Rouse relaxations as long as the mean square displacement is smaller than the tube diameter squared. The Rouse model is not included in Eqs. (1.79) to (1.82)! However, if the fit is restricted to times $t > \tau_e$, the results remain the same. The agreement at times $t < \tau_e$ is somewhat accidental.

What about the slow creep process? As mentioned above, this kind of motion is too slow for NSE, but easily detectable by rheology. Fig. 1.26 shows the loss modulus $G''(\omega)$ versus frequency in a double logarithmic plot. Following Eq. (1.84) for the creep process $G''_{rep}(\omega) \propto \omega^{-1/2}$ is expected and clearly observed at frequencies higher than the inverse disentanglement time τ_d.

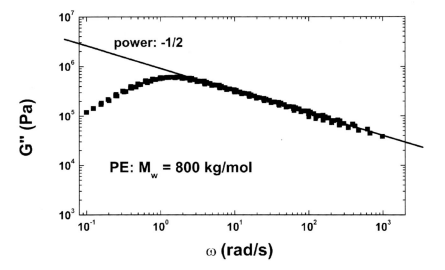

Fig. 1.26 Loss modulus $G''(\omega)$ of polyethylene with $M_w = 800$ kg mol^{-1} (reference temperature $T = 509$ K) measured at the rheometer in Jülich. The solid line illustrates the expectation for the creep process at frequencies higher than the peak frequency ω_d (Eq. 1.84).

Is Everything Fine?
We have shown above that the dynamics in polymer melts is determined by a balance of viscous and entropic forces. A thermally activated fluctuation may be envisaged as a slightly stretched chain. The entropy tries to bend the chain back but has to battle against friction. This dynamics is described in the Rouse model, representing the environment of a test chain by a heat bath that causes thermal activation.

Another approach to the chain dynamics is the mean square displacement Φ_n of a chain segment. Due to the fact that one test segment n is linked to other segments, Φ_n can be described by a sub-Fickian diffusion; the respective time dependence is predicted in the framework of the Rouse model to be proportional to $t^{1/2}$. Both the predicted relaxational structure factor of a labeled test chain as well as the segment mean square displacement have been corroborated by NSE experiments: the former by analyzing the single-chain dynamic structure factor obtained by measuring a few labeled chains in a deuterated matrix for different Q-values, and the latter by an incoherent NSE experiment on fully protonated polymers addressing the self-correlation function. As long as the Q-value is not too high, i.e. we do not observe local details of a segment, and as long as the chains are short, so that they do not entangle (or the time is short enough that they do not know that they are entangled), the Rouse model gives an excellent description of the short-time dynamics in polymer melts.

When the chains become longer, topological chain–chain interactions come into play. We have shown above that the concept of a virtual tube representing the entanglements is very successful in describing the dynamics of long-chain systems. This holds again for the single-chain dynamic structure factor, showing characteristic plateaus at long times where further relaxation is hindered due to topological constraints. Here the reptation model gives a perfect description of the structure factors over the entire range of times and Q-values and allows us to extract the diameter of the virtual tube as the only free parameter of the model. This also holds for the segmental mean square displacement, where the tube concept predicts a transition from a power law $\propto t^{1/2}$ to $\propto t^{1/4}$ at a time constant τ_e. This has indeed been observed for polymers in incoherent NSE experiments addressing the self-correlation function. After a proper evaluation and taking into account non-Gaussian effects in the local reptation regime, the extracted tube diameter is in quantitative agreement with the results from the single-chain dynamic structure factor.

Finally, though not observable in NSE, it has been shown that the slow creep process is traceable in high-molecular-weight polymer melts by analyzing the loss modulus $G''(\omega)$.

This would be the last page of this chapter about polymer dynamics if there were not some observations that are difficult to understand just adopting

Rouse and reptation theory. A few of the many known examples may be listed here:

- The viscosity η in the reptation regime is expected to be proportional to the longest relaxation time. In the reptation concept, this time is the disentanglement time $\tau_d \propto M^3$. Therefore, $\eta \propto M^3$ is anticipated. It has been known for a long time that the $\eta \propto M^3$ dependence is reached only at very high molecular weights, but there exists an intermediate range of molecular weights, where, though the polymer melt is in the well-entangled regime, a power law $\eta \propto M^{3+\alpha}$ with $\alpha \approx 0.4$ is found. Furthermore, it is known from rheological data that the same holds for the disentanglement time τ_d, which in polyisoprene has recently been found to be $\tau_d \propto M^{3.32}$ with a transition to an M^3 dependence at very high molecular weights (Abdel-Goad et al. 2004).

- As derived in Section 1.5, we can write $\Phi_D(\tau_d) = 6D\tau_d \propto Nl^2$ and so, with $\tau_d \propto N^3$ for the diffusion coefficient, $D \propto 1/N^2$. Therefore, for long-chain polymers we expect $D \propto M^{-2}$ following the reptation model. In polybutadiene, Lodge (1999) found $D \propto M^{-2.3}$.

- Finally, and in contrast to what has been presented in Fig. 1.26 for high-molecular-weight polyethylene, in polyisoprene (PI), for example, it has been found that the loss modulus exhibits a -0.25 power-law dependence at frequencies $\omega > \omega_d$ (Abdel-Goad et al. 2004), indicating the existence of additional relaxation processes (McLeish 2002; Likhtman and McLeish 2002). This is illustrated in Fig. 1.27. Note that the total molecular weight of this polyisoprene sample ($M_w = 1000$ kg mol^{-1}) is higher than that of the PE sample presented in Fig. 1.26. However, the number of entanglements $Z = N/N_e$ is higher for PE ($Z_{PE} \approx 400$) than for PI ($Z_{PI} \approx 170$).

It was proposed some time ago that, in addition to the relaxation processes described in the Rouse and reptation theories, there should exist additional degrees of freedom. The main reason for this is the fact that the polymer chains have a finite length. Starting with a chain that is confined in a tube for time $t = t_0$, one may wonder if at a time $t = t_1$ chain segments close to the "open ends" of the tube are as confined as the center part of the chain. Furthermore, it is obvious that not only the test chain itself relaxes but also all chains in the environment that are building the tube: the tube itself is not a fixed object in time!

The former effect – contour length fluctuations (CLF) – will be explored and discussed in the next section; the latter one – constraint release (CR) – will be taken up after that.

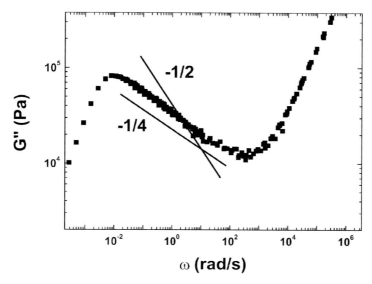

Fig. 1.27 Loss modulus $G''(\omega)$ of polyisoprene with $M_\mathrm{w} = 1000$ kg mol^{-1} at a reference temperature $T = 300$ K measured at the rheometer in Jülich (Abdel-Goad et al. 2004). Figure taken from Wischnewski et al. (2004).

1.7
Limiting Mechanisms for Reptation I: CLF

The finite length of a chain plays an important role due to the existence of chain ends. Their influence is more pronounced in polymer melts of intermediate chain length than in very long-chain systems where the relative weight of the end segments is negligible compared to the total number of segments. We have seen that for PE with $M_\mathrm{w} = 190$ kg mol^{-1} the reptation model gives an excellent description of the single-chain dynamic structure factor. We have also seen that the resulting tube diameter is in agreement with the value extracted from measurements of the segment mean square displacement. Obviously we are in the long-chain limit for the PE system with $M_\mathrm{w} = 190$ kg mol^{-1}. To investigate systematically the chain-length dependence of the chain dynamics, we have measured a series of PE samples with $M_\mathrm{w} = 36$, 24.7, 17.2, 15.2, and 12.4 kg mol^{-1} (Wischnewski et al. 2002).

The experiments were performed at the IN15 spectrometer at the ILL, Grenoble, at $T = 509$ K. The experimental results were corrected for background and resolution. Fig. 1.28 displays the spectra obtained for different molecular weights. The Q-values correspond to: $Q = 0.03$ Å^{-1} (squares), 0.05 Å^{-1} (circles), 0.077 Å^{-1} (triangles up), 0.096 Å^{-1} (diamonds), 0.115 Å^{-1} (triangles down), and 0.15 Å^{-1} (crosses). Filled symbols refer to a wavelength of the incoming neutrons $\lambda = 0.8$ nm, and open symbols to $\lambda = 1.5$ nm.

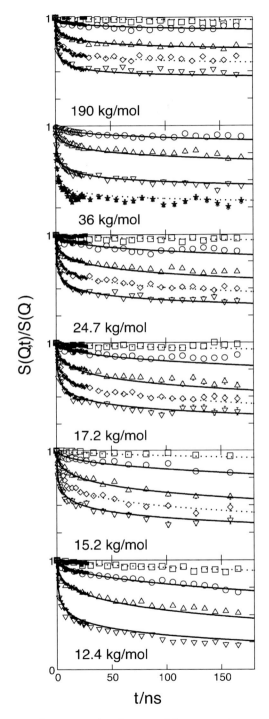

Fig. 1.28 NSE spectra from PE melts of various M_w (see text) (Wischnewski et al. 2002).

For the highest molecular weight $M_w = 190$ kg mol^{-1} and also for $M_w = 36$ kg mol^{-1} the spectra are characterized by an initial fast decay reflecting the unconstrained dynamics at early times followed by local reptation and finally very pronounced plateaus of $S(Q,t)$ at later times signifying the tube constraints. The plateau values for the two samples are practically identical.

Inspecting the results for smaller M_w, we realize that: (1) the dynamic structure factor decays to lower values (at $Q = 1.15$ nm^{-1} the value is ≈ 0.5 for $M_w = 190$ kg mol^{-1}, ≈ 0.4 for $M_w = 24.7$ kg mol^{-1}, and nearly 0.2 for $M_w = 12.4$ kg mol^{-1}); and (2) the long-time plateaus start to slope more and more the smaller M_w becomes, with $M_w = 12.4$ kg mol^{-1} nearly losing the two-relaxation-step character of $S(Q,t)$. Obviously the chain is disentangling from the tube and constraints are successively removed.

Without any theory we can conclude that the spectra change depending on the molecular weight in the sense that the decay of $S(Q,t)$ is stronger and the plateaus are not that pronounced for lower molecular weights. Thereby we have to keep in mind that even the lowest-molecular-weight chain with $M_w = 12.4$ kg mol^{-1} and $N = 849$ still exhibits $N/N_e \approx 6$ entanglements (following Eq. 1.69, $N_e = d^2/l^2 = 4.8^2$ nm$^2/0.17$ nm$^2 = 136$).

Now we will analyze the data quantitatively using the reptation model (Eqs. 1.79 to 1.82). This is demonstrated in Fig. 1.29, where the solid lines show the result for the lowest ($M_w = 12.4$ kg mol^{-1}) and the highest ($M_w = 190$ kg mol^{-1}) molecular weight. The tube diameter is again the only free parameter. For $M_w = 12.4$ kg mol^{-1}, the tube diameter is significantly larger than for the long-chain melt (see Fig. 1.29, $d(M_w=12.4$ kg mol$^{-1}) = 6.0$ nm, $d(M_w=190$ kg mol$^{-1}) = 4.8$ nm), reflecting a reduction of the topological constraints. The dotted lines in the lower part of Fig. 1.29 result from the assumption of a constant $d = 4.8$ nm inserted in the reptation model; it is obvious that this significantly underestimates the amount of relaxation in the low-molecular-weight sample. The data for all molecular weights were evaluated by fitting them with the reptation model. All the parameters were kept constant except for the tube diameter. Table 1.1 shows the resulting d.

Tab. 1.1 Tube diameters for different molecular-weight PE melts as obtained by a fit with the reptation model.

M_w (kg mol^{-1})	d (nm)
190	4.8
36	4.6
24.7	5.4
17.2	5.3
15.2	5.4
12.4	6.0

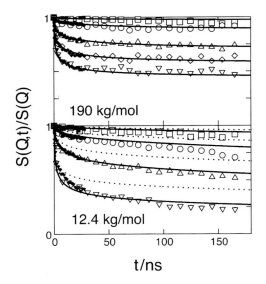

Fig. 1.29 PE data for $M_w = 190$ and 12.4 kg mol^{-1} (Wischnewski et al. 2002). Solid lines represent a fit with the reptation model yielding $d(M_w{=}190\,\text{kg mol}^{-1}) = 4.8$ nm and $d(M_w{=}12.4\,\text{kg mol}^{-1}) = 6.0$ nm. For the dotted lines, the tube diameter was fixed to 4.8 nm.

The significant increase of the tube diameter with decreasing molecular-weight samples is puzzling. There is no doubt that the reptation picture requires a constant tube diameter as long as the chains are in the well-entangled regime, which is the case for our samples. It is evident, in particular considering the worse fit quality at low molecular weights, that the relaxation is determined not only by reptation but also by an additional process that leads to an *apparently* increasing tube diameter with decreasing molecular weight. As long as we stick to the assumption of a fixed tube, it is also evident that, since for the inner segments there is no difference for a long or short chain, this has to be related to the different relative weight of the chain ends with respect to the total length of the chain.

It is known from broad crossover phenomena, like the molecular-weight dependence of the melt viscosity, that limiting mechanisms exist which affect the confinement and thereby limit the reptation process (Doi and Edwards 1986). Contour length fluctuations (CLF) were proposed as one candidate for such processes. The idea is quite simple. The chain ends undergo the same Rouse-like fluctuations as the inner segments. For an end segment, this leads to the effect that at a given time it may be immersed a bit into the tube. When the segment comes out again, it may have a different direction, not aligned to the original tube curvature. In other words, when one chain end is immersed in the tube, the part of the tube that is not occupied is lost, so the effective

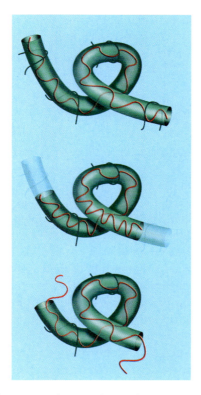

Fig. 1.30 Schematic illustration of contour length fluctuations. Rouse-like motions of end segments lead to the effect that, at a given time, the chain ends are immersed into the tube (central picture). The part of the tube that is not occupied is lost. When they come out again, they may have a different direction, not aligned to the original tube curvature.

tube becomes shorter. The fluctuating chain ends lead to a *time-dependent effective tube length*, i.e. to decreasing constraints with increasing time. This is illustrated in Fig. 1.30.

If we recall the creep process, it is evident that the two relaxation processes are somewhat competitive. Both describe the escape of the chain from the tube confinement: one by diffusion along the tube profile, and the other by destruction of the tube starting from both ends. For times shorter than the Rouse time, which following Eqs. (1.48) and (1.58) is given by $\tau_R = N^2/(\pi^2 W)$, it was recently shown that the effect of reptation on escape from the tube is negligible in comparison to tube length fluctuations (Likhtman and McLeish 2002). This time range corresponds to the times that are accessed in our experiment. Even for the shortest chain $M_w = 12.4$ kg mol^{-1} we get $\tau_R \approx 300$ ns, which is more than the maximum time reached by the high-resolution IN15 NSE spectrometer. This means that for all the NSE experiments we have to consider CLF rather than the creeping process. In this regime the fraction of monomers released from the tube due to contour length fluctuations has a very simple form:

$$\Psi(t) = \frac{1.5}{Z}\left(\frac{t}{\tau_e}\right)^{1/4} \quad (1.85)$$

where $Z = N/N_e$ is again the number of entanglements per chain.

It is straightforward to show that the scaling with t and Z in Eq. (1.85) emerges exactly from the properties of a linear Rouse chain in a tube. In fact, all the approaches to contour length fluctuations so far (Doi and Edwards 1986; des Cloizeaux 1990) predict for $t < \tau_R$ a simple expression for $\Psi(t)$ like Eq. (1.85) but with different prefactors ($\approx$1.2 by Doi and Edwards, and 0.67 by des Cloizeaux). The discrepancy is explained by the different mathematical approximations used. A series of carefully designed one-dimensional stochastic simulations solved the first passage problem for the single Rouse chain in a tube, by taking the limit of the time step tending to zero and the number of monomers going to infinity. These simulations confirmed Eq. (1.85) and gave a result for the prefactor of 1.5 ± 0.02 (Likhtman and McLeish 2002). Without discussing the details, we point to the fact that taking into account the contour length fluctuations for the calculation of the mechanical relaxation functions reproduces the -0.25 power-law dependence in the loss modulus as presented in Fig. 1.27 for chains with not too high numbers of entanglements Z (Likhtman and McLeish 2002).

To incorporate this result in the structure factor calculations, the approximate approach of Clarke and McLeish (1993) was used. We assume that after time t all monomers between 0 and $s(t)$ and between $1 - s(t)$ and 1 have escaped from the tube, where $s(t) = \Psi(t)/2$. Note that here $s(t)$ is a non-dimensional variable between 0 and 0.5, while $s(t)L$ is a section of the contour length following the profile of the primitive path.

This statement contains two approximations: first it assumes that the fraction of the chain that escapes from the tube is the same from each end of the chain, and second it ignores distributions of $s(t)$, replacing it by the single average value. If we make these two assumptions, the rest of the calculation is straightforward. We first use the fact that $R(s,t)$ is a Gaussian variable and therefore

$$\langle \exp[iQ(R(s,t) - R(s',0))] \rangle$$
$$= \exp\left[-\tfrac{1}{2}\sum_{\alpha=x,y,z} Q_\alpha^2 \langle [R_\alpha(s,t) - R_\alpha(s',0)]^2 \rangle\right] \quad (1.86)$$

and then we note that

$$\langle [R_\alpha(s,t) - R_\alpha(s',0)]^2 \rangle$$
$$= \frac{dL}{3} \begin{cases} |s-s'| & \text{for } s(t) < s < 1-s(t) \text{ or } s(t) < s' < 1-s(t) \\ |2s(t)-s-s'| & \text{for } s < s(t) \text{ and } s' < s(t) \\ |2-s-s'-2s(t)| & \text{for } s > 1-s(t) \text{ and } s' > 1-s(t) \end{cases} \quad (1.87)$$

where $L = Zd$ is the contour length. Replacing the summation in the dynamic structure factor (Eq. 1.11) in Section 1.3.1 by integration, we get

$$S_{\text{esc}}(Q,t) = \int_0^1 ds \int_0^1 ds' \exp(-2\mu|s-s'|)$$
$$+ 2\int_0^{s(t)} ds \int_0^{s(t)} ds' \{\exp[-2\mu(2s(t)-s-s')]$$
$$- \exp(-2\mu|s-s'|)\} \quad (1.88)$$

where again $\mu = Q^2 N l^2 / 12$. The integrals can be easily evaluated and we obtain

$$S_{\text{esc}}(Q,t) = \frac{N}{2\mu^2}[2\mu + e^{-2\mu} + 2 - 4\mu s(t) - 4e^{-2\mu s(t)} + e^{-4\mu s(t)}] \quad (1.89)$$

The lines in Fig. 1.28 were obtained by fitting the data with Eqs. (1.79) to (1.81) and adding instead of the creep term (Eq. 1.82) the above-derived expression for CLF (Eq. 1.89). A group of three common Q-values for all molecular weights is displayed by solid lines to facilitate an easy comparison of the molecular-weight dependence of the curves. Additional Q-values only available for some M_w values are represented by dotted lines.

If we compare the experimental spectra with the model prediction, we generally find good agreement. The gradually increasing decay of $S(Q,t)$ with decreasing M_w is described very well with respect to both the magnitude of the effect and the shape of $S(Q,t)$. We further note that, in particular for smaller M_w, the weighted error between fit and data is significantly smaller compared to the fit with the pure reptation model.

Fig. 1.31 compares the tube diameters as obtained from the fit with the reptation model (Table 1.1) and obtained by replacing the creep term by the

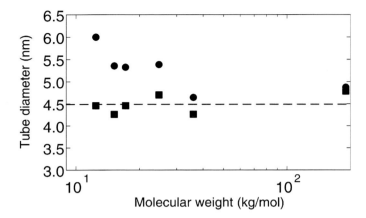

Fig. 1.31 Tube diameters for PE samples with different molecular weights: from pure reptation fit (circles); and when the creep term is replaced by the expression for contour length fluctuations (squares).

concept of contour length fluctuations. At the highest molecular weight, contour length fluctuations are insignificant and both lines of fit yield the same d. At $M_w = 36$ kg mol^{-1}, a slight difference appears, which increases strongly with decreasing length. At $M_w = 12.4$ kg mol^{-1}, the difference in the fitted tube diameters between the two approaches rises to nearly 50%, emphasizing the strong effect of the contour length fluctuation in loosening the grip of the entanglements on a given chain.

However, the main point is that, in contrast to the significant increase of d with decreasing M_w if pure reptation is assumed, if CLF is included the tube diameter stays constant. It is important to note that CLF has been included without introducing any new parameter. The data are described over the entire range of time, Q-values, and molecular weights with one single parameter, the tube diameter d.

Thus, the comparison between the experimental chain-length-dependent dynamic structure factor and theoretical predictions clearly shows that, in the time regime $t \leq \tau_R$, contour length fluctuations are the leading mechanism that limits the chain confinement inherent to the reptation picture. Even for chain lengths corresponding to only ≈6 entanglements, the tube diameter appears to be a well-defined quantity, assuming the same value as for asymptotically long chains. The confinement is lifted from the chain ends inwards, while the chain center remains confined in the original tube.

To conclude this section we underline the consequences of CLF for the macroscopic properties of polymer melts.

1. Since the structure factor $S_{\text{esc}}(Q,t)$ is directly related to the tube survival probability function $\mu_{\text{rep}}(t)$ (Eq. 1.83) and therefore to the relaxation function, it is evident that the modifications of $S_{\text{esc}}(Q,t)$ as described in this section are reflected in a different ω dependence of $G''(\omega)$. It has been shown that CLF introduces an $\omega^{-1/4}$ regime into the spectrum of the loss modulus. In fact, this power law has already been illustrated in Fig. 1.27 for polyisoprene.

2. The power law for chain diffusion, which has been found to deviate from the prediction of the reptation model, as well as the observed exponent of $3+\alpha$ with $\alpha \approx 0.4$ for the molecular-weight dependence of the viscosity are attributed to the CLF (Doi and Edwards 1986; McLeish 2002).

It is the virtue of the NSE experiments that they provided the first experimental proof of the CLF mechanism quantitatively and on a microscopic scale in space and time.

1.8
Limiting Mechanisms for Reptation II: CR

Up to now we have made a crude approximation by assuming that the topological confinement represented by a virtual tube is constant in time. The basic message of the tube concept is that the confinement that a test chain (e.g. a labeled one) experiences can be represented by a tube built by the adjacent chains. The choice of the test chain is arbitrary, and we assume that the dynamics of all chains in the system is the same (at least as long as we consider one-component systems with a narrow distribution of molecular weights). Having this in mind, it is evident that the tube itself is in motion and the respective constraints possess a finite lifetime. For the effect of contour length fluctuations, which leads to a time-dependent effective tube length, it was sufficient to account for the dynamics of the test chain, because CLF is an escape mechanism of one single chain. There is no need to move away from the single-chain picture and, as we have seen above, there is not even a need to introduce any new parameter. In contrast to CLF, the fact that constraints for the test chain are time-dependent is still true if we freeze the test chain: after some time, reptation or CLF of all the other chains would lead to an effectively free test chain. Here, we have left the single-chain picture and arrived at the more complicated many-chain problem.

In the past decades quite a number of concepts have been developed to account for this additional relaxation process, called "constraint release" (CR). One idea was introduced by Graessley and Struglinski (1986). The lifetime of each constraint building a tube is of the order of the disentanglement time τ_{d}.

If a constraint is released, the tube can move locally. This leads to a Rouse-like motion of the tube itself. In the calculation of the stress relaxation function $G(t)$, this Rouse relaxation of the tube can be accounted for by an additional relaxation function of the Rouse type with time constant τ_d. It turns out, on calculating the reptation part and this additional factor, that the relaxed stress is of the same order of magnitude (in the limit of very long chains). This led to the idea to account for CR by introducing the reptation part squared in the relaxation function. The concept was denoted as "double reptation". In the double reptation concept only one relaxation time was used to describe the constraint release (CR) effect, an oversimplification that has to be adjusted by replacing τ_d by a distribution of relaxation times representing the lifetimes of constraints. It is evident that it makes a difference if the constraint for a test chain is built by the end or the middle part of an adjacent chain.

A self-consistent theory that takes into account the distribution of relaxation times or mobilities for different chain segments has been introduced by Rubinstein and Colby (1988). Likhtman and McLeish (2002) used this formalism to simulate the contribution of constraint release, $R(t)$, to the stress relaxation function. One additional factor (c_ν) to adjust the strength of CR was introduced. Their calculation shows that for times $t < \tau_\text{R}$ the relaxation by CR is similar to the relaxation by CLF $[1 - \Psi(t)$ with $\Psi(t)$ as defined in Eq. (1.85)]. In fact, one obtains $R(t)$ by replacing the prefactor in Eq. (1.85) and including c_ν:

$$R(t) = 1 - \frac{1.8}{Z}\left(\frac{c_\nu t}{\tau_\text{e}}\right)^{1/4} \tag{1.90}$$

For times longer than the Rouse time, $R(t>\tau_\text{R})$ is described by the lateral relaxation of the tube, which again is considered as a Rouse chain, but with a significantly longer characteristic relaxation time than τ_R. Altogether, the relaxation is determined by five processes. For short times, Rouse modes inside the tube dominate; and for longer times, local reptation (longitudinal modes with a wavelength longer than the tube diameter d) starts to play a role. The escape of the chain from the tube is dominated either by CLF at times $t < \tau_\text{R}$ or by reptation for $t > \tau_\text{R}$. In the same time regime CR also contributes significantly to the relaxation spectrum. Likhtman and McLeish (2002) achieved a consistent description of experimental rheological data by applying the relaxation function in Eq. (1.90) for the CR mechanism.

Fig. 1.32 shows the calculated effect of all relaxation processes on the loss modulus $G''(\omega)$. The dotted line shows the prediction for pure reptation and Rouse modes. Here, a power law of $-1/2$ in the reptation regime (see Eq. 1.84) and a power of $+1/2$ in the high-frequency Rouse regime are predicted. Taking into account the relaxation by CLF results in the dashed line. Note the change to a $\propto \omega^{-1/4}$ behavior at frequencies above the pure reptation regime, i.e. for

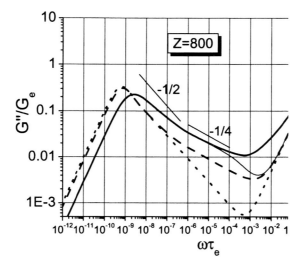

Fig. 1.32 Calculated contributions to $G''(\omega)/G_e$ for a polymer with $Z = 800$ entanglements (Likhtman and McLeish 2002). G_e is the plateau modulus. Lines: pure reptation and Rouse modes (dotted line); adding relaxation by CLF (dashed line); adding relaxation due to CR effect (thin solid line); calculation with all contributions (thick solid line) (see text for further explanation).
Figure courtesy of A. E. Likhtman and T. C. B. McLeish.

$\omega > 1/\tau_R$. Adding the relaxation due to the CR effect yields the thin solid line. Finally, the thick solid line illustrates the result of a calculation with all contributions to the relaxation function, including the longitudinal modes.

It may be anticipated at this point that a description of CR on a microscopic scale, i.e. the contribution of CR to the dynamic structure factor as measured by means of NSE, is not available. It is doubtful if this "many-chain" problem can be accounted for in the framework of the quite simple "single-chain" and tube concept. However, is there any chance of observing the CR relaxation on a microscopic scale? Yes, that can be done! We study by NSE a labeled chain in matrices with different chain lengths. The test chain is always the same, but the adjacent chains, which build the constraints of the test chain, have different lengths, i.e. different capabilities to build the tube. One could also say the time scale for the tube will vary with the matrix chain length.

What do we expect? A long chain in a matrix of the same molecular weight will show the well-known plateaus at high Fourier times, as the dynamics is dominated by the local reptation process in the time window of NSE. With decreasing molecular weight of the matrix chains, we expect a faster relaxation of the test chain because the shorter chains build topological constraints with shorter lifetimes. Finally, for very short matrix chains we expect pure Rouse behavior for the long test chain since the adjacent chains cannot build any tube.

80 | *1 Polymer Dynamics in Melts*

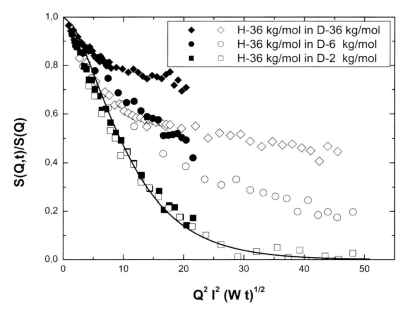

Fig. 1.33 Plot of $S(Q,t)/S(Q)$ measured by NSE (IN15, Grenoble) in a Rouse scaling plot for a protonated test chain with $M_w = 36$ kg mol^{-1} in a deuterated matrix of 36, 6, and 2 kg mol^{-1} each for two Q-values: 0.77 nm^{-1} (filled symbols), and 1.15 nm^{-1} (open symbols). The solid line represents a fit with the Rouse model for the sample with a 2 kg mol^{-1} matrix (Zamponi 2004).

Fig. 1.33 shows the result of such an experiment. It depicts the NSE data of three samples. The test chain is a protonated PE chain with $M_w = 36$ kg mol^{-1}. The matrix is varied between $M_w = 36$ and 2 kg mol^{-1}. The dynamic structure factor is represented in a Rouse scaling as introduced in Section 1.5.1. In this representation, Rouse curves fall on one master curve for different Q-values. For clarity, only two different Q-values are shown for each sample: $Q = 0.77$ nm^{-1} (filled symbols) and 1.15 nm^{-1} (open symbols). For the 2 kg mol^{-1} matrix the dynamic structure factors merge onto one master curve for the two different Q-values. This holds also for other Q-values that are not shown. The dynamic structure factor is very close to the Rouse expectation, though the friction coefficient (the solid line represents a fit with the Rouse model) is higher than expected for a monodisperse 2 kg mol^{-1} sample by a factor of about 2. Note that in this experiment a long chain with a molecular weight of 36 kg mol^{-1} behaves in a Rouse-like way because the short matrix chains are not able to build a tube! The two samples with longer matrix chains show the typical splitting for the two presented Q-values. Furthermore, it is obvious that the relaxation is significantly faster for the sample with the 6 kg mol^{-1} matrix compared to the monodisperse system of protonated 36 kg mol^{-1} chains in a deuterated 36 kg mol^{-1} matrix due to the onset of the CR effect.

The NSE data show clearly the effect of CR. However, we abstain from the introduction of theoretical concepts that could describe the data by means of an oversimplified phenomenological "ad hoc" assumption. It is possible, for instance, to get a satisfactory description of the NSE data by multiplying a tube – Rouse – structure factor with the known reptation structure factor. This adds at the same time a free parameter to the theory, the Rouse time of the tube. However, the significance of such a model is not very strong. Additional degrees of freedom may always be accounted for by some kind of Rouse function, if an additional parameter to be varied is provided. We may at this point stay with the experimental observation of CR, and with the fact that a quantitative description is lacking.

It is probably not a coincidence that, with the description of these obvious limits of the quite simple tube model and its modifications to account for higher-order relaxation processes, we have also reached the end of this chapter about polymer dynamics. A short summary and outlook is given in the next and last section.

1.9
Summary and Outlook

In the previous sections we have tried to develop a consistent picture of the mesoscopic chain dynamics in polymer melts. The systems were always limited to linear homopolymers, the simplest architecture and the simplest chemical composition in this class of soft matter systems.

We started with the dynamics at very short times and demonstrated that in this time regime the segments do not know if they are part of a long entangled system or of a short chain system. This leads to the fact that the dynamics of these two extreme cases cannot be discriminated at very short times. The Rouse model gives a satisfactory description of the segmental motion in this time regime. It is based on the assumption that fluctuations of the observed test chain are activated by a heat bath and that the relaxation of these fluctuations is balanced by viscous forces due to friction and entropic forces. In addition to the internal modes, the chain can undergo a diffusion which is sub-Fickian due to the correlation of different segments within a chain. If the static parameters of the chain are known (segment length, molecular weight), the Rouse model needs only one parameter, namely the effective segmental friction coefficient, to describe the single-chain dynamic structure factor over a wide range of Q-values and Fourier times. It also predicts the time dependence of the segmental mean square displacement, which also has been observed experimentally. If the Q-values become too large, so that the system's specific local properties come into play, the Rouse model starts to fail.

We have shown that this dynamic behavior continues to longer times if there are no constraints present in the system, i.e. the chains are short and do not entangle. If they are longer, the Rouse model starts to fail at longer times – longer than τ_e – when the mean square displacement of the segments reaches a special value. In the famous tube concept, this length scale is identified with the diameter of a virtual tube representing the topological constraints in the system. It has been illustrated that in the limit of very long polymer chains the tube model gives a perfect description of experimental data over the entire range of Q-values and Fourier times accessible by coherent NSE. The only parameter that has to be added to the Rouse parameters is the additional length scale in terms of the tube diameter. It has also been demonstrated by measuring the self-correlation function that the time dependence of the segmental mean square displacement changes due to the topological confinement. The observed behavior is in perfect agreement with the prediction of the reptation model if non-Gaussian behavior is taken into account.

In the intermediate regime of molecular weights, the chains are long enough to build entanglements but not so long that the tube concept gives a satisfactory description of the experimental data. At times longer than τ_e the single-chain dynamic structure factor does not show a fully developed plateau, but a slope that becomes more pronounced with decreasing molecular weight. This points to additional relaxation processes, which may be called secondary in the sense that they are additional to the main relaxation processes accounted for in the tube concept. The observation that they become more pronounced with decreasing chain length points to the fact that they have their origin in the finite length of the chains. Fluctuating chain ends can escape from the confinement by destroying the tube from both ends, a mechanism that is much more important than the creep process for intermediate chain length in the time regime of NSE. This CLF process explains the experimental observation that the plateaus at high Fourier times start to slope if the molecular weight decreases. The CLF process can be calculated and integrated into the reptation dynamic structure factor, giving an excellent description of experimental data over a wide range of Fourier times and Q-values (like before) and additionally for a wide range of molecular weights. Since the underlying process that causes finally CLF is still local reptation, there is no need to introduce new parameters. The tube diameter is still the only variable and it stays constant as expected within the reptation concept for all measured molecular weights if CLF is taken into account.

Finally, the fact that the tube is build by chains that have the same dynamics as the chain that is arrested in the tube leads to a finite lifetime of the tube itself, and thereby to an additional "escape" mechanism, not by the test chain itself, but by a release of the confinement with increasing time. This CR effect can be accounted for in the calculation of the stress relaxation function as measured by rheology. However, though clearly observed in NSE experiments, a quantitative description of NSE data by a modified $S(Q,t)$ is still not available.

As mentioned in the introduction, the goal of polymer science is finally to clarify the relation between macroscopic properties and the architecture and composition of a polymer system. To reach this goal, all relaxation processes that are relevant for the macroscopic behavior have to be accessed, and the interaction between them has to be revealed. This could enable the production of "made-to-measure" polymer systems, i.e. with specific macroscopic properties by selectively manipulating the architecture and composition of the system.

It is evident that the experiments shown here are no more than a first step. A well-disposed reader may agree that the results presented give a more or less consistent picture of the dynamical behavior in polymer melts, though the quantitative provision for CR is not well developed. It is evident that the observed facts may serve as a basis for the understanding of more complicated architectures. The retraction mechanism of an arm that is part of a star polymer or of an H-shaped polymer may not be too different from the CLF mechanism. (Just regard the linear chain as a two-arm star). After the four arms of an H-polymer have retracted completely into the tube of the backbone, the creep of the polymer may be described with the same structure factor as the creep process of a linear chain, but with an unequal larger friction coefficient due to the relaxed arms, which serve as a kind of barb.

However, the dynamics of complicated architectures is more manifold compared to linear chains, the number of processes may be larger, and the interaction between them more complicated. Going one step further by mixing different architectures or mixing systems with different chemical structures results in a mixture of all these relaxation processes. The superposition and interaction of these relaxations follow rules that are up to now widely unknown. How does friction arise in a chemically heterogeneous environment? How can topological confinement of a chain be described if the tube is built by a blend that consists of two components with significantly different tube diameters?

Nevertheless, irrespective of how complicated a system may be, the opportunity remains to label small fragments of one component of such a system and thereby to manipulate the visibility of different processes for neutrons. In combination with the continuously growing opportunities in neutron scattering, chemistry, computing, and complementary experimental techniques like NMR, dielectric spectroscopy or rheology, this may lead to significant progress in this field of research within the coming years.

References

Abdel-Goad, M., Pyckhout-Hintzen, W., Kahle, S., Allgaier, J., Richter, D., and Fetters, L. J., 2004, *Macromolecules* **37**, 8135.
Blanchard, A., 2004, *PhD thesis*, University of Münster.
Boothroyd, A. T., Rennie, A. R., and Boothroyd, C. B., 1991, *Europhys. Lett.* **15**(7), 715.
Clarke, N. and McLeish, T., 1993, *Macromolecules* **26**, 5264.
de Gennes, P.-G., 1981, *J. Phys. (Paris)* **42**, 735.
des Cloizeaux, J., 1990, *Macromolecules* **23**, 4678.
Doi, M. and Edwards, S. F., 1986, *The Theory of Polymer Dynamics*. Clarendon Press, Oxford.
Fatkullin, N. and Kimmich, R., 1995, *Phys. Rev. E* **52**(3), 3273.
Ferry, J., 1970, *Viscoelastic Properties of Polymers*. John Wiley & Sons, New York.
Fischer, E., Kimmich, R., Beginn, U., Moller, M., and Fatkullin, N., 1999, *Phys. Rev. E* **59**(4), 4079.
Flory, P., 1953, *Principles of Polymer Chemistry*. Cornell University Press, Ithaca, NY.
Gotro, J. T. and Graessley, W. W., 1984, *Macromolecules* **17**, 2767.
Graessley, W. W. and Struglinski, M. J., 1986, *Macromolecules* **19**, 1754.
Hsieh, H. L. and Quirk, R. P., 1996, *Anionic Polymerization: Principles and Practical Applications*. Marcel Dekker, New York.
Kamigaito, M., Ando, T., and Sawamoto, M., 2001, *Chem. Rev.* **101**, 3689.
Kirste, R., Kruse, W., and Schelten, J., 1973, *Makromol. Chem.* **162**, 299.
Likhtman, A. E. and McLeish, T. C. B., 2002, *Macromolecules* **35**, 6332.
Lodge, T. P., 1999, *Phys. Rev. Lett.* **83**, 3218.
Marshall, W. and Lovesey, S. W., 1971, *Theory of Thermal Neutron Scattering*, p. 65ff. Clarendon Press, Oxford.
Matyjaszewski, K. and Sawamoto, M., 1996, Controlled/living carbocationic polymerization, chap. 4 in *Cationic Polymerizations: Mechanisms, Synthesis, and Applications*, ed. K. Matyjaszewski. Marcel Dekker, New York.
McLeish, T. C. B., 2002, *Adv. Phys.* **51**, 1379.
Mezei, F., 1980, in *Neutron Spin Echo Proceedings*, Vol. 128, ed. F. Mezei, p. 3. Springer, Berlin.
Milner, S. and McLeish, T., 1998, *Phys. Rev. Lett.* **81**, 725.

Monkenbusch, M., Schaetzler, R., and Richter, D., 1997, *Nucl. Instrum. Meth. Phys. Res. A* **399**(2–3), 301.

Monkenbusch, M., Ohl, M., Richter, D., Pappas, C., Zsigmond, G., Lieutenant, K., and Mezei, F., 2005, *J. Neutron Res.* **13**, 63.

Montes, H., Monkenbusch, M., Willner, L., Rathgeber, S., Fetters, L., and Richter, D., 1999, *J. Chem. Phys.* **110**, 10188.

Odian, G., 1991, *Principles of Polymerization*, 3rd edn. John Wiley & Sons, New York.

Onogi, S., Masuda, T., and Kitagawa, K., 1970, *Macromolecules* **3**, 109.

Paul, W., Smith, G., Yoon, D., Farago, B., Rathgeber, S., Zirkel, A., Willner, L., and Richter, D., 1998, *Phys. Rev. Lett.* **80**(11), 2346.

Richter, D., Butera, R., Fetters, L., Huang, J., Farago, B., and Ewen, B., 1992, *Macromolecules* **25**, 6156.

Richter, D., Farago, B., Butera, R., Fetters, L., Huang, J., and Ewen, B., 1993, *Macromolecules* **26**, 795.

Richter, D., Willner, L., Zirkel, A., Farago, B., Fetters, L. J., and Huang, J., 1994, *Macromolecules* **27**, 7437.

Richter, D., Monkenbusch, M., Allgeier, J., Arbe, A., Colmenero, J., Farago, B., Bae, Y., and Faust, R., 1999, *J. Chem. Phys.* **111**, 6107.

Rubinstein, M. and Colby, R. H., 1988, *J. Chem. Phys.* **89**, 5291.

Schleger, P., Ehlers, G., Kollmar, A., Alefeld, B., Barthelemy, J., Casalta, H., Farago, B., Giraud, P., Hayes, C., Lartigue, C., Mezei, F., and Richter, D., 1999, *Physica B* **266**, 49.

Squires, G., 1978, *Introduction to the Theory of Thermal Neutron Scattering*. Cambridge University Press, Cambridge.

van Hove, L., 1954, *Phys. Rev.* **95**, 249.

Wischnewski, A., Monkenbusch, M., Willner, L., Richter, D., Likhtman, A., McLeish, T., and Farago, B., 2002, *Phys. Rev. Lett.* **8805**, 058301.

Wischnewski, A., Monkenbusch, M., Willner, L., Richter, D., and Kali, G., 2003, *Phys. Rev. Lett.* **90**, 058302.

Wischnewski, A., Zamponi, M., Monkenbusch, M., Willner, L., Pyckhout-Hintzen, W., Richter, D., Likhtman, A. E., McLeish, T. C. B., and Farago, B., 2004, *Physica B* **350**, 193.

Young, R. N., Quirk, R. P., and Fetters, L. J., 1984, *Adv. Polym. Sci.* **54**, 1.

Zamponi, M., 2004, *PhD thesis*, University of Münster.

2
Self-Consistent Field Theory and Its Applications

Mark W. Matsen

Abstract

Self-consistent field theory (SCFT) has emerged as the "state of the art" for modeling the equilibrium phase behavior of structured polymer melts and concentrated polymer solutions. It is a microscopic theory that treats the polymers with the coarse-grained Gaussian model and performs the statistical mechanics using the mean-field approximation. Both simplifications can be rigorously justified for high-molecular-weight polymers, and indeed the theory is well documented for accurately modeling experimental results to the point of being among the most successful theories in soft condensed matter physics. Furthermore, the combination of coarse-graining and mean-field theory produces a computationally efficient algorithm that is both highly predictive and extraordinarily versatile. This chapter provides a detailed derivation and justification of SCFT at the level appropriate for a graduate student. The theory is then demonstrated by a range of explicit applications involving polymer brushes, polymer blends, and block copolymers. The chapter concludes by discussing emerging strategies for going beyond mean-field theory to incorporate fluctuation corrections.

2.1
Introduction

Polymers refer to a spectacularly diverse class of macromolecules formed by covalently linking together small molecules, *monomers*, to form long chains. Fig. 2.1 sketches an assortment of typical architectures, the simplest of which is the linear homopolymer formed from a single strand of identical monomers. There are also copolymer molecules consisting of two or more chemically dis-

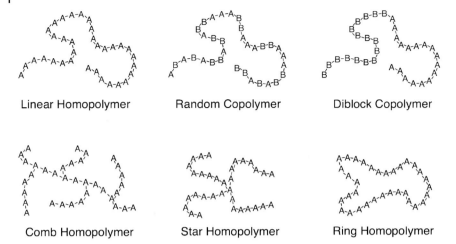

Fig. 2.1 Selection of polymer architectures. Molecules consisting of a single monomer type (i.e. A) are referred to as homopolymer, and those with two or more types (i.e. A and B) are called copolymers.

tinct monomer units, the most common example being the random copolymer, where there is no particular pattern to the monomer sequence. In other cases, the different monomers are grouped together in long intervals to form block copolymer molecules; Fig. 2.1 shows the simplest, a diblock copolymer, formed from one block of A-type units attached to another of B-type units. Although linear chains are the most typical, an almost endless variety of other more elaborate architectures, such as combs, stars, and rings, are possible. The monomers themselves also come in a rich variety, some of the more standard of which are listed in Fig. 2.2. The difference in chemistries can have a profound effect on the molecular interactions and chain flexibility. Some monomers produce rigid rod-like polymers, others produce molecules with liquid-crystalline behavior, a few become ionic in solution, and some are even electrically conducting. The monomer type also has a strong effect on the liquid-to-solid transition temperature, and influences whether the polymeric material solidifies into a glassy or semicrystalline state.

As a result of their economic production coupled with their rich and varied properties, polymers have become a vital class of materials for the industrial sector. In addition, some of their exotic behaviors are now being realized for more sophisticated applications in, for example, the emerging field of nanotechnology. Consequently, research on polymeric macromolecules has flourished over the past several decades, and now stands as one of the dominant areas in soft condensed matter physics. Through the years of experimental and theoretical investigation, a number of elegant theories have been devel-

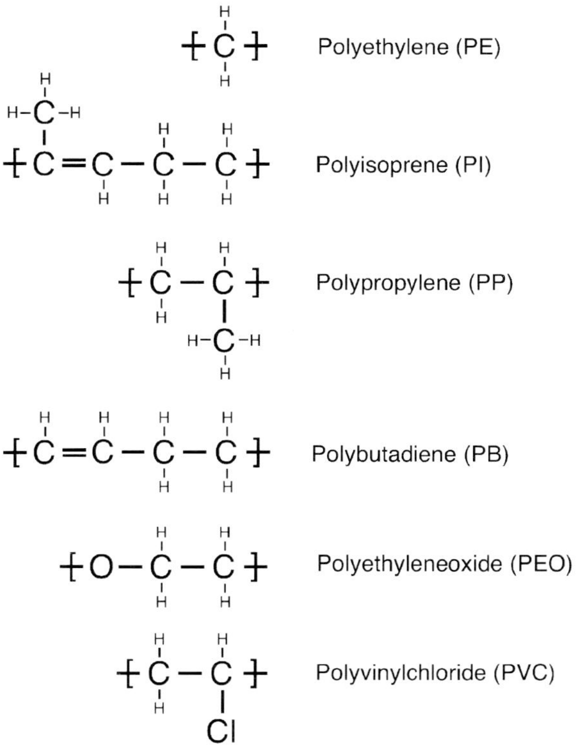

Fig. 2.2 Chemical structures of some common monomers used to form polymer molecules.

oped to describe their behavior, and this chapter is devoted to one of the most successful, namely self-consistent field theory (SCFT) (Edwards 1965).

Considering the difficulty in modeling systems containing relatively simple molecules such as water, one might expect that large polymer molecules would pose an intractable task. However, the opposite is true. The sheer size of these macromolecules (each typically containing something like 10^3 to 10^5 atoms) actually simplifies the problem. As a result of their tremendous molecular weight, the atomic details play a relative small part in their overall behavior, leading to universal properties among the vast array of different polymer types. For example, the characteristic size of a high-molecular-weight polymer in a homogeneous environment scales with the degree of polymerization (i.e. the total number of monomers) to an exponent that is independent of the monomer type. The chemical details of the monomer only affect the proportionality constant. A further advantage of modeling polymers is that their configurations tend to be very open, resulting in a huge degree of interdigitation among the polymers, such that any given molecule is typically in contact with hundreds of others. This has a damping effect on the molecular correlations, which in turn causes mean-field techniques to become highly effective, something that is unfortunately not true of small-molecule systems.

With the advantages favoring mean-field theory, it becomes possible to provide accurate predictions for the equilibrium behavior of polymeric systems. However, polymers are renowned for their slow dynamics and can remain out of equilibrium for long periods of time. In fact, this is effectively a rule for the solid state, where both glassy and semicrystalline polymers become forever trapped in non-equilibrium configurations. This restricts the application of statistical mechanics to the melt (or liquid) state, where the dynamics can be reasonably fast. Even though there are relatively few applications of polymers in their melt state, this is the phase in which materials are processed, and so a thorough understanding of equilibrium melts is paramount.

This chapter provides a basic introduction to SCFT for modeling polymeric melts; for further reading, see Whitmore and Vavasour (1995), Schmid (1998), Fredrickson et al. (2002), and Matsen (2002a). SCFT is a theory with a remarkable track record for versatility and reliability, undoubtedly because of its prudent choice of well-grounded assumptions and approximations. In most applications, the theory employs the simple coarse-grained Gaussian model for polymer chains, and treats their interactions by mean-field theory. Section 2.2 begins by justifying the applicability of the Gaussian model, and then Section 2.3 develops the necessary statistical mechanics for a single chain subjected to an external field, $w(r)$. Following that, a number of useful approximations are introduced for handling special cases such as when $w(r)$ is either weak or strong. This provides the necessary background for discussing SCFT.

As with most theories, the framework of SCFT is best described by demonstrating its application on a representative sample of systems. To this end, we focus on the three examples depicted in Fig. 2.3: polymer brushes, homopolymer interfaces, and block copolymer microstructures. Not only do they represent a varied range of applications, but they are also important systems in their own right. A polymeric brush is formed when chains are grafted to a substrate in such high concentration that they create a dense coating with highly extended configurations. Brushes are simple to prepare and provide a convenient way of modifying the properties of a surface, and can, for example, greatly reduce friction, change the wetting behavior, and affect adhesion properties. Polymeric alloys (or blends) provide an economic method of designing new materials with tailored properties. Because large macromolecules possess relatively little translational entropy, the unlike polymers in an alloy tend to segregate while they are being processed in the melt state, leaving the solid material with a domain structure involving an extensive amount of internal interface. Naturally, the properties of the interface have a pronounced effect on the final material; in particular, a high interfacial tension results in large

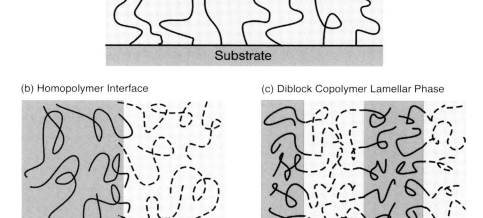

Fig. 2.3 (a) Polymeric brush consisting of polymer chains end-grafted to a flat substrate. (b) Polymer interface separating an A-rich phase (left domain) from a B-rich phase (right domain). (c) Lamellar diblock copolymer microstructure consisting of alternating thin A- and B-rich domains with a repeat period of D.

domains, which is generally undesirable. One way of halting the tendency of domains to coarsen is to chemically bond the unlike polymers together, forming block copolymer molecules. The blocks still segregate into separate domains, but the connectivity of the chain prevents the domains from becoming thicker than the typical size of a single molecule. In fact, the domains tend to form highly ordered periodic geometries, of which the lamellar phase of alternating thin layers pictured in Fig. 2.3(c) is the simplest. This ability to self-assemble into ordered nanoscale morphologies of different geometries is a powerful feature that researchers are now beginning to exploit for various high-tech applications.

2.2
Gaussian Chain

To start off, we consider linear homopolymer chains in a homogeneous environment, and disregard, for the moment, all monomer–monomer interac-

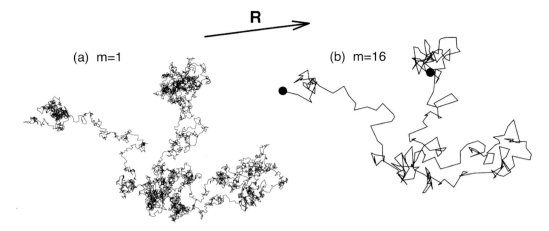

Fig. 2.4 (a) Freely jointed chain with $M = 4000$ fixed-length monomers. (b) Coarse-grained version with $N = 250$ segments, each containing $m = 16$ monomers. The two ends of the polymer are denoted by solid dots, and the end-to-end vector, $\boldsymbol{R}$, is drawn above.

tions including the hard-core ones that prevent the chains from overlapping. In the limit of high molecular weights, all such polymers fall into the same universality class of non-avoiding random walks, and as a consequence they exhibit a degree of common behavior. For instance, their typical size scales with their molecular weight to the universal exponent, $\nu = 1/2$. To investigate this common behavior, there is no need to consider realistic polymer chains involving segments such as those denoted in Fig. 2.2. The fact that the behavior is universal means that it is exhibited by all models, including simple artificial ones.

For maximum simplicity, we select the freely jointed chain, where each monomer has a fixed length, b, and the joint between sequential monomers is completely flexible. Fig. 2.4(a) shows a typical configuration of this model, projected onto the two-dimensional page. It was produced by joining together $M = 4000$ vectors, $\boldsymbol{r}_i$, each generated from the random distribution,

$$p_1(\boldsymbol{r}_i) = \frac{\delta(r_i - b)}{4\pi b^2} \tag{2.1}$$

where $r_i \equiv |\boldsymbol{r}_i|$. The factor in the denominator ensures that the distribution is properly normalized such that

$$\int d\boldsymbol{r}\, p_1(\boldsymbol{r}) = 4\pi \int_0^\infty dr\, r^2 p_1(\boldsymbol{r}) = 1 \tag{2.2}$$

The size of the particular configuration displayed in Fig. 2.4(a) can be characterized by the length of its end-to-end vector,

$$\boldsymbol{R} \equiv \sum_{i=1}^{M} \boldsymbol{r}_i \qquad (2.3)$$

denoted by the arrow at the top of the figure. The length of this vector averaged over all possible configurations, R_0, can then be used as a measure for the typical size of the polymer. Since the simple average, $\langle |\boldsymbol{R}| \rangle$, is difficult to evaluate, it is common practice to use the alternative root mean square (rms) average,

$$R_0 \equiv \sqrt{\langle \boldsymbol{R}^2 \rangle} = \sqrt{\left\langle \sum_{i,j} \boldsymbol{r}_i \cdot \boldsymbol{r}_j \right\rangle} = bM^{1/2} \qquad (2.4)$$

In this case, the formula conveniently simplifies by the fact that $\langle \boldsymbol{r}_i \cdot \boldsymbol{r}_j \rangle = 0$ for all $i \neq j$, because the directions of different steps are uncorrelated. The power-law dependence on the degree of polymerization, M, and the exponent, $\nu = 1/2$, are universal properties of non-avoiding random walks, and would have occurred regardless of the actual model and the measure used to characterize the size. The details of the model and the precise definition of size only affect the proportionality constant, which in this case is simply b.

The underlying reason for the universality of non-avoiding random walks emerges when they are examined on a mesoscopic scale, where the chain is viewed in terms of larger repeat units referred to as *coarse-grained* segments. For this, new distribution functions, $p_m(\boldsymbol{r})$, are defined for the end-to-end vector, $\boldsymbol{r}$, of segments containing m monomers. These distribution functions can be evaluated by using the recursive relation

$$p_m(\boldsymbol{r}) = \int d\boldsymbol{r}_1 \, d\boldsymbol{r}_2 \, p_{m-n}(\boldsymbol{r}_1) p_n(\boldsymbol{r}_2) \delta(\boldsymbol{r}_1 + \boldsymbol{r}_2 - \boldsymbol{r}) \qquad (2.5)$$

where n is any positive integer less than m. By symmetry, each $p_m(\boldsymbol{r})$ depends only on the magnitude, r, of its end-to-end vector, $\boldsymbol{r}$. This fact, combined with the constraint $\boldsymbol{r}_2 = \boldsymbol{r} - \boldsymbol{r}_1$, can be used to reduce the six-dimensional integral in Eq. (2.5) to the two-dimensional one,

$$p_m(\boldsymbol{r}) = 2\pi \int_0^\infty dr_1 \, r_1^2 p_{m-n}(\boldsymbol{r}_1) \int_0^\pi d\theta \, \sin(\theta) p_n(\boldsymbol{r}_2) \qquad (2.6)$$

over the length of $\boldsymbol{r}_1$ and the angle between $\boldsymbol{r}$ and $\boldsymbol{r}_1$. In terms of r_1 and θ, the length of $\boldsymbol{r}_2$ is specified by the cosine law, $r_2^2 = r^2 + r_1^2 - 2rr_1 \cos(\theta)$. Using this relation to substitute θ by r_2, the preceding integral can then be expressed in the slightly more convenient form:

$$p_m(\boldsymbol{r}) = \frac{2\pi}{r} \int_0^\infty dr_1 \, r_1 p_{m-n}(\boldsymbol{r}_1) \int_{|r-r_1|}^{r+r_1} dr_2 \, r_2 p_n(\boldsymbol{r}_2) \qquad (2.7)$$

With the recursion relation in Eq. (2.7), it is now a straightforward exercise to march through calculating $p_m(\mathbf{r})$ for larger and larger segments. Starting with $m = 2$ and $n = 1$, the two-monomer distribution,

$$p_2(\mathbf{r}) = \begin{cases} 1/(8\pi b^2 r) & \text{if } r < 2b \\ 0 & \text{if } 2b < r \end{cases} \qquad (2.8)$$

is obtained from the single-step distribution in Eq. (2.1). One can easily confirm that this distribution is properly normalized as $p_1(\mathbf{r})$ was in Eq. (2.2). Repeating the process with $m = 4$ and $n = 2$ gives the four-monomer distribution,

$$p_4(\mathbf{r}) = \begin{cases} (8b - 3r)/(64\pi b^4) & \text{if } r < 2b \\ (4b - r)^2/(16\pi b^4 r) & \text{if } 2b < r < 4b \\ 0 & \text{if } 4b < r \end{cases} \qquad (2.9)$$

Although the formulas become increasingly complicated, it is straightforward to continue the iterations numerically.

Fig. 2.4(b) displays an equivalent random walk to that in Fig. 2.4(a), but where each segment corresponds to $m = 16$ monomers with a distribution given by $p_{16}(\mathbf{r})$. The act of combining m monomers into larger units, or segments, makes the chain appear smoother and buries microscopic degrees of freedom into the probability distribution, $p_m(\mathbf{r})$. This procedure, where the number of units is reduced from M down to $N \equiv M/m$, is referred to as coarse-graining. It does away with M and b, and replaces them by N and a statistical segment length,

$$a \equiv R_0 N^{-1/2} = bm^{1/2} \qquad (2.10)$$

defined such that $R_0 = aN^{1/2}$.

When the probability distributions, $p_m(\mathbf{r})$, are scaled with respect to the statistical segment length, a, they approach the asymptotic limit

$$p_m(\mathbf{r}) \to \left(\frac{3}{2\pi a^2}\right)^{3/2} \exp\left(-\frac{3r^2}{2a^2}\right) \qquad (2.11)$$

as $m \to \infty$. Fig. 2.5 demonstrates this by comparing distributions of finite m (solid curves) against the asymptotic limit (dashed curve). This convergence to a simple Gaussian distribution is well known in statistics and goes by the name of the *central limit theorem*. The Gaussian (or normal) distribution applies to the average of any large random sample, in this case the

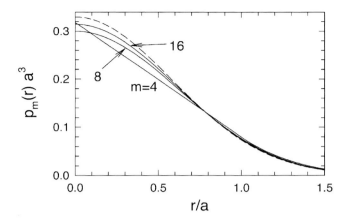

Fig. 2.5 Probability distribution, $p_m(r)$, for a coarse-grained segment with m monomers. The dashed curve denotes the asymptotic limit in Eq. (2.11) for $m \to \infty$.

displacement vectors of m monomers. Even if the lengths of the individual monomers were not fixed, but varied according to some arbitrary distribution, the coarse-grained segments would still develop Gaussian distributions, albeit with a different expression for a. This universality occurs because the Gaussian distribution represents a fixed point of the recursion relation in Eq. (2.5); in other words, if $p_{m-n}(r_1)$ and $p_n(r_2)$ are both Gaussians, then so is $p_m(r)$.

This universality actually extends beyond freely jointed chains to a more general class that can be referred to as Markov chains. An equilibrium configuration of a Markov chain can still be generated by starting at one end and attaching monomers one-by-one according to a given probability distribution. However, in this case, the displacement of the ith monomer, r_i, can have a distribution, $p_1(r_i; r_{i-1})$, that depends on the r_{i-1} of the preceding monomer. In fact, this can be generalized such that the probability of r_i depends on $r_{i-1}, r_{i-2}, \ldots, r_{i-I}$ so long as I remains finite. If this is the case, then sufficiently large coarse-grained segments will still develop a Gaussian distribution with an average chain size obeying $R_0 = aN^{1/2}$.

While the universality class of random walks applies to an impressive range of models, it does not include those with self-avoiding interactions. In that case, the position and orientation of a particular monomer are dependent upon all others in the chain. Nevertheless, such polymer configurations can still be generated with random walks, but those configurations containing overlaps must now be disregarded. The smaller compact configurations are more likely to contain overlapping monomers, and therefore self-avoidance favors larger and more open configurations. Consequently, the size of an isolated polymer in solution scales as $R_0 \propto N^\nu$ with a larger exponent of $\nu \approx 0.6$ (de Gennes 1979). However, when the polymer is in a melt with other polymers, it has to avoid them as well as itself. In this case, the advantage of the more open configurations is lost, and the polymer reverts back to random-walk statistics with the Gaussian probability in Eq. (2.11), although with a modified statistical segment length (Wang 1995).

Fetters et al. (1994) provide a library of experimental results for R_0 as a function of molecular weight and bulk mass densities for various common polymer types, such as those depicted in Fig. 2.2. Given the chemical composition and thus the molecular weight of a monomer, it is then possible to calculate the statistical segment length, a, and the bulk segment density, ρ_0. However, the values obtained depend on the definition of the segment (i.e. the value selected for m). Of course, none of the physical quantities can be affected by such a choice, and indeed this is always the case. If, for example, the number of monomers in each segment is doubled, then N decreases by a factor of 2 and a increases by $\sqrt{2}$, such that the quantity $R_0 = aN^{1/2}$ remains unchanged.

The freedom to redefine segments is linked to the scaling behavior and thus the universality of polymeric systems. However, it can be confusing because the size of N no longer has any absolute meaning, or, in other words, it is impossible to judge whether a polymer is long or short based solely on its value of N. Fortunately, there is a special invariant definition,

$$\mathcal{N} \equiv \left(\frac{R_0^3}{N/\rho_0}\right)^2 = a^6 \rho_0^2 N \qquad (2.12)$$

with a real physical significance that provides a true measure of the polymeric nature of the molecule. It is based on the ratio of the typical volume spanned by a polymer, R_0^3, to its physical volume, N/ρ_0. Thus, small values of $\mathcal{N}$ correspond to compact molecules and large values imply open configurations with many intermolecular contacts. Be aware that some authors assume, without mentioning, that $N = \mathcal{N}$ by innocuously setting the segment volume to $\rho_0^{-1} = a^3$.

2.3
Gaussian Chain in an External Field

The preceding section examined polymers in an ideal homogeneous environment, where no forces were exerted on the chains. In the more interesting situations depicted in Fig. 2.3, the polymers do experience interactions that vary with respect to position, r. Their net effect can generally be represented by a static field, $w(r)$, determined self-consistently from the overall segment distribution in the system. However, we ignore for now the source of the field and start by developing the statistical mechanics for a single polymer acted upon by an arbitrary external field. Although fields are generally real quantities, the derivations that follow hold even when $w(\boldsymbol{r})$ is complex; this fact will be needed in the derivation of self-consistent field theory (SCFT).

We assume that the field varies slowly enough that the chain can be divided into N appropriately sized coarse-grained segments of volume ρ_0^{-1}, each small enough that their local environment is more or less homogeneous, but also large enough that they obey Gaussian statistics with a statistical segment length a. The coarse-grained trajectory of the polymer [e.g. the path in Fig. 2.4(b)] is then specified by a function, $\boldsymbol{r}_\alpha(s)$, where the parameter s ($0 \leq s \leq 1$) runs along the backbone of the polymer such that equal intervals of s have the same molecular weight. (The subscript α will be used to label different molecules when we eventually treat complete systems.) Having specified the trajectory, a dimensionless segment concentration can be defined as

$$\hat{\phi}_\alpha(\boldsymbol{r}) = \frac{N}{\rho_0} \int_0^1 ds\, \delta(\boldsymbol{r} - \boldsymbol{r}_\alpha(s)) \qquad (2.13)$$

The energy of a polymer configuration, $\boldsymbol{r}_\alpha(s)$, turns out to be a local property, meaning that any given interval of the chain, $s_1 \leq s \leq s_2$, can be assigned a definite quantity of energy, $E[\boldsymbol{r}_\alpha; s_1, s_2]$, independent of the rest of the chain. This will become a crucial property in the derivation that follows. The formula for the energy,

$$\frac{E[\boldsymbol{r}_\alpha; s_1, s_2]}{k_B T} = \int_{s_1}^{s_2} ds \left(\frac{3}{2a^2 N} |\boldsymbol{r}'_\alpha(s)|^2 + w(\boldsymbol{r}_\alpha(s)) \right) \qquad (2.14)$$

involves one term that accounts for the Gaussian probability from Eq. (2.11) and another term for the energy of the field. Note that square brackets are used for $E[\boldsymbol{r}_\alpha; s_1, s_2]$ to denote that it is a functional (i.e. a function of a function).

The two primary quantities of the polymer that need to be calculated are the ensemble-averaged concentration, $\phi_\alpha(\boldsymbol{r}) \equiv \langle \hat{\phi}_\alpha(\boldsymbol{r}) \rangle$, and the configurational entropy, S. In order to evaluate these, some partition functions are required. To begin, consider the first sN segments ($0 \leq s \leq 1$) of the chain, constraining the two ends at $\boldsymbol{r}_\alpha(0) = \boldsymbol{r}_0$ and $\boldsymbol{r}_\alpha(s) = \boldsymbol{r}$. The energy of this fragment is $E[\boldsymbol{r}_\alpha; 0, s]$ and thus its partition function is

$$q(\boldsymbol{r}, \boldsymbol{r}_0, s) \propto \int \mathcal{D}\boldsymbol{r}_\alpha \exp\left(-\frac{E[\boldsymbol{r}_\alpha; 0, s]}{k_B T}\right) \delta(\boldsymbol{r}_\alpha(0) - \boldsymbol{r}_0) \delta(\boldsymbol{r}_\alpha(s) - \boldsymbol{r}) \qquad (2.15)$$

This is a functional integral over all configurations, $\boldsymbol{r}_\alpha(t)$, for $0 < t < s$, weighted by the appropriate Boltzmann factor; the delta functions simply exclude those configurations where the two ends are not at r_0 and r. (Those unfamiliar with functional integrals should refer to the Appendix.) In principle, the Boltzmann factor in the partition function should also contain the kinetic energy, and the functional integration should also extend over the momenta of all the monomers. However, the form of the kinetic energy allows the momentum coordinates to be integrated out and included in the proportionality factor (Hong and Noolandi 1981), which in any case has been omitted from Eq. (2.15). A proper evaluation of the proportionality constant would require a detailed account of the exact degrees of freedom available to all the atoms in the polymer molecule, but such information is lost in the coarse-graining procedure. Fortunately, the proportionality constant has no effect on the evaluation of $\phi_\alpha(r)$. Although it does affect S, it does so simply by an additive constant, which is immaterial since we will only be concerned with changes in entropy. Thus, the proportionality relationship in Eq. (2.15) is sufficient for our purposes.

The local property of the polymer energy is manifest mathematically by the equality, $E[r_\alpha; 0, s] = E[r_\alpha; 0, t] + E[r_\alpha; t, s]$ for any $t \in [0, s]$. This implies the recursive relation

$$q(r, r_0, s) = \frac{1}{(a^2 N)^{3/2}} \int dr_1\, q(r, r_1, t) q(r_1, r_0, s-t) \qquad (2.16)$$

which gives the partition function for a fragment of sN segments in terms of the partition functions of two shorter fragments. By including an equality sign in Eq. (2.16), we have effectively chosen a value for the missing proportionality factor in Eq. (2.15); not one that has any actual meaning, but merely the simplest one that renders $q(r, r_0, s)$ dimensionless. This recursion relation also implies

$$q(r, r_0, 0) = (a^2 N)^{3/2} \delta(r - r_0) \qquad (2.17)$$

because Eq. (2.16) must hold for $t \to 0$. Naturally $q(r, r_0, 0) = 0$ when $r \neq r_0$, since $s = 0$ implies that r and r_0 both correspond to the same point on the chain.

The recursion relation, Eq. (2.16), provides a means of building up the partition function for a long chain fragment from those of very short fragments. This reduces the problem to the easier one of finding $q(r, r_0, \epsilon)$, where ϵ is small enough that the field can be treated as a constant for typical chain extensions of $|r - r_0| \lesssim a(\epsilon N)^{1/2}$. In this case, the field simply adds a constant energy of $\epsilon w(r) k_B T$ to each configuration, and thus does not affect the relative probability of its configurations. Therefore, the partition function can be expressed as

$$q(\mathbf{r}, \mathbf{r}_0, \epsilon) \approx \left(\frac{3}{2\pi\epsilon}\right)^{3/2} \exp\left(-\frac{3|\mathbf{r}-\mathbf{r}_0|^2}{2a^2 N\epsilon} - \epsilon w(\mathbf{r})\right) \quad (2.18)$$

which assumes the Gaussian distribution in Eq. (2.11) for a chain of ϵN segments in a homogeneous environment, adjusted for the constant energy of the field and normalized so that it reduces to Eq. (2.17) in the limit $\epsilon \to 0$. For extensions, $|\mathbf{r}-\mathbf{r}_0|$, significantly beyond $a(\epsilon N)^{1/2}$, the partition function approaches zero regardless of whether or not the field energy is well approximated by $\epsilon w(\mathbf{r}) k_B T$. Hence, Eq. (2.18) remains accurate over all extensions, provided that $a(\epsilon N)^{1/2}$ is sufficiently small relative to the spatial variations in $w(\mathbf{r})$.

Now the partition function for large fragments can be calculated iteratively using

$$q(\mathbf{r}, \mathbf{r}_0, s+\epsilon) = \frac{1}{(a^2 N)^{3/2}} \int d\mathbf{r}_\epsilon \, q(\mathbf{r}, \mathbf{r}+\mathbf{r}_\epsilon, \epsilon) q(\mathbf{r}+\mathbf{r}_\epsilon, \mathbf{r}_0, s) \quad (2.19)$$

The first partition function, $q(\mathbf{r}, \mathbf{r}+\mathbf{r}_\epsilon, \epsilon)$, under the integral implies that we only need to know the second one, $q(\mathbf{r}+\mathbf{r}_\epsilon, \mathbf{r}_0, \epsilon)$, accurately for $|\mathbf{r}_\epsilon| \lesssim a(\epsilon N)^{1/2}$. We therefore expand it in the Taylor series

$$q(\mathbf{r}, \mathbf{r}_0, s+\epsilon) = \frac{1}{(a^2 N)^{3/2}} \int d\mathbf{r}_\epsilon \, q(\mathbf{r}, \mathbf{r}+\mathbf{r}_\epsilon, \epsilon)$$
$$\times [1 + \mathbf{r}_\epsilon \cdot \nabla + \tfrac{1}{2}\mathbf{r}_\epsilon \mathbf{r}_\epsilon : \nabla\nabla] q(\mathbf{r}, \mathbf{r}_0, s) \quad (2.20)$$

to second order in $\mathbf{r}_\epsilon$. Here, we have used the dyadic notation (Goldstein 1980) for the second-order term. The expansion now makes it possible to explicitly integrate over $\mathbf{r}_\epsilon$, giving

$$q(\mathbf{r}, \mathbf{r}_0, s+\epsilon) \approx \exp(-\epsilon w(\mathbf{r}))[1 + \tfrac{1}{6} a^2 N\epsilon \nabla^2] q(\mathbf{r}, \mathbf{r}_0, s) \quad (2.21)$$

$$\approx [1 + \tfrac{1}{6} a^2 N\epsilon \nabla^2 - \epsilon w(\mathbf{r})] q(\mathbf{r}, \mathbf{r}_0, s) \quad (2.22)$$

to first order in ϵ. Alternatively, a Taylor series expansion in s gives

$$q(\mathbf{r}, \mathbf{r}_0, s+\epsilon) \approx \left[1 + \epsilon \frac{\partial}{\partial s}\right] q(\mathbf{r}, \mathbf{r}_0, s) \quad (2.23)$$

By direct comparison, it immediately follows that

$$\frac{\partial}{\partial s} q(\mathbf{r}, \mathbf{r}_0, s) = [\tfrac{1}{6} a^2 N \nabla^2 - w(\mathbf{r})] q(\mathbf{r}, \mathbf{r}_0, s) \quad (2.24)$$

which is the modified diffusion equation normally used to evaluate $q(\mathbf{r}, \mathbf{r}_0, s)$. Likewise, a partial partition function

$$q(\mathbf{r}, s) \propto \int \mathcal{D}\mathbf{r}_\alpha \exp\left(-\frac{E[\mathbf{r}_\alpha; 0, s]}{k_\mathrm{B} T}\right) \delta(\mathbf{r}_\alpha(s) - \mathbf{r}) \quad (2.25)$$

can be defined for a chain fragment with a free $s = 0$ end. The fact that

$$q(\mathbf{r}, s) = \frac{1}{(a^2 N)^{3/2}} \int \mathrm{d}\mathbf{r}_0 \, q(\mathbf{r}, \mathbf{r}_0, s) \quad (2.26)$$

implies that it also satisfies Eq. (2.24), given that the diffusion equation is linear. Furthermore, Eq. (2.17) implies the initial condition $q(\mathbf{r}, 0) = 1$.

In a completely analogous way, complementary partition functions are defined for the last $(1-s)N$ segments of the chain. The one for a fixed $s = 1$ end is defined as

$$q^\dagger(\mathbf{r}, \mathbf{r}_0, s) \propto \int \mathcal{D}\mathbf{r}_\alpha \exp\left(-\frac{E[\mathbf{r}_\alpha; s, 1]}{k_\mathrm{B} T}\right) \delta(\mathbf{r}_\alpha(s) - \mathbf{r})\delta(\mathbf{r}_\alpha(1) - \mathbf{r}_0) \quad (2.27)$$

and satisfies the diffusion equation

$$\frac{\partial}{\partial s} q^\dagger(\mathbf{r}, \mathbf{r}_0, s) = -[\tfrac{1}{6} a^2 N \nabla^2 - w(\mathbf{r})] q^\dagger(\mathbf{r}, \mathbf{r}_0, s) \quad (2.28)$$

subject to the condition $q^\dagger(\mathbf{r}, \mathbf{r}_0, 1) = (a^2 N)^{3/2} \delta(\mathbf{r} - \mathbf{r}_0)$. For an unconstrained end, the complementary partition function becomes

$$q^\dagger(\mathbf{r}, s) = \frac{1}{(a^2 N)^{3/2}} \int \mathrm{d}\mathbf{r}_0 \, q^\dagger(\mathbf{r}, \mathbf{r}_0, s) \quad (2.29)$$

satisfying Eq. (2.28) with $q^\dagger(\mathbf{r}, 1) = 1$. The full partition function, $\mathcal{Q}[w]$, for a polymer in the field $w(\mathbf{r})$, is then related to the integral of the two partial partition functions. In the case where both ends of the polymer are free,

$$\mathcal{Q}[w] = \int \mathrm{d}\mathbf{r} \, q(\mathbf{r}, s) q^\dagger(\mathbf{r}, s) \propto \int \mathcal{D}\mathbf{r}_\alpha \exp\left(-\frac{E[\mathbf{r}_\alpha; 0, 1]}{k_\mathrm{B} T}\right) \quad (2.30)$$

Again, the square brackets in $\mathcal{Q}[w]$ denote that it is a functional that depends on the function $w(\mathbf{r})$.

With the partition functions in hand, the average segment distribution, $\phi_\alpha(\mathbf{r})$, can now be evaluated. Here the calculation is shown for an unconstrained polymer, but the procedure is essentially the same if one or both ends are constrained. Following standard statistical mechanics, the average is computed by weighting each configuration by the appropriate Boltzmann factor:

$$\phi_\alpha(\mathbf{r}) = \frac{1}{\mathcal{Q}[w]} \int \mathcal{D}\mathbf{r}_\alpha \, \hat{\phi}_\alpha(\mathbf{r}) \exp\left(-\frac{E[\mathbf{r}_\alpha; 0, 1]}{k_B T}\right)$$

$$= \frac{N}{\rho_0 \mathcal{Q}[w]} \int_0^1 ds \int \mathcal{D}\mathbf{r}_\alpha \, \delta(\mathbf{r} - \mathbf{r}_\alpha(s)) \exp\left(-\frac{E[\mathbf{r}_\alpha; 0, 1]}{k_B T}\right)$$

$$= \frac{N}{\rho_0 \mathcal{Q}[w]} \int_0^1 ds \, q(\mathbf{r}, s) q^\dagger(\mathbf{r}, s) \tag{2.31}$$

This equation provides the practical means of calculating the average segment concentration, but there will also be instances where we need to identify it by the functional derivative

$$\phi_\alpha(\mathbf{r}) = -\frac{N}{\rho_0} \frac{\mathcal{D}\ln(\mathcal{Q}[w])}{\mathcal{D}w(\mathbf{r})} \equiv -\frac{N}{\rho_0} \lim_{\epsilon \to 0} \frac{\ln(\mathcal{Q}[w + \epsilon\delta]) - \ln(\mathcal{Q}[w])}{\epsilon} \tag{2.32}$$

The equivalence of this functional derivative to Eq. (2.31) follows almost immediately from the alternative expression

$$\mathcal{Q}[w] \propto \int \mathcal{D}\mathbf{r}_\alpha \exp\left(-\frac{3}{2a^2 N} \int_0^1 ds \, |\mathbf{r}_\alpha'(s)|^2 - \frac{\rho_0}{N} \int d\mathbf{r}_1 \, w(\mathbf{r}_1) \hat{\phi}_\alpha(\mathbf{r}_1)\right) \tag{2.33}$$

for the partition function. Note that $\mathcal{Q}[w + \epsilon\delta]$ simply refers to the partition function evaluated with $w(\mathbf{r}_1)$ replaced by $w(\mathbf{r}_1) + \epsilon\delta(\mathbf{r}_1 - \mathbf{r})$. (See the Appendix for further information on functional differentiation.)

Lastly, we calculate the entropy, S, or equivalently the entropic free energy, $f_e \equiv -TS$, which, for reasons that will become apparent, is also called the elastic free energy of the polymer. In accord with standard statistical mechanics, the entropic energy is

$$\frac{f_e}{k_B T} = -\ln\left(\frac{\mathcal{Q}[w]}{\mathcal{V}}\right) - \frac{\rho_0}{N} \int d\mathbf{r} \, w(\mathbf{r}) \phi_\alpha(\mathbf{r}) \tag{2.34}$$

The logarithm of $\mathcal{Q}[w]$ gives the total free energy of the polymer in the field, while the integral removes the average internal energy acquired from the field, leaving behind the entropic energy. The extra factor of $\mathcal{V}$ in the logarithm simply adds a constant to f_e such that the entropic energy is measured relative to the homogeneous state; it is easily confirmed that Eq. (2.34) gives $f_e = 0$ for the case of a uniform field.

Since the force exerted on the segments is given by the gradient of the field, an additive constant to the field should have no effect on any physical quantities. Indeed, this is the case. If a constant, w_0, is added to the field such that $w(\mathbf{r}) \Rightarrow w(\mathbf{r}) + w_0$, the partition functions transform as

$$q(\mathbf{r}, s) \Longrightarrow q(\mathbf{r}, s) \exp(-sw_0) \tag{2.35}$$

$$q^\dagger(\mathbf{r}, s) \Longrightarrow q^\dagger(\mathbf{r}, s) \exp(-(1-s)w_0) \tag{2.36}$$

$$\mathcal{Q}[w] \Longrightarrow \mathcal{Q}[w] \exp(-w_0) \tag{2.37}$$

When these substitutions are entered into Eq. (2.31), all the exponential factors cancel, leaving $\phi_\alpha(\mathbf{r})$ unaffected. Likewise, w_0 cancels out of Eq. (2.34) for f_e, using the fact that the average segment concentration is $\bar{\phi}_\alpha = N/\rho_0 V$. This invariance will generally be used to set the spatial average of the field, $\bar{w}$, to zero.

2.4
Strong-Stretching Theory (SST): the Classical Path

In circumstances where the field energy becomes large relative to the thermal energy, $k_\text{B}T$, the polymer is effectively restricted to those configurations close to the ground state, referred to as the *classical path* for reasons that will become apparent in the following section. This opens up the possibility of low-temperature series approximations, but one first needs to locate the ground-state trajectory by minimizing

$$E[\mathbf{r}_\alpha; 0, 1] = \int_0^1 ds\, \mathcal{L}(\mathbf{r}_\alpha(s), \mathbf{r}'_\alpha(s)) \tag{2.38}$$

where

$$\mathcal{L}(\mathbf{r}_\alpha(s), \mathbf{r}'_\alpha(s)) \equiv \frac{3}{2a^2 N}|\mathbf{r}'_\alpha(s)|^2 + w(\mathbf{r}_\alpha(s)) \tag{2.39}$$

This is a standard *calculus of variations* problem, which, as demonstrated in the Appendix, is equivalent to solving the Euler–Lagrange equation

$$\frac{d}{ds}\left(\frac{\partial \mathcal{L}}{\partial r'_\alpha}\right) - \frac{\partial \mathcal{L}}{\partial r_\alpha} = 0 \tag{2.40}$$

This is a slight generalization of Eq. (2.331), due to the fact that $\mathbf{r}_\alpha(s)$ is a vector rather than a scalar function. Nevertheless, the differentiation of $\mathcal{L}$ in

Eq. (2.40) is easily performed and leads to the differential equation

$$\frac{3}{a^2 N} \bm{r}''_\alpha(s) - \nabla w(\bm{r}_\alpha(s)) = 0 \tag{2.41}$$

Solving this vector equation subject to possible boundary conditions for $\bm{r}_\alpha(0)$ and $\bm{r}_\alpha(1)$ provides the classical path.

Although Eq. (2.41) is sufficient, we can derive an alternative equation that provides a useful relation between the local chain extension, $|\bm{r}'_\alpha(s)|$, and the field energy, $w(\bm{r}_\alpha(s))$. This is done by "dotting" Eq. (2.41) with $\bm{r}'_\alpha(s)$ to give

$$\frac{3}{a^2 N} \bm{r}'_\alpha(s) \cdot \bm{r}''_\alpha(s) - \bm{r}'_\alpha(s) \cdot \nabla w(\bm{r}_\alpha(s)) = 0 \tag{2.42}$$

which, in turn, can be rewritten as

$$\frac{\mathrm{d}}{\mathrm{d}s}\left(\frac{3|\bm{r}'_\alpha(s)|^2}{2a^2 N}\right) - \frac{\mathrm{d}}{\mathrm{d}s} w(\bm{r}_\alpha(s)) = 0 \tag{2.43}$$

Performing an integration reduces this to the first-order differential equation

$$\frac{3|\bm{r}'_\alpha(s)|^2}{2a^2 N} - w(\bm{r}_\alpha(s)) = \text{constant} \tag{2.44}$$

In cases of appropriate symmetry, the classical trajectory can be derived from this scalar equation alone.

Once the classical path is known, the average segment concentration is given by

$$\phi_\alpha(\bm{r}) = \frac{N}{\rho_0} \int_0^1 \mathrm{d}s\, \delta(\bm{r} - \bm{r}_\alpha(s)) \tag{2.45}$$

and the entropic free energy is

$$\frac{f_e}{k_B T} = \frac{3}{2a^2 N} \int_0^1 \mathrm{d}s\, |\bm{r}'_\alpha(s)|^2 \tag{2.46}$$

This entropic energy is often referred to as the elastic or stretching energy, because it has the equivalent form to the stretching energy of a thin, elastic thread.

2.5
Analogy with Quantum/Classical Mechanics

Every so often, mathematics and physics are bestowed with a powerful mapping that equates two diverse topics. In one giant leap, years of accumulated knowledge on one topic can be instantly transferred to another. Here, we encounter such an example by the fact that a Gaussian chain in an external field maps onto the quantum mechanics of a single particle in a potential. For those without a background in quantum mechanics, the section can be skimmed over without seriously compromising one's understanding of the remaining sections.

The mapping becomes apparent when examining the path-integral formalism of quantum mechanics introduced by Feynman and Hibbs (1965). With incredible insight, they demonstrate that the quantum-mechanical wavefunction, $\Psi(\boldsymbol{r}, t)$, for a particle of mass m in a potential $U(\boldsymbol{r})$, is given by the path integral

$$\Psi(\boldsymbol{r}, t) = \int \mathcal{D}\boldsymbol{r}_p \, \exp(iS/\hbar) \delta(\boldsymbol{r}_p(t) - \boldsymbol{r}) \tag{2.47}$$

over all possible trajectories $\boldsymbol{r}_p(t)$, where the particle terminates at position $\boldsymbol{r}$ at time t. (Note that $i \equiv \sqrt{-1}$.) In this definition, each trajectory contributes a phase factor given by the classical action

$$S \equiv \int dt \, [\tfrac{1}{2} m |\boldsymbol{r}'_p(t)|^2 - U(\boldsymbol{r}_p(t))] \tag{2.48}$$

which, in this case, is kinetic energy minus potential energy integrated along the particle trajectory. Naturally, Feynman justified this by showing that Eq. (2.47) is equivalent to the usual time-dependent Schrödinger equation

$$i\hbar \frac{\partial}{\partial t} \Psi(\boldsymbol{r}, t) = \left[-\frac{\hbar^2}{2m} \nabla^2 + U(\boldsymbol{r}) \right] \Psi(\boldsymbol{r}, t) \tag{2.49}$$

in much the same way as we derived the diffusion equation for $q(\boldsymbol{r}, s)$ in Section 2.3.

Taking a closer look, the mathematics of this quantum-mechanical problem is exactly that of our polymer in an external field. A precise mapping merely requires the associations:

$$t \iff s \tag{2.50}$$

$$m \iff 3/a^2 N \tag{2.51}$$

$$U(\boldsymbol{r}) \iff -w(\boldsymbol{r}) \tag{2.52}$$

$$S \iff E[\boldsymbol{r}_\alpha; 0, s]/k_B T \tag{2.53}$$

$$\boldsymbol{r}_p(t) \iff \boldsymbol{r}_\alpha(s) \tag{2.54}$$

$$\hbar \iff -i \tag{2.55}$$

$$\Psi(\boldsymbol{r}, t) \iff q(\boldsymbol{r}, s) \tag{2.56}$$

Indeed, these substitutions transform Eq. (2.47) for the wavefunction into Eq. (2.25) for the partition function of sN polymer segments, and likewise the Schrödinger equation (Eq. 2.49) transforms into the diffusion equation (Eq. 2.24).

Extending this analogy further, the strong-stretching theory (SST) developed in Section 2.4 is found to be equivalent to classical mechanics. In Feynman's formulation, the connection between quantum mechanics and classical mechanics becomes exceptionally transparent. When the action, S, of a particle is large relative to $\hbar$, the path integral becomes dominated by the extremum of S, because all paths close to the extremum contribute the same phase factor, and thus add constructively. [This is the basis of Fermat's principle in optics by which the full wave treatment of light can be reduced to classical ray optics (Hecht and Zajac 1974).] However, the procedure of finding the extremum of S is precisely the Lagrangian formalism of classical mechanics (Goldstein 1980). Thus, in the limit of large S, the particle follows the classical-mechanical trajectory determined by Newton's equation of motion (i.e. $m\boldsymbol{a} = \boldsymbol{F}$)

$$m\boldsymbol{r}_p''(t) = -\nabla U(\boldsymbol{r}_p(t)) \tag{2.57}$$

Equally familiar in classical mechanics is the equation for the conservation of energy,

$$\tfrac{1}{2}m|\boldsymbol{r}_p'(t)|^2 + U(\boldsymbol{r}_p(t)) = \text{constant} \tag{2.58}$$

In terms of the analogy, classical mechanics corresponds to the case where $E[\boldsymbol{r}_\alpha; s_1, s_2]/k_\text{B}T$ is large relative to unity, which is the condition for the strong-stretching theory (SST) discussed in the preceding section. Indeed, Eq. (2.57) matches precisely with Eq. (2.41) for the ground-state polymer trajectory, as does Eq. (2.58) with Eq. (2.44). It is this analogy from which the ground-state trajectory acquires its name, the *classical path*.

2.6
Mathematical Techniques and Approximations

Quantum mechanics has experienced a massive activity of research since its introduction midway through the last century. The accumulated knowledge is staggering, and the fact that our problem maps onto it means that we can make significant progress without having to reinvent the wheel. This section demonstrates several useful calculations that draw upon what are now standard techniques in quantum mechanics.

2.6.1
Spectral Method

One of the most widely used and fruitful techniques in quantum mechanics is the spectral method for solving the time-dependent wavefunction, $\Psi(\mathbf{r},t)$. Here, the method is adapted to the evaluation of the partial partition functions, $q(\mathbf{r},s)$ and $q^{\dagger}(\mathbf{r},s)$, for a homopolymer in an external field, $w(\mathbf{r})$. The first step is to solve the analog of the time-independent Schrödinger equation,

$$[\tfrac{1}{6}a^2 N \nabla^2 - w(\mathbf{r})] f_i(\mathbf{r}) = -\lambda_i f_i(\mathbf{r}) \tag{2.59}$$

for all the eigenvalues, λ_i, and eigenfunctions, $f_i(\mathbf{r})$, where $i = 0, 1, 2, \ldots$. For reasons that will be discussed at the end of this section, the eigenvalues are ordered from smallest to largest. As in quantum mechanics, the eigenfunctions are orthogonal and can be normalized to satisfy

$$\frac{1}{V} \int d\mathbf{r}\, f_i(\mathbf{r}) f_j(\mathbf{r}) = \delta_{ij} \tag{2.60}$$

where V is the total volume of the system. Furthermore, the eigenfunctions form a complete basis, which implies that any function of position, $g(\mathbf{r})$, can be expanded as

$$g(\mathbf{r}) = \sum_{i=0}^{\infty} g_i f_i(\mathbf{r}) \tag{2.61}$$

The coefficients g_i of the expansion are determined by multiplying Eq. (2.61) by $f_j(\mathbf{r})$, integrating over $\mathbf{r}$, and using the orthonormal condition from Eq. (2.60). This gives the formula

$$g_i = \frac{1}{V} \int d\mathbf{r}\, f_i(\mathbf{r}) g(\mathbf{r}) \tag{2.62}$$

Once the orthonormal basis functions have been obtained, it is a trivial matter to evaluate $q(\mathbf{r},s)$. One starts by expanding it as

$$q(\mathbf{r},s) = \sum_{i=0}^{\infty} q_i(s) f_i(\mathbf{r}) \tag{2.63}$$

and substituting the expansion into Eq. (2.24). Given the inherent property of the eigenfunctions (Eq. 2.59), the diffusion equation reduces to

$$\sum_{i=0}^{\infty} q_i'(s) f_i(\mathbf{r}) = -\sum_{i=0}^{\infty} \lambda_i q_i(s) f_i(\mathbf{r}) \tag{2.64}$$

With the same steps used to derive Eq. (2.62), it follows that

$$q_i'(s) = -\lambda_i q_i(s) \tag{2.65}$$

for all $i = 0, 1, 2, \ldots$. Differential equations do not come any simpler. Inserting the solution into Eq. (2.63) gives the final result:

$$q(\boldsymbol{r}, s) = \sum_{i=0}^{\infty} c_i \exp(-\lambda_i s) f_i(\boldsymbol{r}) \tag{2.66}$$

where the constant

$$c_i \equiv q_i(0) = \frac{1}{\mathcal{V}} \int d\boldsymbol{r}\, f_i(\boldsymbol{r}) q(\boldsymbol{r}, 0) \tag{2.67}$$

is determined by the initial condition. Similarly, the expression for the complementary partition function is

$$q^\dagger(\boldsymbol{r}, s) = \sum_{i=0}^{\infty} c_i^\dagger \exp(-\lambda_i(1-s)) f_i(\boldsymbol{r}) \tag{2.68}$$

where

$$c_i^\dagger = \frac{1}{\mathcal{V}} \int d\boldsymbol{r}\, f_i(\boldsymbol{r}) q^\dagger(\boldsymbol{r}, 1) \tag{2.69}$$

In practice, it is impossible to complete the infinite sums in Eqs. (2.66) and (2.68), but the eigenvalues are appropriately ordered such that the terms become less and less important. Therefore, the sums can be truncated at some point, $i = M$, where the number of terms retained, $M + 1$, is guided by the desired level of accuracy.

2.6.2
Ground-State Dominance

The previous expansion for $q(\boldsymbol{r}, s)$ in Eq. (2.66) becomes increasingly dominated by the first term involving the *ground-state* eigenvalue, λ_0, the further s is from the end of the chain. For a high-molecular-weight polymer, the approximations

$$q(\boldsymbol{r}, s) \approx c_0 \exp(-\lambda_0 s) f_0(\boldsymbol{r}) \tag{2.70}$$

and

$$q^\dagger(\boldsymbol{r}, s) \approx c_0^\dagger \exp(-\lambda_0(1-s)) f_0(\boldsymbol{r}) \tag{2.71}$$

become accurate over the vast majority of the chain. In this case, it follows that the partition function of the entire chain is

$$\mathcal{Q}[w] \approx \mathcal{V} c_0 c_0^\dagger \exp(-\lambda_0) \tag{2.72}$$

and that the segment concentration is

$$\phi_\alpha(r) \approx \frac{N}{\rho_0 \mathcal{V}} f_0^2(r) \tag{2.73}$$

These simplifications permit us to derive an analytical Landau–Ginzburg free-energy expression for the entropic energy, f_e, of the polymer as a function of its concentration profile, $\phi_\alpha(r)$. Our derivation will assume that concentration gradient

$$\nabla \phi_\alpha(r) = \frac{2N}{\rho_0 \mathcal{V}} f_0(r) \nabla f_0(r) \tag{2.74}$$

is zero at the extremities (i.e. boundaries) of our system, $\mathcal{V}$.

The first step of the derivation is to multiply Eq. (2.59) for $i = 0$ by $f_0(r)$ and integrate over $\mathcal{V}$, producing the result

$$\frac{\rho_0}{N} \int dr\, w(r) \phi_\alpha(r) = \frac{a^2 N}{6 \mathcal{V}} \int dr\, f_0(r) \nabla^2 f_0(r) + \lambda_0 \tag{2.75}$$

Next, this expression for the field energy along with Eq. (2.72) for $\mathcal{Q}[w]$ are inserted into Eq. (2.34), giving the expression

$$\frac{f_e}{k_B T} = -\frac{a^2 N}{6 \mathcal{V}} \int dr\, f_0(r) \nabla^2 f_0(r) - \ln(c_0 c_0^\dagger) \tag{2.76}$$

for the entropic free energy. The logarithmic term can be dropped in the large-N limit. An integration by parts is then performed to re-express the equation as

$$\frac{f_e}{k_B T} = \frac{a^2 N}{6 \mathcal{V}} \int dr\, |\nabla f_0(r)|^2 \tag{2.77}$$

where the boundary term vanishes as a result of the assumption $\nabla \phi_\alpha(r) = 0$ along the edge of $\mathcal{V}$. For the final step of the derivation, the eigenfunction $f_0(r)$ is replaced by the segment concentration $\phi_\alpha(r)$, using Eqs. (2.73) and (2.74). This produces the Landau–Ginzburg free energy expression

$$\frac{f_e}{k_B T} = \frac{a^2 \rho_0}{24} \int dr\, \frac{|\nabla \phi_\alpha(r)|^2}{\phi_\alpha(r)} \tag{2.78}$$

which gives the entropic energy without the need to evaluate the underlying field, $w(\boldsymbol{r})$. The derivation assumes only that the molecular weight (i.e. N) of the polymer is large.

2.6.3
Fourier Representation

Quantum-mechanical problems are often made more tractable by re-expressing them in so-called momentum space. The principle is very similar to that of the spectral method, where quantities are expanded in terms of energy eigenstates. The main difference is that the expansions are now in terms of eigenfunctions, $\exp(\mathrm{i}\boldsymbol{k}\cdot\boldsymbol{r})$, of the momentum operator $\hat{p} \equiv -\mathrm{i}\hbar\nabla$. (Here, $\mathrm{i} \equiv \sqrt{-1}$.) The other difference is that the momentum eigenvalue, $\boldsymbol{k}$, is continuous, which implies that sums are now replaced by integrals and Kronecker delta functions are substituted by Dirac delta functions. For example, the orthogonality condition in Eq. (2.60) becomes

$$\frac{1}{(2\pi)^3}\int \mathrm{d}\boldsymbol{r}\,\exp(\mathrm{i}(\boldsymbol{k}_1-\boldsymbol{k}_2)\cdot\boldsymbol{r}) = \delta(\boldsymbol{k}_1-\boldsymbol{k}_2) \qquad (2.79)$$

Furthermore, the expansion of an arbitrary function, $g(\boldsymbol{r})$, becomes the integral

$$g(\boldsymbol{r}) = \frac{1}{(2\pi)^3}\int \mathrm{d}\boldsymbol{k}\,g(\boldsymbol{k})\exp(\mathrm{i}\boldsymbol{k}\cdot\boldsymbol{r}) \qquad (2.80)$$

and the coefficient of the expansion

$$g(\boldsymbol{k}) \equiv \int \mathrm{d}\boldsymbol{r}\,g(\boldsymbol{r})\exp(-\mathrm{i}\boldsymbol{k}\cdot\boldsymbol{r}) \qquad (2.81)$$

is now a continuous function, referred to as the Fourier transform of $g(\boldsymbol{r})$. Although $g(\boldsymbol{k})$ is complex, its complex conjugate satisfies $g^*(\boldsymbol{k}) = g(-\boldsymbol{k})$, which implies that $|g(\boldsymbol{k})|^2 = g(-\boldsymbol{k})g(\boldsymbol{k})$. Before moving on, we quote a useful identity:

$$\int \mathrm{d}\boldsymbol{r}\,g_1(\boldsymbol{r})g_2(\boldsymbol{r}) = \frac{1}{(2\pi)^3}\int \mathrm{d}\boldsymbol{k}\,g_1(-\boldsymbol{k})g_2(\boldsymbol{k}) \qquad (2.82)$$

for the integral of two functions. It can be proved by substituting the expansions for $g_1(\boldsymbol{r})$ and $g_2(\boldsymbol{r})$ into the left-hand side, identifying the integral over $\boldsymbol{r}$ as the Dirac delta function in Eq. (2.79), and using the sifting property of the delta function from Eq. (2.329).

Now we are ready to Fourier-transform the diffusion equation for $q(\boldsymbol{r},s)$. The first step is to insert the Fourier representations, Eq. (2.80), of $q(\boldsymbol{r},s)$ and

$w(\boldsymbol{r})$ into the diffusion equation (2.24), giving

$$\frac{\partial}{\partial s}\int \mathrm{d}\boldsymbol{k}_2\, q(\boldsymbol{k}_2,s)\exp(\mathrm{i}\boldsymbol{k}_2\cdot\boldsymbol{r})$$

$$=-\frac{a^2 N}{6}\int \mathrm{d}\boldsymbol{k}_2\, k_2^2 q(\boldsymbol{k}_2,s)\exp(\mathrm{i}\boldsymbol{k}_2\cdot\boldsymbol{r})$$

$$-\frac{1}{(2\pi)^3}\int \mathrm{d}\boldsymbol{k}_1\, \mathrm{d}\boldsymbol{k}_2\, w(\boldsymbol{k}_1)q(\boldsymbol{k}_2,s)\exp[\mathrm{i}(\boldsymbol{k}_1+\boldsymbol{k}_2)\cdot\boldsymbol{r}] \quad (2.83)$$

Next the equation is multiplied through by $\exp(-\mathrm{i}\boldsymbol{k}\cdot\boldsymbol{r})$ and integrated over $\boldsymbol{r}$. Using the orthogonality relation in Eq. (2.79), the expression simplifies to

$$\frac{\partial}{\partial s}\int \mathrm{d}\boldsymbol{k}_2\, q(\boldsymbol{k}_2,s)\delta(\boldsymbol{k}_2-\boldsymbol{k})$$

$$=-\frac{a^2 N}{6}\int \mathrm{d}\boldsymbol{k}_2\, k_2^2 q(\boldsymbol{k}_2,s)\delta(\boldsymbol{k}_2-\boldsymbol{k})$$

$$-\frac{1}{(2\pi)^3}\int \mathrm{d}\boldsymbol{k}_1\, \mathrm{d}\boldsymbol{k}_2\, w(\boldsymbol{k}_1)q(\boldsymbol{k}_2,s)\delta(\boldsymbol{k}_1+\boldsymbol{k}_2-\boldsymbol{k}) \quad (2.84)$$

Now we apply the sifting property of the delta function to arrive at the final expression:

$$\frac{\partial}{\partial s}q(\boldsymbol{k},s)=-\frac{k^2 a^2 N}{6}q(\boldsymbol{k},s)-\frac{1}{(2\pi)^3}\int \mathrm{d}\boldsymbol{k}_1\, w(\boldsymbol{k}_1)q(\boldsymbol{k}-\boldsymbol{k}_1,s) \quad (2.85)$$

This equation is then solved for $q(\boldsymbol{k},s)$ subject to the initial condition

$$q(\boldsymbol{k},0)=\int \mathrm{d}\boldsymbol{r}\, q(\boldsymbol{r},0)\exp(-\mathrm{i}\boldsymbol{k}\cdot\boldsymbol{r}) \quad (2.86)$$

For a polymer with a free end,

$$q(\boldsymbol{k},0)=(2\pi)^3\delta(\boldsymbol{k}) \quad (2.87)$$

whereas when the $s=0$ end is fixed at $\boldsymbol{r}_0$,

$$q(\boldsymbol{k},0)=(aN^{1/2})^3\exp(\mathrm{i}\boldsymbol{k}\cdot\boldsymbol{r}_0) \quad (2.88)$$

In either case, the solution to Eq. (2.85) for a homogeneous environment (i.e. $w(\boldsymbol{r})=0$) is readily expressed as

$$q(\boldsymbol{k},s)=q(\boldsymbol{k},0)\exp(-k^2 a^2 Ns/6) \quad (2.89)$$

from which $q(\boldsymbol{r}, s)$ immediately follows. The aim of the next section will be to extend this result to polymers in a weak field.

2.6.4
Random-Phase Approximation

Here, a Landau–Ginzburg free-energy expression is derived for the entropic free energy, f_e, of a polymer molecule in a weakly varying field, $w(\boldsymbol{r})$. As before in Section 2.6.2, this calculation will assume that the two ends of the chain are free, by imposing the initial condition in Eq. (2.87). We will also use the invariance discussed at the end of Section 2.3 to fix the spatial average of the field to $\bar{w} = 0$. Given that $w(\boldsymbol{r})$ only deviates slightly from zero, it is safe to assume that the ensemble-averaged segment concentration, $\phi_\alpha(\boldsymbol{r})$, never strays far from its spatial average,

$$\bar{\phi}_\alpha \equiv \frac{1}{\mathcal{V}} \int \mathrm{d}\boldsymbol{r}\, \phi_\alpha(\boldsymbol{r}) = \frac{N}{\rho_0 \mathcal{V}} \tag{2.90}$$

The first step is to find an approximate solution for $q(\boldsymbol{k}\neq 0, s)$. Since the integral in Eq. (2.85) involves $w(\boldsymbol{k}_1)$, which is small, the multiplicative factor of $q(\boldsymbol{k}-\boldsymbol{k}_1, s)$ need not be particularly accurate. Approximating it by $q(\boldsymbol{k}-\boldsymbol{k}_1, 0)$ from Eq. (2.87) gives

$$\frac{\partial}{\partial s} q(\boldsymbol{k}, s) \approx -x q(\boldsymbol{k}, s) - \frac{1}{(2\pi)^3} \int \mathrm{d}\boldsymbol{k}_1\, w(\boldsymbol{k}_1) q(\boldsymbol{k}-\boldsymbol{k}_1, 0) \tag{2.91}$$

$$= -x q(\boldsymbol{k}, s) - w(\boldsymbol{k}) \tag{2.92}$$

where $x \equiv k^2 a^2 N/6$. Given that $q(\boldsymbol{k}, 0) = 0$ for $\boldsymbol{k} \neq 0$, the solution becomes

$$q(\boldsymbol{k}, s) \approx -w(\boldsymbol{k}) h(x, s) \tag{2.93}$$

where

$$h(x, s) \equiv \frac{1 - \exp(-sx)}{x} \tag{2.94}$$

The second step is to solve for $q(\boldsymbol{k}=0, s)$, in which case Eq. (2.85) reduces to

$$\frac{\partial}{\partial s} q(\boldsymbol{k}, s) = -\frac{1}{(2\pi)^3} \int \mathrm{d}\boldsymbol{k}_1\, w(\boldsymbol{k}_1) q(-\boldsymbol{k}_1, s) \tag{2.95}$$

Since we have set $\bar{w} = 0$, it follows that $w(\boldsymbol{k}_1) = 0$ for $\boldsymbol{k}_1 = 0$, which in turn implies that the approximation from Eq. (2.93) can be inserted inside the integral to give

$$\frac{\partial}{\partial s} q(\mathbf{k}, s) \approx \frac{1}{(2\pi)^3} \int d\mathbf{k}_1 \, w(\mathbf{k}_1) w(-\mathbf{k}_1) h(x_1, s) \qquad (2.96)$$

The solution of this is

$$q(\mathbf{k}, s) \approx q(\mathbf{k}, 0) \left[1 + \frac{1}{2(2\pi)^3 \mathcal{V}} \int d\mathbf{k}_1 \, w(-\mathbf{k}_1) w(\mathbf{k}_1) g(x_1, s) \right] \qquad (2.97)$$

where

$$g(x, s) \equiv \frac{2[\exp(-sx) + sx - 1]}{x^2} \qquad (2.98)$$

Note that we have used the fact that $q(\mathbf{k}=0, 0) = \mathcal{V}$, which follows from Eq. (2.81), and the initial condition $q(\mathbf{r}, 0) = 1$. Now that the Fourier transform, $q(\mathbf{k}, s)$, is known for all $\mathbf{k}$, it can be converted back to

$$q(\mathbf{r}, s) \approx 1 + \frac{1}{2(2\pi)^3 \mathcal{V}} \int d\mathbf{k} \, w(-\mathbf{k}) w(\mathbf{k}) g(x, s)$$

$$- \frac{1}{(2\pi)^3} \int d\mathbf{k} \, w(\mathbf{k}) h(x, s) \exp(i\mathbf{k} \cdot \mathbf{r}) \qquad (2.99)$$

using the inverse Fourier transform, Eq. (2.80). In principle, the second integral in Eq. (2.99) should exclude $\mathbf{k} = 0$, but it can be extended to all $\mathbf{k}$, since $w(\mathbf{k}=0) = 0$. The analogous calculation for the complementary partition function gives

$$q^{\dagger}(\mathbf{r}, s) \approx 1 + \frac{1}{2(2\pi)^3 \mathcal{V}} \int d\mathbf{k} \, w(-\mathbf{k}) w(\mathbf{k}) g(x, 1-s)$$

$$- \frac{1}{(2\pi)^3} \int d\mathbf{k} \, w(\mathbf{k}) h(x, 1-s) \exp(i\mathbf{k} \cdot \mathbf{r}) \qquad (2.100)$$

From the partial partition functions, it follows that the partition for the entire chain is

$$\frac{\mathcal{Q}[w]}{\mathcal{V}} \approx 1 + \frac{1}{2(2\pi)^3 \mathcal{V}} \int d\mathbf{k} \, w(-\mathbf{k}) w(\mathbf{k}) g(x, 1) \qquad (2.101)$$

The Fourier amplitudes of the segment concentration are most easily evaluated using the functional derivative

$$\phi_\alpha(\mathbf{k}) = -\frac{N(2\pi)^3}{\rho_0} \frac{\mathcal{D} \ln(\mathcal{Q}[w])}{\mathcal{D} w(-\mathbf{k})} \qquad (2.102)$$

which is derived similarly to Eq. (2.32) by making the substitution

$$\frac{\rho_0}{N} \int d\mathbf{r}\, w(\mathbf{r}) \hat{\phi}_\alpha(\mathbf{r}) = \frac{\rho_0}{N(2\pi)^3} \int d\mathbf{k}\, w(-\mathbf{k}) \hat{\phi}_\alpha(\mathbf{k}) \qquad (2.103)$$

in Eq. (2.33). Note that the differentiation cannot be used for $\mathbf{k} = 0$, because $\mathcal{Q}[w]$ has been calculated under the assumption $w(\mathbf{k}=0) = 0$. Nevertheless, we already know that $\phi_\alpha(\mathbf{k}=0) = \bar{\phi}_\alpha \mathcal{V} = N/\rho_0$. For $\mathbf{k} \neq 0$, we have

$$\phi_\alpha(\mathbf{k}) = -\bar{\phi}_\alpha w(\mathbf{k}) g(x, 1) \qquad (2.104)$$

It is now just a simple matter of substituting our results into Eq. (2.34) to obtain an expression for the entropic free energy. First, the logarithm of $\mathcal{Q}[w]$ is evaluated with the Taylor series approximation, $\ln(1+x) \approx x$, applied to Eq. (2.101). Then the field energy is rewritten in terms of Fourier coefficients as in Eq. (2.103), and lastly all occurrences of $w(\mathbf{k})$ are eliminated in favor of $\phi(\mathbf{k})$ using Eq. (2.104). The final result is

$$\frac{f_e}{k_B T} = \frac{\rho_0}{2N\bar{\phi}_\alpha (2\pi)^3} \int_{\mathbf{k} \neq 0} d\mathbf{k}\, \frac{\phi_\alpha(-\mathbf{k})\phi_\alpha(\mathbf{k})}{g(x,1)} \qquad (2.105)$$

to second order in $|\phi_\alpha(\mathbf{k})|$. As in Section 2.6.2, this is a Landau–Ginzburg free-energy expression that provides the entropic free energy, f_e, for a specified segment profile, $\phi_\alpha(\mathbf{r})$, without the need to solve for the underlying field, $w(\mathbf{r})$.

This time, the derivation assumes small variations in $\phi_\alpha(\mathbf{r})$, whereas the previous expression in Eq. (2.78) instead assumed that N is large. If we now make this additional assumption that the molecules are large relative to the wavelengths of all the concentration fluctuations (i.e. $x \gg 1$), then $g(x,1) \approx 2/x$ and Eq. (2.105) reduces to

$$\frac{f_e}{k_B T} \approx \frac{a^2 \rho_0}{24 \bar{\phi}_\alpha} \int d\mathbf{r}\, |\nabla \phi_\alpha(\mathbf{r})|^2 \qquad (2.106)$$

where we have made use of the identity

$$\frac{1}{(2\pi)^3} \int d\mathbf{k}\, k^2 \phi_\alpha(-\mathbf{k}) \phi_\alpha(\mathbf{k}) = \int d\mathbf{r}\, |\nabla \phi_\alpha(\mathbf{r})|^2 \qquad (2.107)$$

Hence, Eq. (2.105) becomes equivalent to Eq. (2.78), provided that $\phi_\alpha(\mathbf{r}) \approx \bar{\phi}_\alpha$, which indeed was the assumption used to arrive at Eq. (2.105).

2.7
Polymer Brushes

To proceed further with the theory, it is useful to discuss a specific example, and for that we now examine the polymer brush depicted in Fig. 2.3(a), where n polymers are grafted to a planar substrate of area $\mathcal{A}$. (Brushes are often examined in solvent environments, but for simplicity we will treat only the dry-brush case.) The grafting is enforced by requiring the z component of the trajectories to satisfy $z_\alpha(1) = 0$, for $\alpha = 1, 2, \ldots, n$. Pinning the $s = 1$ end of each molecule to $z = 0$ tends to cause the total segment concentration

$$\hat{\phi}(\mathbf{r}) \equiv \sum_{\alpha=1}^{n} \hat{\phi}_\alpha(\mathbf{r}) \tag{2.108}$$

to accumulate near the substrate, exceeding the bulk segment density ρ_0 permitted by the hard-core repulsive interactions. To correct for this, a static field, $w(\mathbf{r})$, is introduced to mimic the hard-core interactions and force the chains to extend outwards from the substrate in order to avoid overcrowding. More specifically, the field is selected such that it produces a uniform bulk segment concentration, $\hat{\phi}(\mathbf{r}) = 1$, up to the brush edge at $z = L$, determined by the total quantity of polymer, $\mathcal{A}L = \mathcal{V} \equiv nN/\rho_0$.

We assume that the grafting is uniform and sufficiently thick that the brush exhibits translational symmetry, from which it follows that the field $w(z)$ varies only in the z direction. This, in turn, allows us to separate the energy of a polymer fragment as

$$E[\mathbf{r}_\alpha; s_1, s_2] = E_\parallel[x_\alpha; s_1, s_2] + E_\parallel[y_\alpha; s_1, s_2] + E_\perp[z_\alpha; s_1, s_2] \tag{2.109}$$

where

$$E_\parallel[x_\alpha; s_1, s_2] = \frac{3}{2a^2 N} \int_{s_1}^{s_2} ds \, [x'_\alpha(s)]^2 \tag{2.110}$$

$$E_\perp[z_\alpha; s_1, s_2] = \int_{s_1}^{s_2} ds \left(\frac{3}{2a^2 N} [z'_\alpha(s)]^2 + w(z_\alpha(s)) \right) \tag{2.111}$$

This separation implies that the $x_\alpha(s)$ and $y_\alpha(s)$ components of the trajectory are unaffected by the field and continue to perform simple, unperturbed random walks. Consequently, we need only to consider the one-dimensional problem involving the z component of the trajectory, $z_\alpha(s)$.

Rather than starting with the full self-consistent field theory (SCFT), we first present the analytical strong-stretching theory (SST) (Semenov 1985) for thick brushes in which the chains are, effectively, restricted to their classical trajectories. In this limit, a powerful analogy with classical mechanics (Milner et al. 1988) allows us to deduce that $w(z)$ is parabolic, without having to invoke the incompressibility condition, $\hat{\phi}(\mathbf{r}) = 1$. Following that, the effect of fluctuations about the classical path are examined using both the path-integral formulation and the diffusion equation, but still assuming the parabolic potential. We then conclude by describing the SCFT for determining the proper $w(z)$ in the presence of chain fluctuations, and presenting direct comparisons with the preceding SST predictions.

2.7.1
SST for a Brush: the Parabolic Potential

For very thick brushes (i.e. $L/aN^{1/2} \gg 1$), the polymer chains are generally so extended that fluctuations about the classical paths of lowest energy can be ignored to a first approximation. We therefore start by identifying the classical trajectory of a chain with its $s = 0$ end extended to $z = z_0$. The one-dimensional version of Eq. (2.44) implies that the classical path must satisfy

$$\frac{3[z'_\alpha(s)]^2}{2a^2 N} - w(z_\alpha(s)) = \text{constant} \qquad (2.112)$$

Normally, the unknown field, $w(z)$, would be determined using the incompressibility condition, but in this particular situation it can be deduced by exploiting an analogy with classical mechanics (Milner et al. 1988). On physical grounds, the tension of the chain should vanish at the free end (i.e. $z'_\alpha(0) = 0$), just as it would for a heavy length of rope hanging in a gravitational potential. Add to this the fact that the trajectory must finish at the substrate [i.e. $z_\alpha(1) = 0$] regardless of the starting point [i.e. $z_\alpha(0) = z_0$], and we are ready to exploit the analogy with classical mechanics developed in Section 2.5. The analogy maps this polymer problem onto that of a particle that starts from rest and reaches $z = 0$ in exactly one unit of time, regardless of the starting point. As any undergraduate student will know, this is the property of a pendulum (or simple harmonic oscillator) with a period of 4, for which they would have no trouble working out the potential, $U(z)$. Then, from the correspondence in Eq. (2.52), it immediately follows that

$$w(z) = -\frac{3\pi^2 z^2}{8a^2 N} \qquad (2.113)$$

for which the corresponding trajectory is

$$z_\alpha(s) = z_0 \cos(\pi s/2) \qquad (2.114)$$

Inserting the trajectory into Eq. (2.111), the total energy, $E_\perp[z_\alpha; 0, 1]$, of the chain is found to be precisely zero. In reality, there should be a small dependence on z_0 related to the distribution, $g(z_0)$, of the free end, but that is lost in this simple treatment. In the more complete theory (Matsen 2002b) where $g(z_0) \propto \exp(-E_\perp[z_\alpha; 0, 1]/k_B T)$, there is a small entropic tension at the free end of the chain produced by the tendency for a broad $g(z_0)$ distribution, which incidentally negates the argument for the parabolic potential. However, let us stick with the simpler theory for now.

In the simpler version of the theory, $g(z_0)$ has to be determined by requiring the overall segment concentration to be uniform. When a chain with a given end-point z_0 is unable to fluctuate about its classical path, the sNth segment has the delta (i.e. infinitely narrow) distribution

$$\rho(z; z_0, s) = aN^{1/2}\delta(z - z_\alpha(s)) \tag{2.115}$$

Summing this distribution over all the segments gives the concentration

$$\phi(z; z_0) \equiv \int_0^1 ds\, \rho(z; z_0, s) \tag{2.116}$$

$$= \begin{cases} 2aN^{1/2}/(\pi\sqrt{z_0^2 - z^2}) & \text{if } z < z_0 \\ 0 & \text{if } z > z_0 \end{cases} \tag{2.117}$$

of an entire chain with its free end at $z = z_0$. From this, the average concentration over all the chains becomes

$$\phi(z) = \frac{L}{a^2 N} \int_0^1 dz_0\, g(z_0)\phi(z; z_0) \tag{2.118}$$

By requiring $\phi(z) = 1$ over the thickness of the brush, this integral equation can be inverted giving the end-segment distribution

$$g(z_0) = \frac{z_0 a N^{1/2}}{L\sqrt{L^2 - z_0^2}} \tag{2.119}$$

Although the inversion (Netz and Schick 1998) is complicated, the above solution is easily confirmed.

Now that $g(z_0)$ is known, the calculation of the average stretching energy, f_e, of the brush is straightforward. First, the stretching energy of an individual chain is

$$\frac{f_e(z_0)}{k_B T} = \frac{3}{2a^2 N}\int_0^1 ds\, [z'_\alpha(s)]^2 = \frac{3\pi^2 z_0^2}{16 a^2 N} \tag{2.120}$$

Weighting this with respect to $g(z_0)$ gives an average stretching energy per chain of

$$\frac{f_e}{k_BT} = \frac{1}{aN^{1/2}} \int_0^L dz_0 \, g(z_0) \frac{f_e(z_0)}{k_BT} = \frac{\pi^2 L^2}{8a^2 N} \qquad (2.121)$$

Although the predictions of SST are not particularly accurate under realistic experimental conditions, the value of simple analytical expressions cannot be overstated. Comparisons with SCFT in Section 2.7.6 will show that SST does properly capture the general qualitative trends.

Before moving on, we present a useful trick for evaluating f_e that avoids the difficult step of calculating $g(z_0)$. It derives from the fact that $E_\perp[z_\alpha; 0, 1] = 0$, which implies that, for each and every individual chain, the stretching energy equals minus the field energy. Thus,

$$\frac{f_e(z_0)}{k_BT} = -\frac{1}{aN^{1/2}} \int_0^L dz \, w(z) \phi(z; z_0) \qquad (2.122)$$

Substituting this into the integral of Eq. (2.121) gives

$$\frac{f_e}{k_BT} = -\frac{1}{L} \int_0^L dz \, w(z) \phi(z) = \frac{3\pi^2}{8a^2 NL} \int_0^L dz \, z^2 \qquad (2.123)$$

where we have made use of Eq. (2.118). The trivial integral of z^2 gives the previous result in Eq. (2.121). An important advantage of this formula is that it easily extends to brushes grafted to curved substrates, a fact that we will use in Section 2.9.4 for block copolymer morphologies.

2.7.2
Path-Integral Formalism for a Parabolic Potential

While the classical trajectory is the most probable, there are, nevertheless, nearby trajectories that make a significant contribution to the partition function. To calculate their effect, we expand about the lowest energy path in a Fourier sine series

$$z_\alpha(s) = z_0 \cos(\pi s/2) + \sum_{n=1}^{\infty} z_n \sin(n\pi s) \qquad (2.124)$$

which constrains the two ends of the chain to $z = 0$ and $z = z_0$. This allows the integrals to be performed analytically, giving

$$\frac{E_\perp[z_\alpha; 0, 1]}{k_BT} = \sum_{n=1}^{\infty} k_n z_n^2 \qquad (2.125)$$

where
$$k_n = \frac{3\pi^2(4n^2-1)}{16a^2 N} \quad (2.126)$$

As required, $k_n > 0$ for all n, implying that fluctuations about the classical path (i.e. $z_n \neq 0$) increase the energy of the chain, the more so for the higher harmonics.

Now we can evaluate the partition function, integrating over all possible paths, or equivalently all possible amplitudes, z_n, of each harmonic, n. The result is

$$q(0, z_0, 1) \propto \prod_{n=1}^{\infty} \int_{-\infty}^{\infty} dz_n \exp(-k_n z_n^2) = \prod_{n=1}^{\infty} \sqrt{\frac{\pi}{k_n}} = \text{constant} \quad (2.127)$$

It might be disturbing that the product tends to zero as more factors are included, but there is in reality a cutoff at $n \approx M$, since sinusoidal fluctuations in the chain trajectory cease to make any sense when the wavelength becomes smaller than the actual monomer size. What is important is that $q(0, z_0, 1)$ does not actually depend on z_0. So, just as in the classical treatment, the free energy of a chain in the parabolic potential is independent of its extension.

The main effect of the fluctuations is to broaden the single-segment distribution, $\rho(z; z_0, s)$, from the delta function predicted by SST in Eq. (2.115). With fluctuations, the distribution is defined as

$$\rho(z; z_0, s) = aN^{1/2} \int \left(\prod_{n=1}^{\infty} dz_n P_n(z_n) \right) \delta(z - z_\alpha(s)) \quad (2.128)$$

where
$$P_n(z_n) = \sqrt{\frac{k_n}{\pi}} \exp(-k_n z_n^2) \quad (2.129)$$

is the Boltzmann distribution for an nth harmonic fluctuation of amplitude z_n. Using the integral representation

$$\delta(x) = \frac{1}{2\pi} \int_{-\infty}^{\infty} dk \exp(ikx) \quad (2.130)$$

of the Dirac delta function, the single-segment distribution can be rewritten as

$$\rho(z; z_0, s) = \frac{aN^{1/2}}{2\pi} \int dk \exp(ik[z - z_0 \cos(\pi s/2)]) \quad (2.131)$$
$$\times \left(\prod_{n=1}^{\infty} \sqrt{\frac{k_n}{\pi}} \int_{-\infty}^{\infty} dz_n \exp(-k_n z_n^2 - ikz_n \sin(n\pi s)) \right)$$

This allows us to integrate over z_n, which leads to the expression

$$\rho(z;z_0,s) = \frac{aN^{1/2}}{2\pi} \int dk\, \exp\left(ik[z - z_0\cos(\pi s/2)] - \frac{k^2}{4}\sum_{n=1}^{\infty}\frac{\sin^2(n\pi s)}{k_n}\right)$$

(2.132)

The summation can then be eliminated using the Fourier series expansion

$$|\sin(\pi s)| = \frac{2}{\pi} - \frac{4}{\pi}\sum_{n=1}^{\infty}\frac{\cos(2n\pi s)}{4n^2 - 1} = \frac{8}{\pi}\sum_{n=1}^{\infty}\frac{\sin^2(n\pi s)}{4n^2 - 1}$$

(2.133)

With that gone, the final integration over k gives the relatively simple result

$$\rho(z;z_0,s) = \left(\frac{3}{2\sin(\pi s)}\right)^{1/2} \exp\left(-\frac{3\pi[z - z_0\cos(\pi s/2)]^2}{2a^2 N\sin(\pi s)}\right)$$

(2.134)

Hence, the chain fluctuations transform the delta-function distributions (Eq. 2.115) predicted by SST into Gaussian distributions centered about the classical trajectory with a standard deviation (i.e. width) of $\sqrt{a^2 N\sin(\pi s)/3\pi}$. The total concentration of the chain, $\phi(z;z_0)$, has to be performed numerically, but that is a relatively trivial calculation. Fig. 2.6 shows the result for several different extensions compared with the SST prediction from Eq. (2.117). The strong-stretching approximation is inaccurate until $z_0/aN^{1/2} \gtrsim 1$, and even then it is rather poor for $z \approx z_0$. Of course, this should not be surprising, as there is relatively little tension at the free end of the chain.

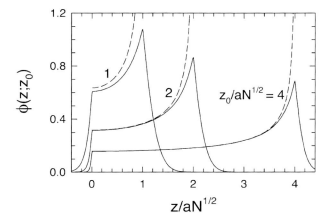

Fig. 2.6 Segment distributions, $\phi(z;z_0)$, of polymers extended to various $z = z_0$ in the parabolic potential, Eq. (2.113). The dashed curves denote the distributions (Eq. 2.117) of the classical paths.

Now we evaluate the average elastic energy, $f_e(z_0)$, of a fluctuating chain extended to $z = z_0$. This is obtained by taking the free energy of a chain in the parabolic potential and subtracting off the average field energy. In mathematical terms,

$$\frac{f_e(z_0)}{k_B T} = -\ln q(0, z_0, 1) - \int dz\, w(z)\phi(z; z_0) \qquad (2.135)$$

Although there is no analytical formula for $\phi(z; z_0)$, Eq. (2.116) allows the field energy to be expressed as

$$\int_0^1 ds \int_0^L dz\, w(z)\rho(z; z_0, s) = -\frac{3\pi^2 z_0^2}{16a^2 N} - \frac{1}{4} \qquad (2.136)$$

which can be evaluated by integrating first over z and then over s. Therefore, the entropic free energy of the chain is

$$\frac{f_e(z_0)}{k_B T} = \frac{3\pi^2 z_0^2}{16a^2 N} + \text{constant} \qquad (2.137)$$

which is precisely the same as predicted by SST (Eq. 2.120), apart from the additive constant. The unknown constant appears because we are unable to specify the proportionality constant in Eq. (2.127), which in turn is because the coarse-graining procedure prevents us from knowing the absolute entropy of the chain; the best we can do is to calculate relative changes in entropy.

2.7.3
Diffusion Equation for a Parabolic Potential

In Section 2.3, we established that the path-integral definition of the partition function is equivalent to the one based on the diffusion equation, just as Feynman's path-integral formalism of quantum mechanics coincides with Schrödinger's differential equation. As in quantum mechanics, the differential equation approach is generally the more practicable method of calculating $q(z, z_0, s)$. Here, we demonstrate that it indeed gives identical results to those obtained in the previous section, where the path-integral approach was used.

Normally the diffusion equation (Eq. 2.24) has to be solved numerically, but there already exists a known analytical solution for parabolic potentials (Merzbacher 1970). In fact, it was originally derived for the wavefunction of a simple harmonic oscillator; once again, we benefit from the analogy with quantum mechanics. Transforming the quantum-mechanical wavefunction in Merzbacher (1970), according to Eqs. (2.50) to (2.56) for the parabolic potential in Eq. (2.113), gives

$$q(z, z_0, s) = \left(\frac{3}{4\sin(\pi s/2)}\right)^{1/2} \exp\left(-\frac{3\pi[(z^2 + z_0^2)\cos(\pi s/2) - 2zz_0]}{4a^2 N \sin(\pi s/2)}\right) \tag{2.138}$$

Although the derivation of this expression is complicated, it is relatively trivial to confirm that it is in fact the proper solution to Eq. (2.24) for the initial condition

$$q(z, z_0, 0) = aN^{1/2}\delta(z - z_0)$$

Substituting $z = 0$ and $s = 1$ into Eq. (2.138), we find that $q(0, z_0, 1) = \sqrt{3/4}$, consistent with the path-integral calculation in Eq. (2.127). In terms of the partition function, the distribution of the sNth segment is given by

$$\rho(z; z_0; s) = \frac{q(0, z, 1-s)q(z, z_0, s)}{q(0, z, 1)} \tag{2.139}$$

The derivation is essentially the same as that for Eq. (2.31). Inserting Eq. (2.138) and simplifying with some basic trigonometric identities gives the identical expression to that in Eq. (2.134), as required by the equivalence of the two approaches. Given this, the remaining results of the preceding section follow. One slight difference is that we now obtain a specific value for the unknown constant in Eq. (2.137), because the initial condition for $q(z, z_0, s)$, in effect, selects a proportionality constant for Eq. (2.127), but not one with any physical significance. Nevertheless, this provides us with a standard reference by which to compare calculations.

2.7.4
Self-Consistent Field Theory (SCFT) for a Brush

The parabolic potential used up to now is only approximate. The argument for it assumes that the tension at the free chain end is zero, but there is in fact a small entropic force that acts to broaden the end-segment distribution, $g(z_0)$ (Matsen 2002b). Here, the self-consistent field theory (SCFT) is described for determining the field in a more rigorous manner. The theory starts with the partition function for the entire system, which is written as

$$Z \propto \int \left(\prod_{\alpha=1}^{n} \tilde{\mathcal{D}}\boldsymbol{r}_\alpha \, \delta(z_\alpha(1) - \epsilon)\right) \delta[1 - \hat{\phi}] \tag{2.140}$$

where the tildes, $\tilde{\mathcal{D}}\boldsymbol{r}_\alpha \equiv \mathcal{D}\boldsymbol{r}_\alpha P[\boldsymbol{r}_\alpha]$, denote that the functional integrals are weighted by

$$P[\boldsymbol{r}_\alpha] = \exp\left(-\frac{3}{2a^2 N}\int_0^1 ds\, |\boldsymbol{r}'_\alpha(s)|^2\right) \tag{2.141}$$

so as to account for the internal entropy of each coarse-grained segment. The functional integrals over each $r_\alpha(s)$ are, in principle, restricted to the volume of the system, $\mathcal{V} = \mathcal{A}L$. To ensure that the chains are properly grafted to the substrate, there are Dirac delta functions constraining the $s = 1$ end of each chain to $z = \epsilon$; we will eventually take the limit $\epsilon \to 0$. Furthermore, there is a Dirac delta functional that constrains the overall concentration, $\hat{\phi}(r)$, to be uniform over all $r \in \mathcal{V}$.

Although the expression for Z is inherently simple, its evaluation is far from trivial. Progress is made by first replacing the delta functional by the integral representation

$$\delta[1 - \hat{\phi}] \propto \int \mathcal{D}W \, \exp\left(\frac{\rho_0}{N} \int d\mathbf{r} \, W(\mathbf{r})[1 - \hat{\phi}(\mathbf{r})]\right) \qquad (2.142)$$

This expression is equivalent to Eq. (2.335) derived in the Appendix, but with $k(x)$ substituted by $-i\rho_0 W(\mathbf{r})/N$. The constants ρ_0 and N have no effect on the limits of integration, but the i implies that $W(\mathbf{r})$ must be integrated in the complex plane along the imaginary axis. Inserting Eq. (2.142) into Eq. (2.140) and substituting $\hat{\phi}(\mathbf{r})$ by Eqs. (2.13) and (2.108) allows the integration over the chain trajectories to be performed. This results in the revised expression

$$Z \propto \int \mathcal{D}W \, (\mathcal{Q}[W])^n \exp\left(\frac{\rho_0}{N} \int d\mathbf{r} \, W(\mathbf{r})\right) \qquad (2.143)$$

where

$$\mathcal{Q}[W] \propto \int \tilde{\mathcal{D}}\mathbf{r}_\alpha \exp\left(-\int_0^1 ds \, W(\mathbf{r}_\alpha(s))\right) \delta(z_\alpha(1) - \epsilon) \qquad (2.144)$$

is the partition function of a single polymer in the external field, $W(\mathbf{r})$. For convenience, the partition function of the system is re-expressed as

$$Z \propto \int \mathcal{D}W \, \exp\left(-\frac{F[W]}{k_B T}\right) \qquad (2.145)$$

where

$$\frac{F[W]}{nk_B T} \equiv -\ln\left(\frac{\mathcal{Q}[W]}{\mathcal{A}aN^{1/2}}\right) - \frac{1}{\mathcal{V}} \int d\mathbf{r} \, W(\mathbf{r}) \qquad (2.146)$$

There are a couple of subtle points that need to be mentioned. First, the present calculation of Z allows the grafted ends to move freely in the $z = \epsilon$ plane, when actually they are grafted to particular spots on the substrate. As long as the chains are densely grafted, the only real consequence of treating the ends as a two-dimensional gas is that $Q[W]$ becomes proportional to $\mathcal{A}$, the area available to each chain end. To correct for this, the $Q[W]$ in Eq. (2.146) is divided by $\mathcal{A}$, which is necessary for $F[W]$ to be a proper extensive quantity that scales with the size of the system. Alternatively, we could imagine that the chain ends are free to move, but then the molecules would become indistinguishable and Z would have to be divided by a factor of $n!$ to avoid the Gibbs paradox (Reif 1965). Either way, we arrive at the same expression for $F[W]$ apart from additive constants. That brings us to the second point, that additive constants to $F[W]$ are of no consequence, since they can be absorbed into the proportionality constant omitted in Eq. (2.145). This fact has been used to insert an extra factor of $aN^{1/2}$ into Eq. (2.146) to make the argument of the logarithm dimensionless.

Even though $W(\mathbf{r})$ takes on imaginary values, $Q[W]$ is solved by exactly the same methods derived in Section 2.3, and thus $F[W]$ is readily computed. However, the functional integral in Eq. (2.145) cannot be performed exactly. To proceed further, we use the fact that $F[W]$ is an analytic functional, which means that the path of integration can be deformed without altering the integral. A standard trick to solve such integrals, the method of steepest descent (Carrier et al. 1966), is to deform the path from the imaginary axis to one that passes through a point where

$$\frac{\mathcal{D} F[W]}{\mathcal{D} W(\mathbf{r})} = 0 \qquad (2.147)$$

in a direction such that the imaginary part of $F[W]$ remains constant. At such a point, the real part of $F[W]$ has a saddle shape, and consequently it is denoted as a *saddle point*. Changing the path of integration in this way concentrates the non-zero contribution of the integral to the neighborhood of the saddle point, allowing for a convenient expansion, the first term of which gives

$$Z \approx \exp\left(-\frac{F[w]}{k_\mathrm{B} T}\right) \qquad (2.148)$$

where $w(\mathbf{r})$ is the solution to Eq. (2.147). Thus, $F[w]$ becomes the SCFT approximation to the free energy of the system. It follows from the functional differentiation in Eq. (2.147) that $w(\mathbf{r})$ must be chosen such that

$$\phi(\mathbf{r}) = 1 \qquad (2.149)$$

where

$$\phi(\mathbf{r}) \equiv -\mathcal{V}\frac{\mathcal{D}\ln(Q[w])}{\mathcal{D} w(\mathbf{r})} \qquad (2.150)$$

is the segment concentration of n polymers in the external field, $w(\mathbf{r})$ (see Eq. 2.32).

The quantity $\phi(\mathbf{r})$ can also be identified as the SCFT approximation for $\langle\hat{\phi}(\mathbf{r})\rangle$, but the explanation for this is often glossed over. It is justified by starting with the formal definition

$$\langle\hat{\phi}(\mathbf{r})\rangle \equiv \frac{1}{Z}\int\left(\prod_{\alpha=1}^{n}\tilde{\mathcal{D}}\mathbf{r}_\alpha\,\delta(z_\alpha(1)-\epsilon)\right)\hat{\phi}(\mathbf{r})\delta[1-\hat{\phi}] \qquad (2.151)$$

where here Z is evaluated with the missing proportionality constant set to unity. Of course, the Dirac delta functional implies that $\langle\hat{\phi}(\mathbf{r})\rangle = 1$, but let us instead evaluate it in the manner of SCFT. With steps equivalent to those that lead to Eq. (2.145), we arrive at

$$\langle\hat{\phi}(\mathbf{r})\rangle = \frac{1}{Z}\int \mathcal{D}W\,\phi(\mathbf{r})\exp\left(-\frac{F[W]}{k_\mathrm{B}T}\right) \qquad (2.152)$$

Here, $\phi(\mathbf{r})$ is the concentration of n polymers subjected to the fluctuating field $W(\mathbf{r})$, but, since the integrand is dominated by $W(\mathbf{r}) \approx w(\mathbf{r})$, the concentration can be evaluated at the saddle point. This allows $\phi(\mathbf{r})$ to be moved outside the functional integration, leading immediately to our intended identity,

$$\langle\hat{\phi}(\mathbf{r})\rangle \approx \frac{\phi(\mathbf{r})}{Z}\int \mathcal{D}W\,\exp\left(-\frac{F[W]}{k_\mathrm{B}T}\right) = \phi(\mathbf{r}) \qquad (2.153)$$

With the saddle-point approximation, the calculation now becomes tractable. The partition function for a chain in the field $w(z)$, with its $s=1$ end fixed at $z=\epsilon$, is

$$\frac{\mathcal{Q}[w]}{\mathcal{V}} = \frac{1}{L}\int_0^L dz\, q(z,s)q^\dagger(z,s) \qquad (2.154)$$

where the integral can be evaluated with any $0 \le s \le 1$. The partial partition function, $q(z,s)$, is obtained by solving the diffusion equation (Eq. 2.24) with the initial condition $q(z,0)=1$, and similarly the complementary function, $q^\dagger(z,s)$, is calculated with Eq. (2.28) subject to $q^\dagger(z,1) = aN^{1/2}\delta(z-\epsilon)$. These quantities also provide the concentration

$$\phi(z) = \frac{\mathcal{V}}{\mathcal{Q}[w]}\int_0^1 ds\, q(z,s)q^\dagger(z,s) \qquad (2.155)$$

which must be equated to unity, by an appropriate adjustment of the field $w(z)$. Once the field is determined self-consistently, the free energy of the brush, $F[w]$, can be evaluated using Eq. (2.146).

2.7.5
Boundary Conditions

In the bulk region of a melt, the long-range attractive interactions favoring a high density are balanced against short-range hard-core interactions to produce a more or less uniform segment concentration, ρ_0. The details of this mechanism are easily avoided by invoking the incompressibility assumption, $\hat{\phi}(\boldsymbol{r}) = 1$. This is no longer valid near the substrate ($z = 0$) nor the air interface ($z = L$). At any surface, the dimensionless concentration $\hat{\phi}(\boldsymbol{r})$ must drop to zero. Although the decay in concentration is generally very sharp, it cannot occur as a true step function. Instead, there must be a continuous surface profile, $\phi_{\rm s}(z)$, that switches from 1 to 0 over a narrow region next to the surface.

The calculation of $\phi_{\rm s}(z)$ requires an involved treatment that takes a detailed account of the molecular interactions. (Although we will be explicitly referring to the $z = 0$ surface, everything we say applies equally to the $z = L$ surface.) In a proper treatment, the partial partition functions would be solved under the boundary conditions $q(0, s) = q^\dagger(0, s) = 0$, which prevent the polymers from crossing the $z = 0$ plane and ensure that $\phi_{\rm s}(z) \to 0$ as $z \to 0$. For discussion purposes, we will assume that the profile is not so sharp as to invalidate our coarse-grained model. It will certainly be sharp enough for the ground-state dominance approximation to hold, which implies that the surface field is given by

$$w_{\rm s}(z) = \frac{a^2 N}{6} \frac{\nabla^2 \sqrt{\phi_{\rm s}(z)}}{\sqrt{\phi_{\rm s}(z)}} \qquad (2.156)$$

However, one further relation between $w_{\rm s}(z)$ and $\phi_{\rm s}(z)$ is required in order to solve for the surface profile. This is where the details of the molecular interactions must enter.

Provided that we are not concerned with knowing the details of the surface, such as $\phi_{\rm s}(z)$ and the resulting surface tension, it is possible to proceed without a full treatment of the surface profile. This involves implementing the alternative boundary conditions

$$\frac{\partial}{\partial z} q(0, s) = \frac{\partial}{\partial z} q^\dagger(0, s) = 0 \qquad (2.157)$$

and solving the SCFT for $w(z)$ ignoring the deviation from $\phi(z) = 1$ at the surface. The full solution would only differ in the narrow region next to the substrate where $\phi_{\rm s}(z) < 1$. More specifically, if we were given $\phi_{\rm s}(z)$, the proper solution would be obtained by making the substitutions

$$q(z, s) \implies q(z, s) \sqrt{\phi_{\rm s}(z)} \qquad (2.158)$$

$$q^\dagger(z,s) \Longrightarrow q^\dagger(z,s)\sqrt{\phi_s(z)} \tag{2.159}$$

$$w(z) \Longrightarrow w(z) + w_s(z) \tag{2.160}$$

The added factor of $\sqrt{\phi_s(z)}$ would restore the proper boundary condition for the partial partition functions, and would transform the concentration profile from $\phi(z) = 1$ to $\phi(z) = \phi_s(z)$. One can also confirm that the transformation in the field is exactly as required for the transformed partition functions to satisfy the diffusion equations (Eqs. 2.24 and 2.28); the proof of this requires that the pre-transformed partition functions satisfy

$$\left[\frac{\partial}{\partial z} q(z,s)\right]\left[\frac{\partial}{\partial z}\sqrt{\phi_s(z)}\right] = \left[\frac{\partial}{\partial z} q^\dagger(z,s)\right]\left[\frac{\partial}{\partial z}\sqrt{\phi_s(z)}\right] = 0 \tag{2.161}$$

Indeed, it is safe to assume that the boundary conditions in Eq. (2.157) extend over the narrow region where $\phi_s(z)$ has a non-zero gradient.

For most problems such as the brush, we simply assume that there is no variation in the profile $\phi_s(z)$ of each surface, rendering them as inert features that conveniently decouple from the behavior of the remaining system. Thus, we can continue to ignore the details of the molecular interactions and solve the field subject to the incompressibility condition $\phi(z) = 1$, with the partial partition functions satisfying reflecting boundary conditions (Eq. 2.157) at $z = 0$ and $z = L$.

For the case of the polymer brush, there is still one more issue to deal with at the $z = 0$ substrate. The grafting of the chain ends in a perfect plane, which for the moment is at $z = \epsilon$, produces a delta function in the field (Likhtman and Semenov 2000). Our intention is to incorporate this singularity into the boundary condition, restoring $w(z)$ to a continuous, finite function. We must, however, begin with the true field

$$w_t(z) = \tfrac{1}{6}\beta a^2 N \delta(z-\epsilon) + w(z) \tag{2.162}$$

containing the singularity at a small infinitesimal distance ϵ from the substrate. The diffusion equation (Eq. 2.24) requires the delta function in $w_t(z)$ to be balanced by an infinity in $\nabla^2 q(\epsilon, s)$. In other words, the delta function produces a sharp kink in $\partial q(z,s)/\partial z$ at $z = \epsilon$, and similarly in $\partial q^\dagger(z,s)/\partial z$. The strength of the kink is determined by integrating the diffusion equation from $z = 0$ to 2ϵ, which gives

$$\frac{a^2 N}{6}\left[\frac{\partial}{\partial z}q(2\epsilon,s) - \frac{\partial}{\partial z}q(0,s)\right] - \frac{\beta a^2 N}{6} q(\epsilon,s) \approx 0 \tag{2.163}$$

Our original reflecting boundary condition implies that $\partial q_z(0,s)/\partial z = 0$. Dropping this term from Eq. (2.163) and then taking the limit $\epsilon \to 0$ yields the new boundary condition

$$\frac{\partial}{\partial z}q(0,s) = \beta q(0,s) \tag{2.164}$$

Naturally, the complementary partition function $q^\dagger(z,s)$ satisfies the same boundary condition at $z=0$. The constant β is determined by the requirement that $w(z)$ does not develop a delta function at $z=0$.

2.7.6
Spectral Solution to SCFT

There are two main ways to solve the SCFT: one is a direct real-space method (Matsen 2004), where the diffusion equations are solved using finite differences for the derivatives (the Crank–Nicolson algorithm); and the other involves a spectral method (Matsen and Schick 1994a) much like that discussed in Section 2.6.1. The spectral method is particularly efficient for lower grafting densities, while the real-space one is better suited to thick brushes. Note that there are also semi-spectral algorithms (Rasmussen and Kalosakas 2002) that combine the advantages of the two methods. Here, we demonstrate the spectral method.

Since the field is unknown, the functional expansions will be performed in terms of Laplacian eigenfunctions, $f_i(z)$, defined by

$$\nabla^2 f_i(z) = -\frac{\lambda_i}{L^2} f_i(z) \tag{2.165}$$

subject to the boundary conditions $f_i'(0) = \beta f_i(0)$ and $f_i'(L) = 0$. Note that the L^2 in Eq. (2.165) has been included so that λ_i becomes dimensionless. The solution to this eigenvalue problem is

$$f_i(z) = C_i \cos\left[\sqrt{\lambda_i}(1 - z/L)\right] \tag{2.166}$$

where the boundary conditions require

$$\sqrt{\lambda_i} \tan\left(\sqrt{\lambda_i}\right) = \beta L \tag{2.167}$$

With the normalization constant set to

$$C_i = \sqrt{\frac{4\sqrt{\lambda_i}}{2\sqrt{\lambda_i} + \sin(2\sqrt{\lambda_i})}} \tag{2.168}$$

the basis functions obey the orthonormal condition in Eq. (2.60). Once the basis functions are determined, there are a couple of useful quantities to calculate. The first is the symmetric tensor

$$\Gamma_{ijk} \equiv \frac{1}{L}\int dz\, f_i(z)f_j(z)f_k(z) \qquad (2.169)$$

and the second is the set of coefficients for the identity function

$$\mathcal{I}_i \equiv \frac{1}{L}\int_0^L dz\, f_i(z) = C_i\frac{\sin(\sqrt{\lambda_i})}{\sqrt{\lambda_i}} \qquad (2.170)$$

defined such that $\sum_i \mathcal{I}_i f_i(z) = 1$.

With the preamble complete, we are now ready to solve for the partial partition function. The procedure is slightly more involved than in Section 2.6.1, because the field is now omitted from Eq. (2.165), but that is the price we pay to have simple basis functions. Nevertheless, the procedure is much the same. The expansions

$$q(z,s) = \sum_i q_i(s)f_i(z) \quad \text{and} \quad w(z) = \sum_k w_k f_k(z)$$

are first inserted into the diffusion equation (Eq. 2.24), giving

$$\sum_i \frac{dq_i(s)}{ds} f_i(z) = -\frac{a^2 N}{6L^2}\sum_i \lambda_i q_i(s) f_i(z) - \sum_{ik} w_k q_i(s) f_i(z) f_k(z) \qquad (2.171)$$

By multiplying through by $f_j(z)$, integrating over r, and using the orthonormal condition, this can be rewritten as a system of first-order linear ordinary differential equations

$$\frac{d}{ds}q_i(s) = \sum_j A_{ij} q_j(s) \qquad (2.172)$$

subject to the initial conditions $q_i(0) = \mathcal{I}_i$. The coefficients of the matrix $\mathbf{A}$ are given by

$$A_{ij} \equiv -\frac{\lambda_i a^2 N}{6L^2}\delta_{ij} - \sum_k w_k \Gamma_{ijk} \qquad (2.173)$$

Note that the boundary conditions on $q(z,s)$ are already accounted for by those on $f_i(z)$.

The solution to Eq. (2.172) is simply

$$q_i(s) = \sum_j T_{ij}(s) q_j(0) \qquad (2.174)$$

where $\mathbf{T}(s) \equiv \exp(\mathbf{A}s)$ is referred to as a transfer matrix. To evaluate it, we perform a matrix diagonalization

$$\mathbf{A} = \mathbf{U}\mathbf{D}\mathbf{U}^{-1} \qquad (2.175)$$

where the elements of the diagonal matrix, $d_k \equiv D_{kk}$, are the eigenvalues of $\mathbf{A}$ and the respective columns of $\mathbf{U}$ are the corresponding normalized eigenvectors. Since $\mathbf{A}$ is symmetric, all the eigenvalues will be real and $\mathbf{U}^{-1} = \mathbf{U}^T$. We can then express $\mathbf{T}(s) = \mathbf{U}\exp(\mathbf{D}s)\mathbf{U}^T$, from which it follows that

$$T_{ij}(s) = \sum_k \exp(sd_k) U_{ik} U_{jk} \qquad (2.176)$$

Similarly, the solution for the complementary function,

$$q^\dagger(z,s) \equiv \sum_i q_i^\dagger(s) f_i(z),$$

is given by

$$q_i^\dagger(s) = \sum_j T_{ij}(1-s) q_j^\dagger(1) \qquad (2.177)$$

where

$$q_i^\dagger(1) = \frac{1}{L} \int_0^L dz\, q^\dagger(z,1) f_i(z) = C_i \cos\left(\sqrt{\lambda_i}\right) \frac{a N^{1/2}}{L} \qquad (2.178)$$

This result is obtained by first integrating with $q^\dagger(z,1) = aN^{1/2}\delta(z-\epsilon)$, and then taking the limit $\epsilon \to 0$. Once the partial partition functions are evaluated, the partition function for the entire chain is obtained by

$$\frac{\mathcal{Q}[w]}{\mathcal{V}} = \sum_i q_i(s) q_i^\dagger(s) = \sum_i \exp(d_i) \bar{q}_i(0) \bar{q}_i^\dagger(1) \qquad (2.179)$$

where we have made the convenient definitions

$$\bar{q}_i(0) \equiv \sum_j q_j(0) U_{ji} \quad \text{and} \quad \bar{q}_i^\dagger(1) \equiv \sum_j q_j^\dagger(1) U_{ji}.$$

The coefficients of the average polymer concentration can be expressed as

$$\phi_i = \frac{\mathcal{V}}{\mathcal{Q}[w]} \sum_{jk} I_{jk} \Gamma_{ijk} \qquad (2.180)$$

where the components of the matrix **I** are defined as

$$I_{jk} \equiv \int_0^1 ds\, q_j(s) q_k^\dagger(s) \tag{2.181}$$

$$= \sum_{m,n} \left[\frac{\exp(d_m) - \exp(d_n)}{d_m - d_n} \right] U_{jm} U_{kn} \bar{q}_m(0) \bar{q}_n^\dagger(1) \tag{2.182}$$

Note that, in the instances where $d_m = d_n$, the factor in the square brackets reduces to $\exp(d_m)$.

Given the formula to evaluate the segment concentration, the coefficients of the field w_i can now be adjusted to satisfy the incompressibility condition $\phi(z) = 1$, which is expressed in terms of coefficients as

$$\phi_i = \mathcal{I}_i \tag{2.183}$$

In practice, the functional expansion has to be truncated at some value $i = M$, where M is chosen based on an acceptable error tolerance. In principle, the first $(M+1)$ components w_i would be adjusted to satisfy the first $(M+1)$ conditions for ϕ_i. However, the components of Eq. (2.183) are not completely independent, due to the fact that SCFT is not affected by an additive constant to $w(z)$ (see the discussion at the end of Section 2.3). To rectify this problem, the first condition, $\phi_0 = \mathcal{I}_0$, can be replaced by

$$\sum_i \mathcal{I}_i w_i = 0 \tag{2.184}$$

which sets the spatial average of $w(z)$ to zero. Now the set of equations are independent, and they can be solved by standard numerical techniques. There are various techniques that can be used (Thompson et al. 2004); we generally use the Broyden algorithm, which is a quasi-Newton–Raphson method. This requires a reasonable initial guess for $w(z)$, but that can be provided by the SST estimate in Eq. (2.113). Fig. 2.7 shows the field $w(z)$ obtained from SCFT (solid curves) for several different brush thicknesses compared against the SST result (dashed curve) from Eq. (2.113). As expected, the SST prediction becomes increasingly accurate as $L \to \infty$.

We have glossed over the fact that β was introduced in order to prevent a delta function from developing in $w(z)$ at $z = 0$. The signature of a delta function is that the field coefficients w_i fail to converge to zero in the limit of large i. To prevent this, we could include the extra condition $w_M = 0$ in our Broyden iterations and use it to determine β. The estimate, $\beta \approx 3L/a^2 N$, by Likhtman and Semenov (2000) provides a suitable initial guess. In reality,

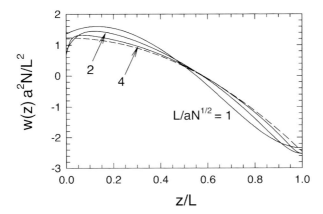

Fig. 2.7 Self-consistent field $w(z)$ plotted for several brush thicknesses L (solid curves). The dashed curve denotes the SST prediction in Eq. (2.113), shifted vertically such that its average is zero.

this initial guess turns out to be very accurate, and since the spectral method can cope with a small delta function, we simply use the initial guess provided by Likhtman and Semenov (2000).

Although the end-segment distribution function $g(z_0)$ is not a direct by-product of SCFT, it is easily evaluated. The formula for it,

$$g(z_0) = \frac{\mathcal{V}}{\mathcal{Q}[w]} q^\dagger(z_0, 0) \tag{2.185}$$

is derived in the same manner as Eq. (2.155) for $\phi(z)$. Fig. 2.8 shows the SCFT prediction for $g(z_0)$ (solid curves) calculated for several different brush thicknesses compared against the SST prediction (dashed curve) from Eq. (2.119). As with the field, SST becomes increasingly accurate as $L \to \infty$.

The free energy of the brush is given by

$$\frac{F}{nk_\mathrm{B}T} = -\ln\left(\frac{\mathcal{Q}[w]}{\mathcal{A}aN^{1/2}}\right) - \frac{\beta a^2 N}{6L} \tag{2.186}$$

The second term comes from the delta-function contribution to the true potential $w_\mathrm{t}(z)$ in Eq. (2.162), which was removed from $w(z)$ by the use of the boundary condition in Eq. (2.164). The resulting energy is plotted in Fig. 2.9. Included with a dotted curve is an approximation,

$$\frac{F}{nk_\mathrm{B}T} \approx \frac{\pi^2 L^2}{8a^2 N} - \ln\left(\frac{\sqrt{3}\,L}{2aN^{1/2}}\right) + 0.1544$$

$$- 0.64\left(\frac{L}{aN^{1/2}}\right)^{-2/3} - 0.09\left(\frac{L}{aN^{1/2}}\right)^{-4/3} \tag{2.187}$$

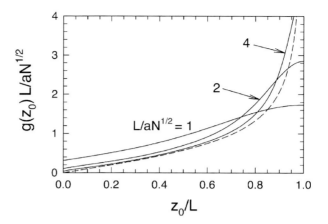

Fig. 2.8 End-segment distribution $g_0(z_0)$ calculated by SCFT for several brush thicknesses L (solid curves). The dashed curve denotes the SST prediction from Eq. (2.119).

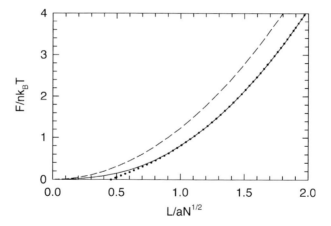

Fig. 2.9 Free energy F per chain versus brush thickness L calculated by SCFT (solid curve). The dashed curve shows the SST prediction from Eq. (2.121), and the dotted curve denotes the improved approximation in Eq. (2.187), which on this scale is indistinguishable from SCFT for $L/aN^{1/2} \gtrsim 1$.

based on a theoretical expansion by Likhtman and Semenov (2000). The two coefficients of 0.64 and 0.09 were obtained by a previous fit to SCFT (Matsen 2004) at $L/aN^{1/2} \gg 1$, but, as Fig. 2.9 clearly shows, the fit remains accurate down to reasonably low values of L. Notice that the dominant term in Eq. (2.187) is precisely the SST prediction from Eq. (2.121).

2.8 Polymer Blends

The polymer brush treated in the previous section provided a good illustration of how SCFT works. The procedure began with an expression for the partition function of the system, Z, expressed in terms of functional integrals over the path of each molecule. An integral representation was used to replace a Dirac delta functional, so that the functional integrations over each polymer could be performed. However, this left a functional integral over a field, which could not be evaluated and required a saddle-point approximation. Although these basic steps remain common to all SCFT calculations, the simple brush misses out some common features of the more elaborate systems. Most notably, the brush contained only a single molecular species, which in turn was composed of only one monomer type. This section demonstrates the natural extension to multicomponent systems.

Let us now consider a blend of n_A A-type homopolymers with n_B B-type homopolymers, and for simplicity assume that both species have the same degree of polymerization, N, and the same statistical segment length, a. The volume of the system is therefore $\mathcal{V} = nN/\rho_0$, where $n = n_A + n_B$ is the total number of molecules in the blend. (For convenience, we follow the standard convention of defining the A and B coarse-grained segments such that they each occupy the same volume, ρ_0^{-1}.) The volume fraction of A-type segments (i.e. composition of the blend) will be denoted by $\phi \equiv n_A N/\mathcal{V}\rho_0$. Again, all polymers are parameterized by s running from 0 to 1, and the configurations of A- and B-type polymers are specified by $\mathbf{r}_{A,\alpha}(s)$ and $\mathbf{r}_{B,\alpha}(s)$, respectively. Given this, the overall A-segment concentration can be expressed as

$$\hat{\phi}_A(\mathbf{r}) = \frac{N}{\rho_0} \sum_{\alpha=1}^{n_A} \int_0^1 ds\, \delta(\mathbf{r} - \mathbf{r}_{A,\alpha}(s)) \qquad (2.188)$$

with an analogous expression specifying the B-segment concentration, $\hat{\phi}_B(\mathbf{r})$.

With the inclusion of two chemically distinct segments, it is no longer possible to sidestep the issue of monomer–monomer interactions. For discussion purposes, assume that the interaction between two monomers, i and j, obeys a simple Lennard-Jones potential

$$u(r_{ij}) = \epsilon \left[\left(\frac{\sigma}{r_{ij}}\right)^{12} - 2\left(\frac{\sigma}{r_{ij}}\right)^6 \right] \qquad (2.189)$$

where r_{ij} is their separation. The incompressibility assumption

$$\hat{\phi}_A(\mathbf{r}) + \hat{\phi}_B(\mathbf{r}) = 1 \qquad (2.190)$$

acts to maintain the spacing between neighboring monomers at the minimum of the potential energy, $r_{ij} = \sigma$. The detailed shape of the potential is therefore no longer needed; we can simply assume that the interaction energy is $-\epsilon$ between neighboring monomers and zero otherwise. In the brush system, all the interactions were the same and thus they added up to a constant energy that could be ignored, assuming the total number of monomer contacts remained more or less fixed. In the blend, however, we must account for the fact that the depth of the potential will differ between the A–A, A–B, and B–B contacts. Noting that the range of the monomer–monomer interactions, σ, is negligible on the coarse-grained scale, the interactions can be treated as contact energies and thus the internal energy can be expressed as

$$U \equiv -\frac{N^2 \epsilon_{AA}}{2\rho_0^2} \sum_{\alpha,\beta} \int ds\, dt\, \delta(\mathbf{r}_{A,\alpha}(s) - \mathbf{r}_{A,\beta}(t))$$

$$-\frac{N^2 \epsilon_{AB}}{\rho_0^2} \sum_{\alpha,\beta} \int ds\, dt\, \delta(\mathbf{r}_{A,\alpha}(s) - \mathbf{r}_{B,\beta}(t))$$

$$-\frac{N^2 \epsilon_{BB}}{2\rho_0^2} \sum_{\alpha,\beta} \int ds\, dt\, \delta(\mathbf{r}_{B,\alpha}(s) - \mathbf{r}_{B,\beta}(t)) \quad (2.191)$$

Using the incompressibility assumption, $\hat{\phi}_A(\mathbf{r}) + \hat{\phi}_B(\mathbf{r}) = 1$, this simplifies to

$$\frac{U[\hat{\phi}_A, \hat{\phi}_B]}{k_B T} = \chi \rho_0 \int d\mathbf{r}\, \hat{\phi}_A(\mathbf{r}) \hat{\phi}_B(\mathbf{r}) \quad (2.192)$$

where the interaction strengths are grouped together into a single dimensionless parameter

$$\chi \equiv \frac{\epsilon_{AA} - 2\epsilon_{AB} + \epsilon_{BB}}{2 k_B T \rho_0} \quad (2.193)$$

Note that the value of χ is inversely proportional to the segment density ρ_0, and thus is not an invariant quantity; one must be careful of this fact when referring to literature values of χ. Equation (2.192) is the simplest and by far the most common expression for the interactions, but SCFT is perfectly capable of accommodating more elaborate treatments (Matsen 2002a), if so desired.

2.8.1
SCFT for a Polymer Blend

The partition function for a polymer blend is hardly any more involved than that for the polymer brush in Section 2.7.4. In this case, it becomes

$$Z \propto \frac{1}{n_A! n_B!} \int \prod_{\alpha=1}^{n_A} \tilde{\mathcal{D}} \boldsymbol{r}_{A,\alpha} \prod_{\beta=1}^{n_B} \tilde{\mathcal{D}} \boldsymbol{r}_{B,\beta} \exp\left(-\frac{U[\hat{\phi}_A, \hat{\phi}_B]}{k_B T}\right) \delta[1 - \hat{\phi}_A - \hat{\phi}_B] \tag{2.194}$$

As before, there is a functional integration over the configuration of each polymer and a Dirac delta functional is used to enforce the incompressibility condition. Of course, the delta functions constraining the chain ends are gone, and there is now a Boltzmann weight accounting for the energy of the segment interactions. Also present are factors of $n_A!$ and $n_B!$ to account for the indistinguishability of the A and B polymers, respectively.

This time, we start by inserting a functional integral over $\delta[\Phi_A - \hat{\phi}_A]$, allowing the operator, $\hat{\phi}_A(\boldsymbol{r})$, to be swapped with the ordinary function, $\Phi_A(\boldsymbol{r})$. This transforms $\delta[1 - \hat{\phi}_A - \hat{\phi}_B]$ into $\delta[1 - \Phi_A - \hat{\phi}_B]$, which, in turn, allows the exchange of $\hat{\phi}_B(\boldsymbol{r})$ with $1 - \Phi_A(\boldsymbol{r})$. The result of this manipulation is

$$Z \propto \frac{1}{n_A! n_B!} \int \mathcal{D}\Phi_A \prod_{\alpha=1}^{n_A} \tilde{\mathcal{D}} \boldsymbol{r}_{A,\alpha} \prod_{\beta=1}^{n_B} \tilde{\mathcal{D}} \boldsymbol{r}_{B,\beta} \exp\left(-\frac{U[\Phi_A, 1-\Phi_A]}{k_B T}\right)$$

$$\times \delta[\Phi_A - \hat{\phi}_A]\delta[1 - \Phi_A - \hat{\phi}_B] \tag{2.195}$$

As with the brush, the two delta functionals are replaced with integral representations analogous to that in Eq. (2.142), which allows us to complete the integrals over the polymer configurations, leading to the result:

$$Z \propto \frac{1}{n_A! n_B!} \int \mathcal{D}\Phi_A \mathcal{D}W_A \mathcal{D}W_B \left(\frac{\rho_0}{N} \mathcal{Q}_A[W_A]\right)^{n_A} \left(\frac{\rho_0}{N} \mathcal{Q}_B[W_B]\right)^{n_B}$$

$$\times \exp\left(-\frac{U[\Phi_A, 1-\Phi_A]}{k_B T} + \frac{\rho_0}{N} \int d\boldsymbol{r} \left[W_A \Phi_A + W_B(1 - \Phi_A)\right]\right) \tag{2.196}$$

where

$$\mathcal{Q}_\gamma[W_\gamma] \propto \int \tilde{\mathcal{D}} \boldsymbol{r}_{\gamma,\alpha} \exp\left(-\int_0^1 ds\, W_\gamma(\boldsymbol{r}_{\gamma,\alpha}(s))\right) \tag{2.197}$$

is the partition function of a single γ-type polymer in the external field $W_\gamma(\boldsymbol{r})$. (For the purpose of future simplification, a factor of $(\rho_0/N)^n$ has been extracted from the unspecified proportionality constant in Eq. 2.196.) Next, the factorials are replaced by the usual Stirling approximations [e.g. $\ln(n_\gamma!) \approx n_\gamma \ln(n_\gamma) - n_\gamma$], giving

$$Z \propto \int \mathcal{D}\Phi_A \mathcal{D}W_A \mathcal{D}W_B \, \exp\left(-\frac{F[\Phi_A, W_A, W_B]}{k_B T}\right) \qquad (2.198)$$

where

$$\frac{F}{nk_B T} = \phi\left[\ln\left(\frac{\phi\mathcal{V}}{\mathcal{Q}_A[W_A]}\right) - 1\right] + (1-\phi)\left[\ln\left(\frac{(1-\phi)\mathcal{V}}{\mathcal{Q}_B[W_B]}\right) - 1\right]$$

$$+ \frac{1}{\mathcal{V}} \int d\mathbf{r} \, [\chi N \Phi_A(1-\Phi_A) - W_A \Phi_A - W_B(1-\Phi_A)] \qquad (2.199)$$

As before with the brush, all the steps given above are exact. The partition functions, $\mathcal{Q}_A[W_A]$ and $\mathcal{Q}_B[W_B]$, can still be calculated by the method outlined above in Section 2.3, and thus $F[\Phi_A, W_A, W_B]$ can be evaluated without difficulty. It is the functional integration of $\exp(-F/k_B T)$ in Eq. (2.198) that poses a problem. So once again, the saddle-point approximation is implemented, which requires us to locate the extremum denoted by the lower-case functions, ϕ_A, w_A, and w_B. Setting to zero the functional derivative of $F[\Phi_A, W_A, W_B]$ with respect to W_A gives the condition

$$\phi_A(\mathbf{r}) = -\mathcal{V}\phi \frac{\mathcal{D} \ln(\mathcal{Q}_A[w_A])}{\mathcal{D} w_A(\mathbf{r})} \qquad (2.200)$$

Equation (2.32) identifies $\phi_A(\mathbf{r})$ as the average concentration of n_A polymers in the external field $w_A(\mathbf{r})$, which is evaluated by

$$\phi_A(\mathbf{r}) = -\frac{\mathcal{V}\phi}{\mathcal{Q}_A[w_A]} \int_0^1 ds \, q_A(\mathbf{r}, s) q_A^\dagger(\mathbf{r}, s) \qquad (2.201)$$

where

$$\mathcal{Q}_A[w_A] = \int d\mathbf{r} \, q_A(\mathbf{r}, s) q_A^\dagger(\mathbf{r}, s) \qquad (2.202)$$

Here, the partial partition functions $q_A(\mathbf{r}, s)$ and $q_A^\dagger(\mathbf{r}, s)$ obey the previous diffusion equations (Eqs. 2.24 and 2.28), but with the field $w_A(\mathbf{r})$. Differentiation of $F[\Phi_A, W_A, W_B]$ with respect to W_B leads to the incompressibility requirement

$$\phi_A(\mathbf{r}) + \phi_B(\mathbf{r}) = 1 \qquad (2.203)$$

where

$$\phi_B(\mathbf{r}) \equiv -\mathcal{V}\phi \frac{\mathcal{D}\ln(\mathcal{Q}_B[w_B])}{\mathcal{D}w_B(\mathbf{r})} \qquad (2.204)$$

is the average concentration of n_B polymers in the external field $w_B(\mathbf{r})$, evaluated in the same manner as $\phi_A(\mathbf{r})$. Following arguments already presented in Section 2.7.4, $\phi_A(\mathbf{r})$ and $\phi_B(\mathbf{r})$ are the SCFT approximations for $\langle \hat{\phi}_A(\mathbf{r}) \rangle$ and $\langle \hat{\phi}_B(\mathbf{r}) \rangle$, respectively. The last remaining differentiation of $F[\Phi_A, W_A, W_B]$ with respect to Φ_A gives the self-consistent field condition

$$w_A(\mathbf{r}) - w_B(\mathbf{r}) = \chi N[1 - 2\phi_A(\mathbf{r})] \qquad (2.205)$$

In principle, the two Eqs. (2.203) and (2.205) should be enough to determine the two functions $w_A(\mathbf{r})$ and $w_B(\mathbf{r})$, but this is not exactly the case. With similar arguments to those at the end of Section 2.3, one can show that $F[\phi_A, w_A, w_B]$ is unaffected by a constant added to either $w_A(\mathbf{r})$ or $w_B(\mathbf{r})$. These degrees of freedom can be removed by fixing their spatial averages to $\bar{w}_A = \bar{w}_B = 0$, and adding an appropriate constant to Eq. (2.205) such that

$$w_A(\mathbf{r}) - w_B(\mathbf{r}) = 2\chi N[\phi - \phi_A(\mathbf{r})] \qquad (2.206)$$

With this revised field equation, the SCFT approximation of the free energy $F[\phi_A, w_A, w_B]$ given by Eq. (2.199) can be re-expressed as

$$\frac{F}{nk_B T} = \frac{F_h}{nk_B T} - \phi \ln\left(\frac{\mathcal{Q}_A[w_A]}{\mathcal{V}}\right) - (1-\phi)\ln\left(\frac{\mathcal{Q}_B[w_B]}{\mathcal{V}}\right)$$
$$- \frac{\chi N}{\mathcal{V}} \int d\mathbf{r}\, [\phi_A(\mathbf{r}) - \phi][\phi_B(\mathbf{r}) - 1 + \phi] \qquad (2.207)$$

where

$$\frac{F_h}{nk_B T} = \phi[\ln \phi - 1] + (1-\phi)[\ln(1-\phi) - 1] + \chi N \phi(1-\phi) \qquad (2.208)$$

2.8.2
Homogeneous Phases and Macrophase Separation

The simplest solution to the self-consistent field equations is that for uniform fields, $w_A(\mathbf{r}) = w_B(\mathbf{r}) = 0$, corresponding to a homogeneous mixture. In this case, the partition functions reduce to $\mathcal{Q}_A[w_A] = \mathcal{Q}_B[w_B] = \mathcal{V}$, and the segment concentrations simplify to $\phi_A(\boldsymbol{r}) = \phi$ and $\phi_B(\boldsymbol{r}) = 1 - \phi$. Inserting these results into Eq. (2.207), we find that $F = F_h$. Indeed, Eq. (2.208) is the well-known Flory–Huggins expression for the free energy of a homogeneous blend of composition ϕ. Fig. 2.10(a) shows the free-energy curve for $\chi N = 3$. Notice that there is a region, between the two inflection points denoted by solid diamond symbols, where the free energy has negative curvature. This property implies that the system is unstable toward macrophase separation, where the melt splits into A- and B-rich phases occupying separate regions of the volume $\mathcal{V}$.

To determine whether or not the blend will macrophase-separate, we need an expression for the combined free energy, $F_s = F_h^{(1)} + F_h^{(2)}$, of two distinct phases of compositions $\phi^{(1)}$ and $\phi^{(2)}$, respectively. If the blend is to macrophase-separate, it has to obey some basic conservation rules. First, the number of molecules in the two phases, $n^{(1)}$ and $n^{(2)}$, must satisfy

$$n^{(1)} + n^{(2)} = n \qquad (2.209)$$

and second, the number of A-type polymers in each phase must comply with

$$n^{(1)}\phi^{(1)} + n^{(2)}\phi^{(2)} = n\phi \qquad (2.210)$$

These conservation laws imply that the fraction of molecules in phase (1) will be

$$\frac{n^{(1)}}{n} = \frac{\phi^{(2)} - \phi}{\phi^{(2)} - \phi^{(1)}} \qquad (2.211)$$

Now we can rewrite the total energy F_s of a macrophase-separated system as

$$\frac{F_s}{nk_BT} = \frac{F_h^{(2)}}{n^{(2)}k_BT} + \left(\frac{F_h^{(1)}}{n^{(1)}k_BT} - \frac{F_h^{(2)}}{n^{(2)}k_BT}\right)\frac{n^{(1)}}{n} \qquad (2.212)$$

which is plotted in Fig. 2.10(a) for the example where $\phi^{(1)} = 0.3$ and $\phi^{(2)} = 0.6$. It is clear from the form of Eqs. (2.211) and (2.212) that F_s versus ϕ is simply a straight line extending from $F_s = F_h^{(1)}$ and $\phi = \phi^{(1)}$ to $F_s = F_h^{(2)}$ and $\phi = \phi^{(2)}$. Naturally, the line must terminate at $\phi^{(1)}$ and $\phi^{(2)}$ since the overall composition ϕ is a volume average over the two individual phases (i.e. $\phi^{(1)} \leq \phi \leq \phi^{(2)}$). In this particular example, the entire line F_s lies below the curve F_h, and thus macrophase separation is preferred at all compositions $0.3 < \phi < 0.6$. The volumes occupied by the two resulting phases would occur in the ratio

$$\frac{n^{(1)}}{n^{(2)}} = \frac{\phi^{(2)} - \phi}{\phi - \phi^{(1)}} \qquad (2.213)$$

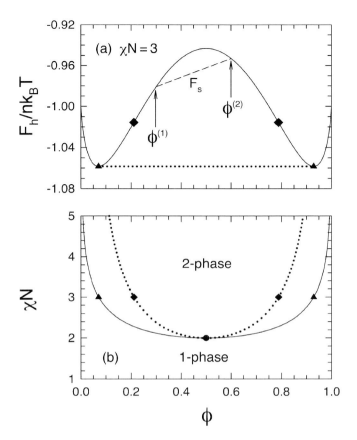

Fig. 2.10 (a) Free energy F_h of a homogeneous blend as a function of composition ϕ, calculated for $\chi N = 3$. The dashed line denotes the free energy F_s for a macrophase-separated blend with compositions $\phi^{(1)} = 0.3$ and $\phi^{(2)} = 0.6$, and the dotted line is the double-tangent construction locating the binodal points denoted by solid triangles. The two solid diamonds denote spinodal points, where the curvature of the free-energy curve switches sign.
(b) Phase diagram showing the binodal (solid curve) and spinodal (dotted curve) curves as a function of χN. The solid triangle and diamond symbols correspond to those in panel (a), and the solid circle denotes a critical point.

which is known as the *lever rule*, on account of its graphical interpretation in terms of the two portions, $(\phi^{(2)} - \phi)$ and $(\phi - \phi^{(1)})$, of the straight line. Of course, the particular compositions, $\phi^{(1)} = 0.3$ and $\phi^{(2)} = 0.6$, do not represent the global free-energy minimum.

The lowest possible energy that can be achieved by macrophase separation is determined by the double-tangent construction denoted with a dotted line in Fig. 2.10(a). At any ϕ along this line, the blend will ultimately macrophase-separate into the compositions at the two ends of the line marked by solid triangles and referred to as *binodal points*. Near these binodal compositions, however, the system is metastable against macrophase separation and can exist in the single mixed state for a considerable time. This is because the positive curvature in F_h causes the free energy initially to increase as $\phi^{(1)}$ and $\phi^{(2)}$ depart from the average composition ϕ. Only when $\phi^{(1)}$ and $\phi^{(2)}$ are sufficiently far apart does the system benefit from a reduction in free energy. The presence of this energy barrier requires the phase separation to proceed by the slow process of nucleation and growth. In contrast, there is no energy barrier at compositions between the *spinodal points* denoted by the solid diamonds, due to the fact that the curvature in the free energy is negative. In this instance, phase separation is relatively fast and occurs by a process referred to as spinodal decomposition.

In this symmetric case where both polymers are the same size, the double-tangent line is horizontal and thus the binodals occur at zero slope in the free energy, which works out to be

$$\chi N = \frac{1}{2\phi - 1} \ln\left(\frac{\phi}{1-\phi}\right) \tag{2.214}$$

The spinodals occur when the second derivative is zero, which corresponds to

$$\chi N = \frac{1}{2\phi(1-\phi)} \tag{2.215}$$

The phase diagram in Fig. 2.10(b) shows the binodals (solid curves) between which the system favors macrophase separation, and the spinodals (dotted curves) between which the separation is more or less immediate. Both sets of curves terminate at a critical point ($\chi N = 2$), below which the interaction strength is too weak to induce phase separation, regardless of the composition.

2.8.3
Scattering Function for a Homogeneous Blend

A beam of radiation (e.g. X-rays or neutrons) will pass undeflected through a perfectly homogeneous blend, $\hat{\phi}_A(r) = \phi$, assuming that the wavelength λ is above the atomic resolution and that absorption is negligible. However, there are always thermal fluctuations that disturb the uniform composition, which in turn scatter the radiation by various angles θ. To predict the resulting pattern of radiation, $I(\theta)$, we must first calculate the free-energy cost, $F[\Phi_A]$, of producing a specified composition fluctuation, $\hat{\phi}_A(r) = \Phi_A(r)$. The fact that we will continue to implement mean-field theory implies that, in reality, the free energy we calculate will be for a fixed $\phi_A(r) \equiv \langle \hat{\phi}_A(r) \rangle$ as opposed to a fixed $\hat{\phi}_A(r)$; this limitation will be discussed further in Section 2.11.

To perform the calculation, we use the Fourier representation developed in Section 2.6.4. In terms of the Fourier transforms of the composition, $\phi_A(\mathbf{k})$ and $\phi_B(\mathbf{k})$, the free energy is approximated by

$$\frac{F}{nk_BT} = \frac{F_h}{nk_BT} + \frac{1}{2\phi(2\pi)^3\mathcal{V}} \int_{\mathbf{k}\neq 0} d\mathbf{k}\, \frac{\phi_A(-\mathbf{k})\phi_A(\mathbf{k})}{g(x,1)}$$

$$+ \frac{1}{2(1-\phi)(2\pi)^3\mathcal{V}} \int_{\mathbf{k}\neq 0} d\mathbf{k}\, \frac{\phi_B(-\mathbf{k})\phi_B(\mathbf{k})}{g(x,1)}$$

$$+ \frac{\chi N}{(2\pi)^3\mathcal{V}} \int_{\mathbf{k}\neq 0} d\mathbf{k}\, \phi_A(-\mathbf{k})\phi_B(\mathbf{k}) \qquad (2.216)$$

The first term is the free energy of a perfectly homogeneous state (Eq. 2.208), the following two integrals represent the loss in configurational entropy (Eq. 2.105) experienced by the A- and B-type polymers, and the final integral accounts for the increased number of A–B contacts (Eq. 2.192 transformed by Eq. 2.82). The incompressibility condition is enforced by setting $\phi_B(\mathbf{k}) = -\phi_A(\mathbf{k})$, which simplifies the free energy to

$$\frac{F}{nk_BT} = \frac{F_h}{nk_BT} + \frac{N}{2(2\pi)^3\mathcal{V}} \int_{\mathbf{k}\neq 0} d\mathbf{k}\, S^{-1}(\mathbf{k})\phi_A(-\mathbf{k})\phi_A(\mathbf{k}) \qquad (2.217)$$

where $S(\mathbf{k})$, referred to as the scattering function, is defined by

$$S^{-1}(\mathbf{k}) = \frac{1}{N\phi(1-\phi)g(\frac{1}{6}k^2a^2N,\,1)} - 2\chi \qquad (2.218)$$

The scattering function for a 50/50 blend is plotted in Fig. 2.11 at several values of χN within the one-phase region of the phase diagram in Fig. 2.10(b). The inverse of $S(\mathbf{k})$ determines the free-energy cost of producing a compositional fluctuation of wavelength $D = 2\pi/k$. In the one-phase region, $S^{-1}(\mathbf{k}) > 0$ for all $\mathbf{k}$, implying that all fluctuations cost energy and are thus suppressed. However, as the two-phase region is entered, $S^{-1}(\mathbf{k})$ becomes negative, starting with the longest wavelengths. For small $\mathbf{k}$,

$$S^{-1}(\mathbf{k}) \approx \frac{1 + \frac{1}{18}k^2a^2N}{N\phi(1-\phi)} - 2\chi \qquad (2.219)$$

which implies that the blend becomes unstable to long-wavelength fluctuations at

$$\chi N = \frac{1}{2\phi(1-\phi)} \qquad (2.220)$$

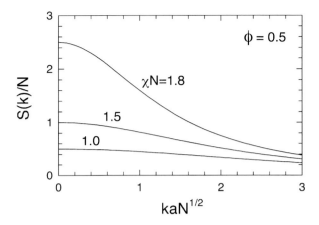

Fig. 2.11 Scattering function $S(\boldsymbol{k})$ for binary blends of symmetric composition, $\phi = 0.5$, plotted for several degrees of segregation, χN.

This is precisely the spinodal line (Eq. 2.215) calculated in the preceding section.

The quantity $S(\boldsymbol{k})$ is called the scattering function because of its direct relevance to the pattern $I(\theta)$ produced by the scattering of radiation. First, there is a one-to-one correspondence between k and θ due to the fact that a periodic compositional fluctuation of wavevector $\boldsymbol{k}$ predominantly scatters

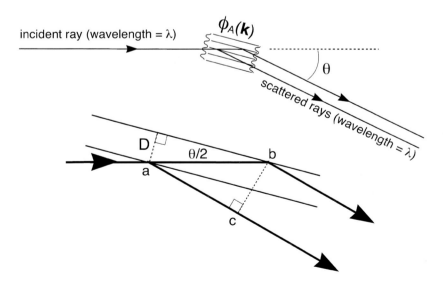

Fig. 2.12 Diagram showing scattered rays of wavelength λ due to a composition fluctuation $\phi_A(\boldsymbol{k})$ of period $D \equiv 2\pi/k$. The sketch at the bottom shows the condition for a first-order Bragg reflection, where the path difference $\overline{ab} - \overline{ac}$, equals one wavelength λ.

radiation by a particular angle θ determined by the Bragg condition outlined in Fig. 2.12. First-order constructive interference requires that the path difference $\overline{\mathrm{ab}} - \overline{\mathrm{ac}}$ equals one wavelength λ of the incident radiation. By simple trigonometry, the path length between points "a" and "b" is $\overline{\mathrm{ab}} = D/\sin(\theta/2)$, where $D = 2\pi/k$ is the period of the composition fluctuation. The alternative path length between "a" and "c" is $\overline{\mathrm{ac}} = \overline{\mathrm{ab}}\cos(\theta) = \overline{\mathrm{ab}}[1 - 2\sin^2(\theta/2)]$. Thus, the Bragg condition relating k and θ is

$$\theta = 2\sin^{-1}\left(\frac{\lambda k}{4\pi}\right) \qquad (2.221)$$

Second, the intensity of the scattered radiation, $I(\theta)$, is proportional to the ensemble average of $|\phi_A(\mathbf{k})|^2$, where the orientation of $\mathbf{k}$ fulfills the Bragg condition. Since the probability of a harmonic fluctuation is given by the Boltzmann factor $\exp(-cS^{-1}(\mathbf{k})|\phi_A(\mathbf{k})|^2)$, where c is independent of $\mathbf{k}$ and $\phi_A(\mathbf{k})$, it follows that $\langle|\phi_A(\mathbf{k})|^2\rangle \propto S(\mathbf{k})$. This, in turn, implies that $I(\theta) \propto S(\mathbf{k})$, and hence the name "scattering function".

2.8.4
SCFT for a Homopolymer Interface

When a blend macrophase-separates, there is inevitably an interface between the two phases as illustrated in Fig. 2.3(b). Here, we calculate its properties and determine the resulting excess free energy per unit area, referred to as the interfacial tension, γ_I. Since the interface is unfavorable, it tends to be flat so as to minimize its total area $\mathcal{A}$. Therefore, we set up a Cartesian coordinate system with the z axis perpendicular to the interface, and we consider a volume $\mathcal{V} = L\mathcal{A}$, where L is sufficiently large that the system attains the properties of the bulk A- and B-rich phases at $z = -L/2$ and $z = L/2$, respectively. Because of the translational symmetry parallel to the interface, there will be no spatial variation in the x and y directions. These facts allow us to impose reflecting boundary conditions at the edges of $\mathcal{V}$.

As for the brush in Section 2.7.6, we again solve this problem using the spectral method. For the symmetry of our current problem, the appropriate basis functions are

$$f_i(z) = \begin{cases} 1 & \text{if } i = 0 \\ \sqrt{2}\cos(i\pi(1 + 2z/L)) & \text{if } i > 0 \end{cases} \qquad (2.222)$$

This is, in fact, the same set of functions as used for the polymer brush in Eq. (2.166), but with $\beta = 0$. In this case, the eigenvalues of the Laplacian become

$$\lambda_i = 4\pi^2 i^2 \tag{2.223}$$

and the symmetric tensor simplifies to

$$\Gamma_{ijk} = (\delta_{i+j,k} + \delta_{|i-j|,k})/\sqrt{2} \tag{2.224}$$

when all the indices are non-zero. If one or more of the indices are zero, then it reduces to a single delta function (e.g. $\Gamma_{ij0} = \delta_{ij}$ when $k=0$).

The SCFT calculation for this system proceeds remarkably like that of the polymer brush, with only a couple of key differences. The first difference occurs because there are now two distinct fields, $w_A(z)$ and $w_B(z)$, acting on the A and B segments, respectively. This results in two different transfer matrices, $\mathbf{T}_A(s) \equiv \exp(\mathbf{A}s)$ and $\mathbf{T}_B(s) \equiv \exp(\mathbf{B}s)$, where the coefficients of $\mathbf{A}$ and $\mathbf{B}$ are

$$A_{ij} \equiv -\frac{2\pi^2 i^2 a^2 N}{3L^2}\delta_{ij} - \frac{w_{A,(i+j)} + w_{A,|i-j|}}{\sqrt{2}} \tag{2.225}$$

$$B_{ij} \equiv -\frac{2\pi^2 i^2 a^2 N}{3L^2}\delta_{ij} - \frac{w_{B,(i+j)} + w_{B,|i-j|}}{\sqrt{2}} \tag{2.226}$$

These matrices are equivalent to that in Eq. (2.173), but with the simplified expressions for λ_i and Γ_{ijk} explicitly inserted. We have also assumed that the spatial averages of $w_A(z)$ and $w_B(z)$ are both zero, which, for the current set of basis functions, implies $w_{A,0} = w_{B,0} = 0$.

The fact that all chain ends are now free is accounted for by the conditions $q_\gamma(\mathbf{r},0) = 1$ and $q_\gamma^\dagger(\mathbf{r},1) = 1$, where γ denotes either an A- or B-type polymer. In terms of the coefficients $q_{\gamma,i}(s)$ and $q_{\gamma,i}^\dagger(s)$, these conditions become

$$q_{\gamma,i}(0) = \delta_{i0} \tag{2.227}$$

$$q_{\gamma,i}^\dagger(1) = \delta_{i0} \tag{2.228}$$

Other than this one minor difference, the calculation of the concentrations coefficients $\phi_{\gamma,i}$ proceeds exactly as before in Eqs. (2.180) and (2.181).

The procedure for solving the self-consistent field equations (Eq. 2.203 and 2.206) changes ever so slightly from the case of the polymeric brush. The summations are still truncated at $i = M$, where M is large enough that the resulting inaccuracy is negligible. This time, however, there is no β to worry about, and the arbitrary constants to the fields have already been taken care of by setting $w_{A,0} = w_{B,0} = 0$. The remaining $2M$ field coefficients are determined by requiring

$$\phi_{A,i} + \phi_{B,i} = 0 \tag{2.229}$$

$$w_{A,i} - w_{B,i} = -2\chi N \phi_{A,i} \tag{2.230}$$

for $i = 1, 2, 3, \ldots, M$. Once these equations are satisfied, the free energy is given by Eq. (2.207), which simplifies to

$$\frac{F}{nk_BT} = \frac{F_h}{nk_BT} - \phi \ln(q_{A,0}(1))$$
$$- (1-\phi)\ln(q_{B,0}(1)) - \chi N \sum_{i=1}^{M} \phi_{A,i}\phi_{B,i} \tag{2.231}$$

This expression is arrived at by the fact that $\mathcal{Q}_\gamma = \mathcal{V}q_{\gamma,0}(1)$, $\phi_{A,0} = \phi$, and $\phi_{B,0} = 1 - \phi$.

Fig. 2.13(a) shows the segment profiles calculated at several degrees of segregation within the two-phase region of Fig. 2.10(b). The interface appears in the center of the system at $z = 0$, because a symmetric blend composition of $\phi = 0.5$ was chosen. A convenient definition for the width over which the profile switches between the A- and B-rich phases is

$$w_I \equiv \frac{\phi_A(L/2) - \phi_A(-L/2)}{\phi'_A(0)} \tag{2.232}$$

As χN increases, the interface becomes narrower as illustrated in Fig. 2.13(b) so as to reduce the overlap between the A- and B-segment profiles. The interfacial tension γ_I associated with the interface is given by the excess free energy per unit area of interface:

$$\gamma_I \equiv \frac{F - F_s}{\mathcal{A}} = \left(\frac{F - F_s}{nk_BT}\right)\left(\frac{L}{aN^{1/2}}\right)\rho_0 a N^{-1/2} k_B T \tag{2.233}$$

As χN increases beyond 2, so γ_I grows monotonically from zero as shown in Fig. 2.14. These calculations can also be extended to curved interfaces (Matsen 1999), if one also wishes to evaluate the bending moduli of the interface.

2.8.5
Interface in a Strongly Segregated Blend

In the limit of large χN, it becomes possible to derive analytical predictions for the interface between the A- and B-rich homopolymer phases. The strong field gradients that develop at the interface cause the ground-state dominance approximation from Section 2.6.2 to become increasingly accurate. This allows the free energy of the blend to be expressed as

$$\frac{F}{nk_BT} = \frac{a^2 N}{24L} \int dz \left(\frac{[\phi'_A(z)]^2}{\phi_A(z)} + \frac{[\phi'_B(z)]^2}{\phi_B(z)}\right)$$

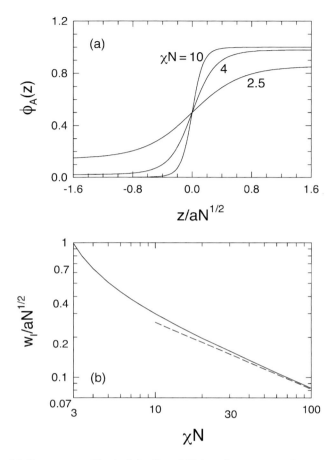

Fig. 2.13 (a) Segment profile $\phi_A(z)$ of an A/B interface at $z = 0$ calculated for several different levels of segregation, χN. (b) Interfacial width w_I plotted logarithmically as a function of segregation. The dashed line denotes the strong segregation prediction in Eq. (2.239).

$$+\frac{\chi N}{L}\int dz\, \phi_A(z)\phi_B(z) \qquad (2.234)$$

where the configurational entropy of the polymers has been approximated by Eq. (2.78). The dependence on the B-polymer concentration is easily eliminated using the incompressibility condition, $\phi_B(z) = 1 - \phi_A(z)$. This leaves one unknown concentration, $\phi_A(z)$, which is determined by minimizing F subject to the boundary conditions $\phi_A(-L/2) = 0$ and $\phi_A(L/2) = 1$. Note that we will ultimately extend L to infinity.

The functional minimization is facilitated by invoking the substitution

$$\phi_A(z) = \sin^2(\Theta(z)) \qquad (2.235)$$

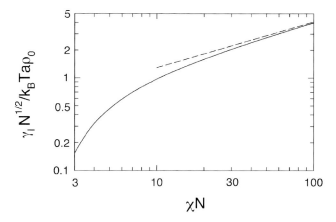

Fig. 2.14 Interfacial tension γ_I plotted logarithmically as a function of segregation, χN. The dashed line denotes the strong segregation prediction in Eq. (2.244).

which recasts the free-energy expression as

$$\frac{F}{nk_BT} = \frac{1}{L}\int dz\,\{\tfrac{1}{6}a^2 N[\Theta'(z)]^2 + \chi N \sin^2(\Theta(z))\cos^2(\Theta(z))\} \quad (2.236)$$

with the new boundary conditions $\Theta(-L/2) = 0$ and $\Theta(L/2) = \pi/2$. This functional has the identical form to that encountered in Section 2.4, where the classical trajectory was calculated for an arbitrary field $w(\boldsymbol{r})$. Applying the equivalent steps here to the Euler–Lagrange equation leads to

$$\tfrac{1}{6}a^2 N[\Theta'(z)]^2 - \chi N \sin^2(\Theta(z))\cos^2(\Theta(z)) = \text{constant} \quad (2.237)$$

The boundary conditions on $\Theta(z)$, coupled with the fact that the profile must become flat in the limit $z \to \pm\infty$, imply that the constant of integration must be zero, and thus

$$w_I \Theta'(z) = 2\sin(\Theta(z))\cos(\Theta(z)) \quad (2.238)$$

where

$$w_I \equiv \frac{2a}{\sqrt{6\chi}} \quad (2.239)$$

Returning to the original function, $\phi_A(z)$, Eq. (2.238) becomes

$$w_I \phi_A'(z) = 4\phi_A(z)[1 - \phi_A(z)] \quad (2.240)$$

which can be rearranged as

$$\frac{d\phi_A}{\phi_A(1-\phi_A)} = \frac{4\,dz}{w_I} \quad (2.241)$$

Integrating this equation, using the method of partial fractions on the right-hand side, and setting the constant of integration such that $\phi_A(0) = 1/2$, results in the predicted profile

$$\phi_A(z) = \tfrac{1}{2}[1 + \tanh(2z/w_I)] \quad (2.242)$$

Referring to the definition in Eq. (2.232), the constant, w_I, can now be interpreted as the interfacial width. This approximate width is plotted as a dashed line in Fig. 2.13(b) alongside the SCFT prediction, and indeed it is reasonably accurate for $\chi N \gtrsim 10$.

Inserting the calculated profile into the free-energy expression (2.234) gives

$$\frac{F}{nk_BT} = \frac{aN}{L}\sqrt{\frac{\chi}{6}} \quad (2.243)$$

where the integral $\int dx \cosh^{-2}(x) = 2$ has been used. Given that the free energy of the two bulk phases is $F_s = 0$ in the strong segregation limit, it follows from Eq. (2.233) that the interfacial tension is

$$\gamma_I = k_B T a \rho_0 \sqrt{\frac{\chi}{6}} \quad (2.244)$$

Fig. 2.14 demonstrates that this well-known estimate becomes reasonably good once $\chi N \gtrsim 10$.

2.8.6
Grand-Canonical Ensemble

Up to this point, the polymer blend has been treated in the canonical ensemble (Hong and Noolandi 1981), where the numbers of A- and B-type polymers are fixed. However, the statistical mechanics of multicomponent blends can be equally well performed in the grand-canonical ensemble (Matsen 1995a), where n_A and n_B are permitted to fluctuate with chemical potentials μ_A and μ_B controlling their respective averages. The grand-canonical partition function, Z_g, is related to the canonical one, Z, by the expression

$$Z_g = \sum_{n_A=0}^{\infty} \sum_{n_B=0}^{\infty} z_A^{n_A} z_B^{n_B} Z \quad (2.245)$$

where $z_A = \exp(\mu_A/k_BT)$ and $z_B = \exp(\mu_B/k_BT)$. When the melt is treated as incompressible such that $n_A + n_B = n$, one of the chemical potentials becomes redundant and the definition can be simplified to

$$Z_g = \sum_{n_A=0}^{\infty} \sum_{n_B=0}^{\infty} z^{n_A} Z \qquad (2.246)$$

with $z = \exp(\mu/k_BT)$. This is evaluated by inserting the expression for Z from Eq. (2.196) and summing over n_A and n_B. The sum over n_A is carried out using the Taylor series expansion

$$\sum_{n_A=0}^{\infty} \frac{1}{n_A!} \left(\frac{\rho_0}{N} z \mathcal{Q}_A[W_A]\right)^{n_A} = \exp\left(\frac{\rho_0}{N} z \mathcal{Q}_A[W_A]\right) \qquad (2.247)$$

with an equivalent expansion for the sum over n_B. With the summations completed, the partition function reduces to

$$Z_g \propto \int \mathcal{D}\Phi_A \, \mathcal{D}W_A \, \mathcal{D}W_B \, \exp\left(-\frac{F_g[\Phi_A, W_A, W_B]}{k_BT}\right) \qquad (2.248)$$

where

$$\frac{F_g}{nk_BT} = -z\frac{\mathcal{Q}_A[W_A]}{V} - \frac{\mathcal{Q}_B[W_B]}{V} \qquad (2.249)$$

$$+ \frac{1}{V} \int d\mathbf{r} \left[\chi N \Phi_A(1-\Phi_A) - W_A\Phi_A - W_B(1-\Phi_A)\right]$$

Just as before, $F_g[\Phi_A, W_A, W_B]$ can be readily evaluated, but the functional integration in Eq. (2.248) cannot be done. The saddle-point approximation is therefore once again applied, producing the identical self-consistent conditions

$$\phi_A(\mathbf{r}) + \phi_B(\mathbf{r}) = 1 \qquad (2.250)$$

$$w_A(\mathbf{r}) - w_B(\mathbf{r}) = \chi N(1 - 2\phi_A(\mathbf{r})) \qquad (2.251)$$

derived in the canonical ensemble. However, $\phi_A(\mathbf{r})$ and $\phi_B(\mathbf{r})$ are now defined by

$$\phi_A(\mathbf{r}) \equiv -\frac{z}{V}\frac{\mathcal{D}\mathcal{Q}_A[w_A]}{\mathcal{D}w_A(\mathbf{r})} = z\int_0^1 ds \, q_A(\mathbf{r},s) q_A^\dagger(\mathbf{r},s) \qquad (2.252)$$

$$\phi_{\rm B}(\bm{r}) \equiv -\frac{1}{\mathcal{V}}\frac{\mathcal{D}\mathcal{Q}_{\rm B}[w_{\rm B}]}{\mathcal{D}w_{\rm B}(\bm{r})} = \int_0^1 ds\, q_{\rm B}(\bm{r},s)q_{\rm B}^\dagger(\bm{r},s) \tag{2.253}$$

but they are still identified, using an analogous argument to that in Section 2.7.4, as the SCFT approximations for $\langle\hat{\phi}_{\rm A}(\bm{r})\rangle$ and $\langle\hat{\phi}_{\rm B}(\bm{r})\rangle$, respectively. The self-consistent field equations are solved as before in the canonical ensemble, but this time additive constants to the fields do affect $F_{\rm g}$. Therefore, we are no longer free to select the spatial averages for $\bar{w}_{\rm A}$ and $\bar{w}_{\rm B}$, and thus Eq. (2.251) is used as is. Once the fields are determined, the grand-canonical free energy is given by $F_{\rm g}[\phi_{\rm A}, w_{\rm A}, w_{\rm B}]$ from Eq. (2.249).

The advantages of the grand-canonical ensemble come into play when dealing with macrophase separation and coexisting phases. Rather than having to perform a double-tangent construction like that in Fig. 2.10(a), coexistence is determined by simply equating the grand-canonical free energies of the two phases. Here we demonstrate the procedure for the binary homopolymer blends treated in Section 2.8.2. Since the A- and B-rich phases are homogeneous, the fields are constant and therefore the diffusion equations for the partial partition functions are easily solved. From their solutions, it follows that

$$\phi_{\rm A}(\bm{r}) = z \exp(-w_{\rm A}) \tag{2.254}$$

$$\phi_{\rm B}(\bm{r}) = \exp(-w_{\rm B}) \tag{2.255}$$

$$\mathcal{Q}_{\rm A}[w_{\rm A}] = \mathcal{V}\exp(-w_{\rm A}) \tag{2.256}$$

$$\mathcal{Q}_{\rm B}[w_{\rm B}] = \mathcal{V}\exp(-w_{\rm B}) \tag{2.257}$$

The incompressibility condition (2.250) is enforced by setting $\phi \equiv \phi_{\rm A}(\bm{r}) = 1 - \phi_{\rm B}(\bm{r})$, which allows the four equations to be rewritten as

$$w_{\rm A} = -\ln\phi + \mu/k_{\rm B}T \tag{2.258}$$

$$w_{\rm B} = -\ln(1-\phi) \tag{2.259}$$

$$\mathcal{Q}_{\rm A}[w_{\rm A}] = \mathcal{V}\phi/z \tag{2.260}$$

$$\mathcal{Q}_{\rm B}[w_{\rm B}] = \mathcal{V}(1-\phi) \tag{2.261}$$

The remaining self-consistent field condition (Eq. 2.251) requires that ϕ satisfy

$$\frac{\mu}{k_{\rm B}T} = \ln\phi - \ln(1-\phi) + \chi N(1-2\phi) \tag{2.262}$$

In general, there will be one solution, $\phi = \phi^{(1)}$, corresponding to a B-rich phase and another, $\phi = \phi^{(2)}$, for an A-rich phase. Their grand-canonical free energies are then given by Eq. (2.249), which, for homogeneous phases, becomes

$$\frac{F_{g,h}}{nk_BT} = \phi[\ln\phi - 1] + (1-\phi)[\ln(1-\phi) - 1] + \chi N\phi(1-\phi) - \frac{\mu\phi}{k_BT} \quad (2.263)$$

Notice that $F_{g,h} = F_h - \mu n_A$, which is the standard relation between the grand-canonical and canonical free energies.

Fig. 2.15 plots the grand-canonical free energies, $F_{g,h}^{(1)}$ and $F_{g,h}^{(2)}$, of the two phases as a function of the chemical potential μ. For purposes of comparison, this is done for the same degree of segregation, $\chi N = 3$, as used in Fig. 2.10(a), where the canonical free energy F_h was plotted as a function of composition ϕ. In the grand-canonical ensemble, the coexistence is associated with the crossing of $F_{g,h}^{(1)}$ and $F_{g,h}^{(2)}$, which, due to symmetry, occurs at $\mu = 0$. It is reasonably straightforward to show that the negative slope of the curves in Fig. 2.15 is equal to the composition of the blend, which confirms that the more horizontal curve, $F_{g,h}^{(1)}$, corresponds to a B-rich phase, while the steeper one, $F_{g,h}^{(2)}$, represents an A-rich phase. According to Eq. (2.262), the compositions of the coexisting phases at $\mu = 0$ satisfy

$$\chi N = \frac{1}{1-2\phi}\ln\left(\frac{1-\phi}{\phi}\right) \quad (2.264)$$

which is precisely the same condition as derived in Section 2.8.2 using the canonical ensemble. Indeed, the two ensembles always provide equivalent

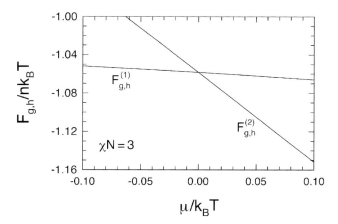

Fig. 2.15 Grand-canonical free energy $F_{g,h}$ of homogeneous blends as a function of chemical potential μ, calculated for $\chi N = 3$. The two curves, $F_{g,h}^{(1)}$ and $F_{g,h}^{(2)}$, correspond to separate B- and A-rich phases, respectively. The B-rich phase is stable at $\mu < 0$ and the A-rich one is favored for $\mu > 0$; they coexist at $\mu = 0$.

results and therefore either one can, in principle, do the job. However, there can be significant computational advantages of using them in conjunction with each other (Matsen 2003a).

2.9
Block Copolymer Melts

One way of preventing the macrophase separation of A- and B-type homopolymers is to bond them together covalently into a single AB diblock copolymer as depicted in Fig. 2.1. In this architecture, the first fN segments ($0 \leq f \leq 1$) of the polymer are type A, and the remaining $(1-f)N$ segments are type B. The A and B segments still segregate at large χN, but the domains remain microscopic in size, and thus the process is referred to as microphase separation. Not only that, the domains within the resulting microstructures take on periodically ordered geometries, such as the lamellar phase depicted in Fig. 2.3(c).

In this section, we consider a melt of n identical diblock copolymer molecules, occupying a fixed volume of $\mathcal{V} = nN/\rho_0$. This time, there is only a single molecular species, and thus one set of functions, $\boldsymbol{r}_\alpha(s)$ with $\alpha = 1, 2, \ldots, n$, is sufficient to specify the configuration of the system. In terms of these trajectories, the dimensionless A-segment concentration is expressed as

$$\hat{\phi}_A(\boldsymbol{r}) = \frac{N}{\rho_0} \sum_{\alpha=1}^{n} \int_0^f ds\, \delta(\boldsymbol{r} - \boldsymbol{r}_\alpha(s)) \qquad (2.265)$$

and that of the B segments is the same, but with the integration extending from $s = f$ to $s = 1$. As before with the homopolymer blend, the segment interactions are described by the same $U[\hat{\phi}_A, \hat{\phi}_B]$ as defined in Eq. (2.192).

2.9.1
SCFT for a Diblock Copolymer Melt

The SCFT for diblock copolymers (Helfand 1975) is remarkably similar to that for the homopolymer blend considered in Section 2.8.1. The partition function for a block copolymer melt,

$$Z \propto \frac{1}{n!} \int \prod_{\alpha=1}^{n} \tilde{\mathcal{D}} \boldsymbol{r}_\alpha \, \exp\left(-\frac{U[\hat{\phi}_A, \hat{\phi}_B]}{k_B T}\right) \delta[1 - \hat{\phi}_A - \hat{\phi}_B] \qquad (2.266)$$

is virtually the same, except that there are now only functional integrals over one molecular type. Proceeding as before, the partition function is converted to the same form as Eq. (2.198) for the polymer blend, but with

$$\frac{F}{nk_BT} = -\ln\left(\frac{\mathcal{Q}[W_A, W_B]}{\mathcal{V}}\right) \qquad (2.267)$$

$$+ \frac{1}{\mathcal{V}} \int d\mathbf{r}[\chi N\Phi_A(1-\Phi_A) - W_A\Phi_A - W_B(1-\Phi_A)]$$

where

$$\mathcal{Q}[W_A, W_B] \propto \int \tilde{\mathcal{D}}\mathbf{r}_\alpha \exp\left(-\int_0^f ds\, W_A(\mathbf{r}_\alpha(s)) - \int_f^1 ds\, W_B(\mathbf{r}_\alpha(s))\right) \qquad (2.268)$$

is identified as the partition function for a single diblock copolymer with its A and B blocks subjected to the fields $W_A(\mathbf{r})$ and $W_B(\mathbf{r})$, respectively. Note that an irrelevant constant of unity has been dropped from Eq. (2.267).

As always in SCFT, the free energy of the melt is approximated by $F[\phi_A, w_A, w_B]$, where the functions ϕ_A, w_A, and w_B correspond to a saddle point obtained by equating the functional derivatives of Eq. (2.267) to zero. The derivative of $F[\Phi_A, W_A, W_B]$ with respect to W_A leads to the condition

$$\phi_A(\mathbf{r}) = -\mathcal{V}\frac{\mathcal{D}\ln(\mathcal{Q}[w_A, w_B])}{\mathcal{D}w_A(\mathbf{r})} \qquad (2.269)$$

which identifies $\phi_A(\mathbf{r})$ as the average A-segment concentration from n diblock copolymers in the fields $w_A(\mathbf{r})$ and $w_B(\mathbf{r})$. Differentiation with respect to W_B leads to the incompressibility constraint

$$\phi_A(\mathbf{r}) + \phi_B(\mathbf{r}) = 1 \qquad (2.270)$$

where

$$\phi_B(\mathbf{r}) \equiv -\mathcal{V}\frac{\mathcal{D}\ln(\mathcal{Q}[w_A, w_B])}{\mathcal{D}w_B(\mathbf{r})} \qquad (2.271)$$

is the analogous B-segment concentration. The remaining functional derivative with respect to Φ_A provides the self-consistent field condition

$$w_A(\mathbf{r}) - w_B(\mathbf{r}) = 2\chi N(f - \phi_A(\mathbf{r})) \qquad (2.272)$$

where again an appropriate constant has been added to allow for $\bar{w}_A = \bar{w}_B = 0$. With that, the free energy reduces to

$$\frac{F}{nk_BT} = \chi Nf(1-f) - \ln\left(\frac{\mathcal{Q}[w_A, w_B]}{\mathcal{V}}\right)$$

$$-\frac{\chi N}{\mathcal{V}}\int d\boldsymbol{r}\,(\phi_A(\boldsymbol{r})-f)(\phi_B(\boldsymbol{r})-1+f) \qquad (2.273)$$

where the first term, $\chi N f(1-f)$, gives the free energy of the disordered state in which all the A and B blocks are homogeneously mixed.

The statistical mechanics for a diblock copolymer in fields $w_A(\boldsymbol{r})$ and $w_B(\boldsymbol{r})$ differs ever so slightly from that of the homopolymer considered in Section 2.3. The full single-polymer partition function is still expressed as

$$\mathcal{Q}[w_A, w_B] = \int d\boldsymbol{r}\, q(\boldsymbol{r},s)q^\dagger(\boldsymbol{r},s) \qquad (2.274)$$

and the partial partition functions still satisfy the same diffusion equations with $q(\boldsymbol{r},0) = q^\dagger(\boldsymbol{r},1) = 1$. The diblock architecture enters simply by the fact that the field $w(\boldsymbol{r})$ in Eqs. (2.24) and (2.28) is substituted by $w_A(\boldsymbol{r})$ for $0 < s < f$ and by $w_B(\boldsymbol{r})$ for $f < s < 1$. Once the partial partition functions are evaluated, the average A-segment concentration is given by

$$\phi_A(\boldsymbol{r}) = \frac{\mathcal{V}}{\mathcal{Q}[w_A, w_B]}\int_0^f ds\, q(\boldsymbol{r},s)q^\dagger(\boldsymbol{r},s) \qquad (2.275)$$

and $\phi_B(\boldsymbol{r})$ is given by the same integral, but with an appropriate change of limits.

2.9.2
Scattering Function for the Disordered Phase

Here a Landau–Ginzburg free-energy functional (Leibler 1980) is calculated for small variations in the A-segment profile, $\phi_A(\boldsymbol{r})$, about the average value of $\bar{\phi}_A = f$, following analogous steps to those in Section 2.8.3 for polymer blends. As before, we assume that the variations in $w_A(\boldsymbol{r})$ and $w_B(\boldsymbol{r})$ are sufficiently small such that $q(\boldsymbol{r}, f)$ is well described by Eq. (2.99) with the field $w_A(\boldsymbol{r})$ and similarly such that $q^\dagger(\boldsymbol{r}, f)$ is given by Eq. (2.100) with the field $w_B(\boldsymbol{r})$. These approximations are then inserted into Eq. (2.274) to give

$$\frac{\mathcal{Q}[w_A, w_B]}{\mathcal{V}} \approx 1 + \frac{1}{2(2\pi)^3 \mathcal{V}} \int d\boldsymbol{k}\,[S_{11} w_A(-\boldsymbol{k})w_A(\boldsymbol{k})$$
$$+ S_{22} w_B(-\boldsymbol{k})w_B(\boldsymbol{k}) + 2S_{12} w_A(-\boldsymbol{k})w_B(\boldsymbol{k})] \qquad (2.276)$$

where $S_{11} \equiv g(x, f)$, $S_{22} \equiv g(x, 1-f)$, and $S_{12} \equiv h(x, f)h(x, 1-f)$. Differentiating $\mathcal{Q}[w_A, w_B]$ with respect to the Fourier transforms of the fields, as we did in Section 2.8.3, gives the Fourier transforms of the concentrations,

$$\phi_A(\boldsymbol{k}) = -S_{11} w_A(\boldsymbol{k}) - S_{12} w_B(\boldsymbol{k}) \qquad (2.277)$$

$$\phi_B(\boldsymbol{k}) = -S_{22}w_B(\boldsymbol{k}) - S_{12}w_A(\boldsymbol{k}) \tag{2.278}$$

for all $\boldsymbol{k} \neq 0$. These equations are then inverted to obtain the expressions

$$w_A(\boldsymbol{k}) = \frac{-S_{22}\phi_A(\boldsymbol{k}) + S_{12}\phi_B(\boldsymbol{k})}{\det(S)} \tag{2.279}$$

$$w_B(\boldsymbol{k}) = \frac{S_{12}\phi_A(\boldsymbol{k}) - S_{11}\phi_B(\boldsymbol{k})}{\det(S)} \tag{2.280}$$

for the fields, where $\det(S) \equiv S_{11}S_{22} - S_{12}^2$. At this point, the incompressibility condition can be used to set $\phi_B(\boldsymbol{k}) = -\phi_A(\boldsymbol{k})$, for all $\boldsymbol{k} \neq 0$. Substituting the expressions for the two fields into that for $\mathcal{Q}[w_A, w_B]$, and then inserting that into the logarithm of Eq. (2.273), gives

$$\frac{F}{nk_BT} = \chi N f(1-f) + \frac{N}{2(2\pi)^3 \mathcal{V}} \int_{\boldsymbol{k} \neq 0} d\boldsymbol{k}\, S^{-1}(\boldsymbol{k})\phi_A(-\boldsymbol{k})\phi_A(\boldsymbol{k}) \tag{2.281}$$

to second order in $|\phi_A(\boldsymbol{k})|$, where

$$S^{-1}(\boldsymbol{k}) = \frac{g(x,1)}{N\det(S)} - 2\chi \tag{2.282}$$

is the inverse of the scattering function for a disordered melt. In a somewhat analogous but considerably more complicated procedure, the scattering functions can also be evaluated for periodically ordered phases (Yeung et al. 1996; Shi et al. 1996). It involves a very elegant method akin to band-structure calculations in solid-state physics, providing yet another example of where SCFT draws upon existing techniques from quantum mechanics.

Fig. 2.16 shows the disordered-state $S(\boldsymbol{k})$ for diblock copolymers of symmetric composition, $f = 0.5$, plotted for a series of χN values. As the interaction strength increases, $S(\boldsymbol{k})$ develops a peak over a sphere of wavevectors at radius $kaN^{1/2} = 4.77$, which corresponds to composition fluctuations of wavelength $D/aN^{1/2} = 1.318$. The peak eventually diverges at $\chi N = 10.495$, at which point the disordered state becomes unstable and switches by a continuous transition to the ordered lamellar phase. With the exception of $f = 0.5$, the spinodal point is preempted by a discontinuous transition to the ordered phase, but an approximate treatment of this requires, at the very least, fourth-order terms in the free-energy expansion of Eq. (2.281) (Leibler 1980). Better still, the next section will show how to calculate the exact mean-field phase boundaries.

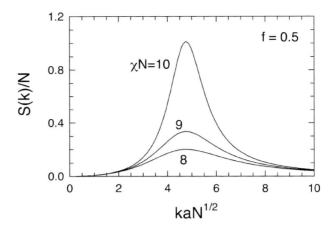

Fig. 2.16 Scattering function $S(\mathbf{k})$ for a disordered diblock copolymer melt of symmetric composition, $f = 0.5$, plotted for several degrees of segregation, χN.

2.9.3
Spectral Method for the Ordered Phases

Now we turn our attention to the ordered block copolymer phases. Due to the periodicity of their domain structures, the spectral method is by far the most efficient method for solving the SCFT. Here, the method is illustrated starting with the simplest microstructure, the one-dimensional lamellar phase pictured in Fig. 2.3(c). Our chosen coordinate system orients the z axis perpendicular to the interfaces, so that there is translational symmetry in the x and y directions. This allows for the relatively simple basis functions

$$f_i(z) = \begin{cases} 1 & \text{if } i = 0 \\ \sqrt{2}\cos(i\pi(1+2z/D)) & \text{if } i > 0 \end{cases} \quad (2.283)$$

where D corresponds to the lamellar period. In fact, these are the precise functions used in Section 2.8.4 for the interface of a binary homopolymer blend, but with L replaced by D. Consequently, the eigenvalues λ_i and the tensor Γ_{ijk} are given by exactly the same Eqs. (2.223) and (2.224). Again, these quantities are used to construct the transfer matrices, $\mathbf{T}_A(s) \equiv \exp(\mathbf{A}s)$ and $\mathbf{T}_B(s) \equiv \exp(\mathbf{B}s)$, where

$$A_{ij} = -\frac{\lambda_i a^2 N}{6D^2}\delta_{ij} - \sum_k w_{A,k}\Gamma_{ijk} \quad (2.284)$$

$$B_{ij} = -\frac{\lambda_i a^2 N}{6D^2}\delta_{ij} - \sum_k w_{B,k}\Gamma_{ijk} \qquad (2.285)$$

The evaluation of the transfer matrices is performed by the same diagonalization procedure detailed in Section 2.7.6. Just as before, the coefficients for the partial partition functions are expressed in terms of the transfer matrices as

$$q_i(s) = \begin{cases} T_{A,i0}(s) & \text{if } s \leq f \\ \sum_j T_{B,ij}(s-f)q_j(f) & \text{if } s > f \end{cases} \qquad (2.286)$$

$$q_i^\dagger(s) = \begin{cases} \sum_j T_{A,ij}(f-s)q_j^\dagger(f) & \text{if } s < f \\ T_{B,i0}(1-s) & \text{if } s \geq f \end{cases} \qquad (2.287)$$

although there are now separate expressions for $s < f$ and $s > f$ due to the field switching from $w_A(\boldsymbol{r})$ to $w_B(\boldsymbol{r})$. The coefficients for the average polymer concentration continue to be expressed as

$$\phi_{\gamma,i} = \frac{1}{q_0(1)}\sum_{jk} I_{\gamma,jk}\Gamma_{ijk} \qquad (2.288)$$

in terms of matrices $\mathbf{I}_\gamma$ defined much as before in Eq. (2.181). For the A-segment concentration,

$$I_{A,jk} \equiv \int_0^f ds\, q_j(s)q_k^\dagger(s) \qquad (2.289)$$

$$= \sum_{m,n}\left[\frac{\exp(fd_{A,m}) - \exp(fd_{A,n})}{d_{A,m} - d_{A,n}}\right]U_{A,jm}U_{A,kn}\bar{q}_m(0)\bar{q}_n^\dagger(f) \qquad (2.290)$$

Note that, for equal eigenvalues, $d_{A,m} = d_{A,n}$, the factor in the square brackets reduces to $f\exp(fd_{A,m})$. The expression for $I_{B,jk}$ is analogous, but with $d_{B,i}$ and $U_{B,ij}$ calculated from the **B** matrix, and with $s = 0$ and f switched to $s = f$ and 1, respectively.

The self-consistent field conditions for $w_{A,i}$ and $w_{B,i}$ remain exactly as in Eqs. (2.229) and (2.230) for the homopolymer blends. We still follow the same practice of setting the spatial average of the fields to zero, which implies $w_{A,0} = w_{B,0} = 0$, and the sums are still truncated at $i = M$, where M is chosen sufficiently large to meet the desired numerical accuracy. Once the field coefficients, $w_{A,i}$ and $w_{B,i}$ for $i = 1, 2, \ldots, M$, are determined, the free energy can be evaluated by

$$\frac{F}{nk_B T} = \chi N f(1-f) - \ln(q_0(1)) - \chi N \sum_{i=1}^{M} \phi_{A,i} \phi_{B,i} \qquad (2.291)$$

where we have exploited the fact that $\mathcal{Q}[w_A, w_B] = \mathcal{V} q_0(1)$, $\phi_{A,0} = f$, and $\phi_{B,0} = 1 - f$. In this block copolymer case, there is the one additional step of minimizing the free energy with respect to the periodicity D of the microstructure.

Fig. 2.17(a) shows the predicted segment profile $\phi_A(z)$ of the lamellar phase for symmetric diblocks, $f = 0.5$, at several different degrees of segregation, χN. The lamellar phase first appears at $\chi N = 10.495$ with a period of $D/aN^{1/2} = 1.318$, consistent with the divergence in $S(\mathbf{k})$ discussed in Section 2.9.2. At weak segregations such as $\chi N \approx 11$, the profile is approximately sinusoidal. By $\chi N = 20$, the centers of the A and B domains become relatively pure, and by $\chi N = 50$, the A–B contacts are restricted to the relatively narrow interfacial regions. The series of profiles clearly demonstrates that the domain spacing D swells and the interfacial width

$$w_I \equiv \frac{\phi_A(D/2) - \phi_A(-D/2)}{\phi_A'(0)} \qquad (2.292)$$

narrows as the segregation increases. Figs. 2.17(b) and (c) show these trends on logarithmic scales so as to highlight the well-known scaling behavior that emerges for strongly segregated microstructures. The dashed line in Fig. 2.17(b) denotes the domain scaling of $D \propto a\chi^{1/6} N^{2/3}$, where the proportionality constant will be derived in the next section using strong segregation theory (SST). At intermediate segregations, the rate of increase is somewhat more rapid, consistent with the fact that the experimentally measured exponent for N tends to be ~ 0.8 (Almdal et al. 1990). The dashed and dotted lines in Fig. 2.17(c) show SST predictions for the interfacial width, which will be discussed in the next section.

The lamellar phase is only observed for $f \approx 0.5$. When the diblock copolymer becomes sufficiently asymmetric, the domains transform into other periodic geometries as depicted in Fig. 2.18. In the cylindrical (C) phase, the shorter minority blocks form cylindrical domains aligned in a hexagonal packing with the longer majority blocks filling the intervening space. In the spherical (S) phase, the minority blocks form spheres that generally pack in a body-centered cubic (bcc) arrangement. Note, however, that there are occasions where the spherical (S_{cp}) phase prefers close-packed (e.g. fcc or hcp) arrangements. In addition to these *classical* phases, two *complex* phases, gyroid (G)

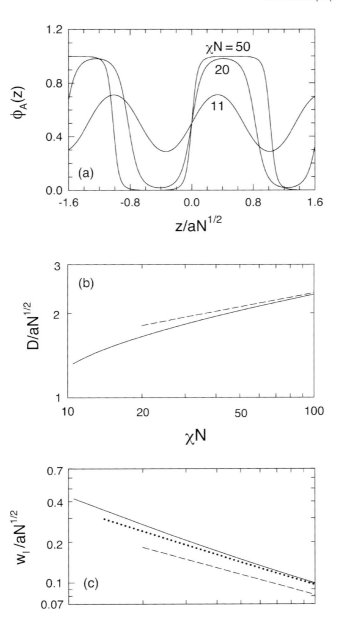

Fig. 2.17 (a) Segment profiles $\phi_A(z)$ from the lamellar phase of symmetric diblock copolymers, $f = 0.5$, plotted for three levels of segregation, χN. (b) Domain spacing D versus segregation plotted logarithmically; the dashed curve denotes the SST prediction from Eq. (2.302). (c) Interfacial width w_I versus segregation plotted logarithmically; the dashed and dotted lines correspond to the SST predictions in Eqs. (2.239) and (2.299), respectively.

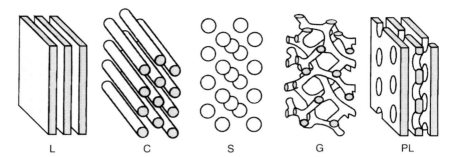

Fig. 2.18 Minority domains from the periodically ordered phases observed in diblock copolymer melts. The first three, lamellar (L), cylindrical (C), and spherical (S), are referred to as the "classical" phases, and the last two, gyroid (G) and perforated-lamellar (PL), are denoted as the "complex" phases. Although the PL phase is often observed, our best experimental evidence indicates that it is only ever metastable (Hajduk et al. 1997).

and perforated-lamellar (PL), have also been observed. The minority domains of the G phase (Hajduk et al. 1994; Schulz et al. 1994) form two interweaving networks composed of three-fold coordinated junctions. The PL phase (Hamley et al. 1993) is much like the classical L phase, but the minority-component lamellas develop perforations through which the majority-component layers become connected. The perforations tend to be aligned hexagonally within the layers and staggered between neighboring layers.

Fortunately, the only part of the spectral method that changes when considering the non-lamellar morphologies are the basis functions, and these changes are entirely contained within the values of λ_i and Γ_{ijk}. Take the C phase, for example. Its hexagonal unit cell is shown in Fig. 2.19 with a coordinate system defined such that the cylinders are aligned in the z direction. The orthonormal basis functions for this symmetry start off as

$$f_0(\boldsymbol{r}) = 1 \tag{2.293}$$

$$f_1(\boldsymbol{r}) = \sqrt{2/3}\,[\cos(2Y) + 2\cos(X)\cos(Y)] \tag{2.294}$$

$$f_2(\boldsymbol{r}) = \sqrt{2/3}\,[\cos(2X) + 2\cos(X)\cos(3Y)] \tag{2.295}$$

$$f_3(\boldsymbol{r}) = \sqrt{2/3}\,[\cos(4Y) + 2\cos(2X)\cos(2Y)] \tag{2.296}$$

$$f_4(\boldsymbol{r}) = \sqrt{4/3}\,[\cos(3X)\cos(Y) + \cos(2X)\cos(4Y)$$
$$+ \cos(X)\cos(5Y)] \tag{2.297}$$

where $X \equiv 2\pi x/D$ and $Y \equiv 2\pi y/\sqrt{3}D$. (The general formula is provided by Henry and Lonsdale (1969) under the space-group symmetry of P6mm.)

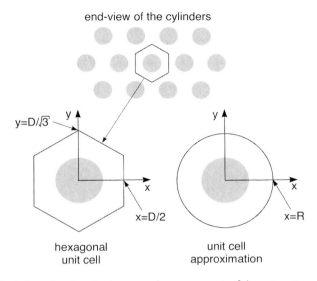

Fig. 2.19 End-view showing the hexagonal arrangement of the minority domains of the cylinder (C) phase. Below to the left is the proper hexagonal unit cell, and to the right is a circular unit-cell approximation (UCA), where the radius R is set by the equal-area condition, $\pi R^2 = \sqrt{3}D^2/2$.

The corresponding eigenvalues of the Laplacian are $\lambda_0 = 0$, $\lambda_1 = 16\pi^2/3$, $\lambda_2 = 16\pi^2$, $\lambda_3 = 64\pi^2/3$, $\lambda_4 = 112\pi^2/3$, and so on. Although there are simply too many elements of Γ_{ijk} to begin enumerating them, it is a trivial matter to generate them by computer. Once they have been evaluated and stored away in a file, the free energy of the C phase is calculated using exactly the same expressions as for the L phase. The same goes for the S and G phases, where their basis functions are catalogued by Henry and Lonsdale (1969) under the space-group symmetries Im$\bar{3}$m and Ia$\bar{3}$d, respectively. For the S_{cp} and PL phases, there are near-degeneracies (Matsen and Bates 1996a) in the arrangement of spheres and the stacking of layers, respectively. As suggested by Matsen and Bates (1996a), this is likely to result in irregular non-periodic arrangements, which is indeed consistent with recent experimental observations (Sakamoto et al. 1997; Zhu et al. 2001). Nevertheless, for the purpose of calculating phase boundaries, it is sufficient to use the symmetries Fm$\bar{3}$m (i.e. fcc packing) and R$\bar{3}$m (i.e. ABCABC... stacking), respectively. The PL phase is also different in that it has two distinct periodicities, which implies that the eigenvalue equation (2.165) for the basis functions must be generalized to

$$\nabla^2 f_i(\boldsymbol{r}) = -\left(\frac{\lambda_{\|,i}}{D_\|^2} + \frac{\lambda_{\perp,i}}{D_\perp^2}\right) f_i(\boldsymbol{r}) \qquad (2.298)$$

where $D_\parallel$ is the in-plane spacing between perforations and $D_\perp$ is the out-of-plane lamellar spacing. However, this distinction only enters in an obvious extension to Eqs. (2.284) and (2.285), and in the need to minimize F with respect to both $D_\parallel$ and $D_\perp$.

Comparing the free energies of the disordered phase and all the ordered phases of Fig. 2.19 generates the phase diagram in Fig. 2.20, which maps the phase of lowest free energy as a function of composition f and segregation χN. This theoretical diagram is in excellent agreement with experiment (Matsen 2002a) apart from a few minor differences. Experimental phase diagrams (Bates et al. 1994) have varying degrees of asymmetry about $f = 0.5$, but this is well accounted for by conformational asymmetry (Matsen and Bates 1997a) occurring due to unequal statistical lengths of the A and B segments. Aside from that, the small remaining differences with experiment are generally attributed to fluctuation effects, which will be discussed later in Section 2.11. According to the theory, the only complex microstructure present in the equilibrium phase diagram is G, and based on extrapolations (Matsen and Bates 1996a) its region of stability pinches off at $\chi N \approx 60$. Although the PL phase never, at any point, possesses the lowest free energy, it is very close to being stable in the G region. This integrates well with recent experiments (Hajduk et al. 1997) demonstrating that occurrences of PL eventually convert to G, given sufficient time. In addition to predicting a phase diagram in agreement with experiment, the theory also provides powerful intuitive explanations for the behavior in terms of spontaneous interfacial curvature and packing frustration (Matsen 2002a).

Many, particularly older, block copolymer calculations employ a unit-cell approximation (UCA), where the hexagonal cell is replaced by a circular cell of equal area as depicted in Fig. 2.19. This creates a rotational symmetry that transforms the two-dimensional diffusion equation into a simpler one-dimensional version. An equivalent approximation is also available for the spherical phase. The diblock copolymer phase diagram calculated (Vavasour and Whitmore 1992) with this UCA is very similar to that in Fig. 2.20, except that it leaves out the G phase as there are no UCAs for the complex phases. Nevertheless, the complex phase region is a small part of the diblock copolymer phase diagram, and thus its omission is not particularly serious. While there is no longer any need to invoke this approximation for simple diblock copolymer melts, there are many other systems with dauntingly large parameter spaces (e.g. blends) that are still computationally demanding. For those that are dominated by the classical phases, the omission of the complex phases is a small price to pay for the tremendous computational advantage gained by implementing the UCA. The spectral method can still be used exactly as described above, but with the expansion performed in terms of Bessel functions (Matsen 2003a).

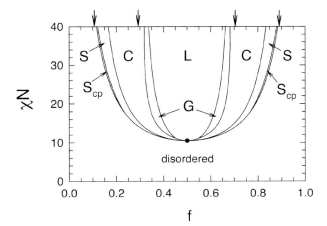

Fig. 2.20 Theoretical phase diagram for diblock copolymer melts showing where the various morphologies in Fig. 2.18 are stable. All phase transitions are discontinuous except for the mean-field critical point denoted by the solid dot. Along part of the order–disorder transition (ODT) is a narrow region, S_{cp}, where spherical domains prefer a close-packed arrangement over the usual bcc one. In the strong segregation limit, the G region pinches off at $\chi N \approx 60$, the ODT approaches the two edges of the diagram, and the C/S and L/C boundaries approach the fixed compositions estimated by the arrows at the top of the diagram.

2.9.4
SST for the Ordered Phases

The free energy of a well-segregated diblock copolymer microstructure has three main contributions: the tension of the internal interface, and the stretching energies of the A and B blocks (Matsen and Bates 1997b). In the large-χN regime, these energies can all be approximated by simple expressions, producing a very useful analytical strong segregation theory (SST) (Semenov 1985). In fact, the interface of a block copolymer melt is the same as that of a homopolymer blend, to a first-order approximation. The derivation in Section 2.8.5 just requires sufficiently steep field gradients at the interface such that the ground-state dominance applies, and this continues to be the case. Therefore, the tension γ_I in Eq. (2.244) and the interfacial width w_I in Eq. (2.239) also apply to block copolymer melts. Indeed, the approximation for the width, shown by the dashed line in Fig. 2.17(c), is reasonable, although not nearly as accurate as it was in Fig. 2.13(b) for the homopolymer blend. A much better approximation,

$$w_I = \frac{2a}{\sqrt{6\chi}} \left[1 + \frac{4}{\pi} \left(\frac{3}{\pi^2 \chi N} \right)^{1/3} \right] \qquad (2.299)$$

denoted by the dotted curve in Fig. 2.17(c), can be derived by accounting for the connectivity of the A and B blocks (Semenov 1993), but it is considerably more complicated. For simplicity, we stick with the simple first-order approximation for γ_I in Eq. (2.244).

The stretching energies of the A and B blocks are approximated by noting that, as χN increases, the junction points become strongly pinned to narrow interfaces, while the domains swell in size. Consequently, the A and B domains can be described as strongly stretched brushes. The lamellar (L) phase consists of four brushes per period D, and it follows from the incompressibility constraint that the thicknesses of the A and B brushes are $fD/2$ and $(1-f)D/2$, respectively. Since all the brushes are flat, their energies are given by Eq. (2.121), and thus the total free energy of the L phase, F_L, can be approximated as

$$\frac{F_L}{nk_BT} = \frac{\gamma_I \Sigma}{k_B T} + \frac{\pi^2 [fD/2]^2}{8a^2[fN]} + \frac{\pi^2[(1-f)D/2]^2}{8a^2[(1-f)N]} \qquad (2.300)$$

where $\Sigma = 2N/D\rho_0$ is the interfacial area per molecule. Inserting the interfacial tension from Eq. (2.244), the expression simplifies to

$$\frac{F_L}{nk_BT} = 2\sqrt{\frac{\chi N}{6}} \left(\frac{D}{aN^{1/2}}\right)^{-1} + \frac{\pi^2}{32}\left(\frac{D}{aN^{1/2}}\right)^2 \qquad (2.301)$$

The equilibrium domain spacing, obtained by minimizing F_L, is then given by

$$\frac{D}{aN^{1/2}} = 2\left(\frac{8\chi N}{3\pi^4}\right)^{1/6} \qquad (2.302)$$

which is shown in Fig. 2.17(b) with a dashed line. This provides the domain-size scaling alluded to earlier. Inserting the equilibrium value of D into Eq. (2.301) provides the final expression,

$$\frac{F_L}{nk_BT} = \tfrac{1}{4}(9\pi^2 \chi N)^{1/3} \qquad (2.303)$$

for the free energy of the lamellar phase.

For non-lamellar phases, the SST becomes complicated, unless the unit-cell approximation (UCA) is implemented. First, this allows the interfacial shape to be determined by symmetry rather than by minimizing the free energy. For instance, the symmetry of the approximate unit cell shown in Fig. 2.19 for the cylindrical (C) phase implies a perfectly circular interface. Given this, the incompressibility condition requires the radius of the interface to be $R_{\rm I} \equiv \sqrt{f}R$, from which it follows that the interfacial area per molecule is

$$\Sigma = \frac{2N\sqrt{f}}{R\rho_0} \qquad (2.304)$$

Second, the symmetry requires that the strongly stretched chains follow straight trajectories in the radial direction, for which Eq. (2.123) is easily generalized. The stretching energy of the γ-type blocks (γ = A or B) becomes

$$\frac{f_{\gamma,\rm e}}{k_{\rm B}T} = \frac{3\pi^2}{8a^2 N_\gamma} \times \frac{1}{\mathcal{V}_\gamma} \int d\mathbf{r}\, z_{\rm d}^2 \qquad (2.305)$$

which involves an average of $z_{\rm d}^2$ over the volume $\mathcal{V}_\gamma$ of the γ domain, where $z_{\rm d}$ is the radial distance to the interface and N_γ is the number of segments in the γ-type block. For a C phase with A-type cylinders and a B-type matrix, the stretching energy of the A blocks becomes

$$\frac{f_{\rm A,e}}{k_{\rm B}T} = \frac{3\pi^2}{4a^2 N f^2 R^2} \int_0^{R_{\rm I}} d\rho\, \rho(\rho - R_{\rm I})^2 = \frac{\pi^2 R^2}{16 a^2 N} \qquad (2.306)$$

and that of the B blocks is

$$\frac{f_{\rm B,e}}{k_{\rm B}T} = \frac{3\pi^2}{4a^2 N (1-f)^2 R^2} \int_{R_{\rm I}}^{R} d\rho\, \rho(\rho - R_{\rm I})^2 = \frac{\pi^2 R^2}{16 a^2 N} \alpha_{\rm B} \qquad (2.307)$$

where

$$\alpha_{\rm B} \equiv \frac{(1-\sqrt{f})^3 (3+\sqrt{f})}{(1-f)^2} \qquad (2.308)$$

Combining the three contributions, the total free energy of the C phase is written as

$$\frac{F_{\rm C}}{n k_{\rm B} T} = 2\sqrt{\frac{\chi N f}{6}} \left(\frac{R}{a N^{1/2}}\right)^{-1} + \frac{\pi^2 (1+\alpha_{\rm B})}{16} \left(\frac{R}{a N^{1/2}}\right)^2 \qquad (2.309)$$

Again, minimization leads to a domain size that scales as $R \sim a\chi^{1/6} N^{2/3}$, which when inserted back into Eq. (2.309) gives

$$\frac{F_{\rm C}}{n k_{\rm B} T} = \tfrac{1}{4}[18\pi^2 (1+\alpha_{\rm B}) \chi N]^{1/3} \qquad (2.310)$$

Comparing $F_{\rm L}$ and $F_{\rm C}$, we find that the C phase has a lower free energy than the L phase for $f < 0.2991$. The symmetry of the phase diagram implies that a

C phase with B-rich cylinders becomes preferred over L when $f > 0.7009$. The analogous calculation for the spherical (S) phase (Matsen and Bates 1997b) predicts C/S boundaries at $f = 0.1172$ and $f = 0.8828$.

There is one problem that has been overlooked. Equation (2.305) provided the stretching energies while avoiding the need to work out the distribution of chain ends. However, had we evaluated the concentration of B ends, we would have encountered a negative distribution alongside the interface. Of course, a negative concentration is unphysical. The proper solution (Ball et al. 1991) has instead a narrow exclusion zone next to the interface that is free of chain ends, and as a consequence the SST field $w_B(r)$ deviates slightly from the parabolic potential. (The classical-mechanical analogy with a simple harmonic oscillator used in Section 2.7.1 only holds if the chain ends have a finite population over the entire domain.) Fortunately, the effect is exceptionally small and can be safely ignored (Matsen and Whitmore 1996).

The only significant inaccuracy comes from the unit-cell approximation (UCA) (Matsen and Whitmore 1996). An improved SST-based calculation (Likhtman and Semenov 1994) with the proper unit cell of the C phase, but still assuming a circular interface and radial trajectories, predicts an L/C phase boundary at $f = 0.293$. Although a relatively minor problem in this instance, the assumption of straight trajectories [in general, the shortest distance to an interface (Likhtman and Semenov 1994)] minimizes the stretching energy without enforcing the connectivity of the A and B blocks. This approximation can, in fact, lead to highly erroneous predictions (Matsen 2003b). One way of enforcing the connectivity is to subdivide the A and B domains into wedges along which the diblocks follow straight paths with a kink at the interface (Olmsted and Milner 1998). Furthermore, the proper interface is not a perfect circle, but is slightly perturbed towards the six corners of the hexagonal unit cell (Matsen and Bates 1997b). Recently Likhtman and Semenov (1997) have presented a method for performing SST calculations that minimizes the free energy with respect to the interfacial shape while maintaining the connectivity of the blocks (the trajectory of each block is still straight and the exclusion zones are still ignored), but the method is highly computational. With this more accurate SST algorithm, they showed that the gyroid (G) phase becomes unstable in the strong segregation limit, consistent with the SCFT extrapolations by Matsen and Bates (1996a), but unfortunately they did not calculate accurate values for the remaining L/C and C/S boundaries. At present, our best SST estimates for these boundaries are $f = 0.294$ and $f = 0.109$, respectively, obtained by adjusting the UCA predictions according to SCFT-based corrections calculated by Matsen and Whitmore (1996). These values are denoted by arrows at the top of Fig. 2.20, and indeed they correspond reasonably well with the SCFT boundaries at finite χN.

2.10
Current Track Record and Future Outlook for SCFT

Although SCFT has only been demonstrated here for three relatively simple systems, it is a fantastically versatile theory that can be applied to multi-component mixtures with any number of species (Hong and Noolandi 1981; Matsen 1995a) and to polymer architectures of virtually any complexity (Matsen and Schick 1994b; Matsen and Schick 1994c). Solvent molecules can also be included in the theory (Naughton and Matsen 2002). Interactions are easily generalized (Matsen 2002a) beyond the simple Flory–Huggins form in Eq. (2.192), and the constraint $\hat{\phi}_A(r) + \hat{\phi}_B(r) = 1$ can be relaxed to allow for some degree of compressibility (Yeung et al. 1994). To include liquid-crystalline interactions (Netz and Schick 1996) or to account for chain stiffness (Morse and Fredrickson 1994; Matsen 1996), the Gaussian model for flexible polymers can be substituted by a worm-like chain model (Takahashi and Yunoki 1967), in which the flexibility can be adjusted. The possibilities are truly limitless.

Not only is SCFT highly versatile, it also has a track record to rival any and all theories in soft condensed matter physics. On the topic of polymer brushes, SCFT has recently resolved (Matsen and Gardiner 2001) a discrepancy between SST (Leibler et al. 1994) and experiment (Reiter and Khanna 2000) on the subtle effect of autophobic dewetting, where a small interfacial tension of entropic origins causes a homopolymer film to dewet from a chemically identical brush. However, to date, the majority of its triumphs have involved the subtleties of block copolymer phase behavior. Most notably, SCFT correctly predicted the complex phase behavior of diblock copolymer melts several years in advance of experiment. At the time that SCFT predicted (Matsen and Schick 1994a) the phase diagram in Fig. 2.20, experiments had reported four stable complex phases: the double-diamond (Thomas et al. 1987), the perforated-lamellar (PL) (Hamley et al. 1993), the modulated-lamellar (Hamley et al. 1993), and the gyroid (G) (Hajduk et al. 1994; Schulz et al. 1994). The existence of the modulated-lamellar phase, based solely on scattering experiments, was the least certain, and strongly conflicted with our theoretical understanding of the complex phase behavior (Matsen and Bates 1996b; Matsen 2002b). It is now accepted and well supported by theory (Yeung et al. 1996) that the observed scattering pattern was simply a result of anisotropic fluctuations from the classical lamellar (L) phase. It has also become apparent that the double-diamond phase was nothing more than a misidentified gyroid phase (Hajduk et al. 1995), attributed to the remarkable similarities of the two microstructures, coupled with the limited resolution of scattering experiments at the time. Conversely, the perforated-lamellar phase is most definitely observed in diblock copolymer melts, but it is now accepted to be a metastable state that, given sufficient time, converts to the gyroid morphology (Hajduk et al. 1997). With that final realization, experiments are now in line with the SCFT prediction that gyroid is the only stable complex phase in diblock copolymer melts.

However, in more elaborate architectures, notably ABC triblock copolymers, a zoo of complex morphologies are possible (Bates and Fredrickson 1999), and SCFT is having equal success in resolving their complex phase behavior (Matsen 1998; Shefelbine et al. 1999; Wickham and Shi 2001). Another impressive accomplishment of SCFT has been the successful treatment (Matsen 1997) of a delicate symmetry-breaking transition in the grain boundary of a lamellar morphology, where a symmetric chevron boundary switches to an asymmetric omega boundary as the angle between two lamellar grains increases (Gido and Thomas 1994). Arguably, the most impressive accomplishment up to now involves the macrophase separation of chemically equivalent small and large diblock copolymers (Hashimoto et al. 1993). The fact that SCFT is the only theory to correctly predict (Matsen 1995b) this effect is impressive in its own right, but the truly stunning achievement emerged later when Papadakis et al. (1998) purposely synthesized diblock copolymers in order to quantitatively test one of the predicted SCFT phase diagrams. The agreement between the theoretical and experimental phase boundaries is virtually perfect, which is particularly remarkable, given the fact that the competing theories fail to predict any boundaries at all. With this level of success, SCFT has unequivocally emerged as the state-of-the-art theory for structured polymeric melts.

The superb track record of SCFT can be attributed to the sound approximations upon which it is built. Although the Gaussian chain model treats polymers as simple microscopic elastic threads, which may seem at first to be highly artificial, it is a well-grounded model for high-molecular-weight polymers as justified in Section 2.2. In this limit of high molecular weight, the separation between the atomic and molecular length scales allows for the effective treatment of the monomer–monomer interactions by the simple Flory–Huggins form $U[\hat{\phi}_A, \hat{\phi}_B]$ in Eq. (2.192) coupled with the incompressibility constraint $\hat{\phi}_A(r) + \hat{\phi}_B(r) = 1$. Beyond these approximations in the underlying model, the only other is the saddle-point approximation, which amounts to mean-field theory. Again, this approximation becomes increasingly accurate for large invariant polymerization indices $\mathcal{N}$, where the individual polymers acquire more and more contacts with their neighboring molecules. The conventional understanding is that SCFT becomes exact in the limit of $\mathcal{N} \to \infty$. Nevertheless, real polymers are of finite molecular weight, and thus there is still a need for further development.

2.11
Beyond SCFT: Fluctuation Corrections

The main shortcoming of SCFT is that it does not properly account for composition fluctuations (Matsen 2002b), where $\hat{\phi}_A(r)$ and $\hat{\phi}_B(r)$ deviate from their ensemble averages, $\langle\hat{\phi}_A(r)\rangle$ and $\langle\hat{\phi}_B(r)\rangle$, respectively. These fluctuations are best known for disordering the weakly segregated phases near the mean-field critical points in Figs. 2.10(b) and 2.20, but they also have some significant effects on strongly segregated melts. In particular, capillary-wave fluctuations broaden the internal interfaces in both polymer blends and block copolymer microstructures. Furthermore, fluctuations are responsible for disordering the lattice arrangement of the close-packed spherical (S_{cp}) phase predicted by SCFT in Fig. 2.20 for diblock copolymer melts (Sakamoto et al. 1997; Matsen 2002b; Wang et al. 2005).

Composition fluctuations are often treated with the use of Landau–Ginzburg free-energy expressions like those derived in Eqs. (2.217), (2.234), and (2.281). The formal procedure involves breaking the calculation of the partition function Z into two separate steps. For the polymer blend considered in Section 2.8.1, the procedure begins with the calculation of

$$Z_{LG}[\Phi_A] \propto \frac{1}{n_A! n_B!} \exp\left(-\frac{U[\Phi_A, 1-\Phi_A]}{k_B T}\right) \qquad (2.311)$$

$$\times \prod_{\alpha=1}^{n_A} \tilde{\mathcal{D}} r_{A,\alpha} \prod_{\beta=1}^{n_B} \tilde{\mathcal{D}} r_{B,\beta} \, \delta[\Phi_A - \hat{\phi}_A]\delta[1 - \Phi_A - \hat{\phi}_B]$$

which provides the Landau–Ginzburg free energy

$$F_{LG}[\Phi_A] = -k_B T \ln(Z_{LG}[\Phi_A])$$

for a fixed composition profile $\Phi_A(r)$. Next, the functional integration

$$Z = \int \mathcal{D}\Phi_A \, Z_{LG}[\Phi_A] \qquad (2.312)$$

is performed over all possible compositions. While it is obvious that these steps reproduce the result of Eq. (2.195), $F_{LG}[\Phi_A]$ is no easier to calculate than the actual free energy, $F = -k_B T \ln Z$, of the system.

To proceed, the Landau–Ginzburg free energy can be approximated by the methods of SCFT. It may seem odd to correct the SCFT prediction for F by using the SCFT prediction for $F_{LG}[\Phi_A]$, but SCFT is better suited to the latter problem, where the composition is not permitted to fluctuate. Still, the SCFT calculation of $F_{LG}[\Phi_A]$ can run into problems, because it only constrains the average concentration of A segments [i.e. $\langle\hat{\phi}_A(r)\rangle = \Phi_A(r)$] rather than the actual concentration [i.e. $\hat{\phi}_A(r) = \Phi_A(r)$]. For well-segregated compositions, where $\Phi_A(r)$ is generally close to either zero or one, there is little distinction between constraining $\hat{\phi}_A(r)$ as opposed to its ensemble average, and therefore the method works well. The treatment of capillary-wave fluctuations in both polymer blends and block copolymer microstructures agrees well with experiment (Shull et al. 1993; Sferrazza et al. 1997). However, for the most widely used application (Fredrickson and Helfand 1987) on the Brazovskii fluctuations (Brazovskii 1975) of weakly segregated block copolymer melts, this Landau–Ginzburg treatment is questionable. It predicts the ODT in Fig. 2.20 to shift upwards, destroying the mean-field critical point and opening up direct transitions between the disordered state and each of the ordered phases (Hamley and Podneks 1997). Although these qualitative predictions agree with both experiment (Bates et al. 1988) and simulation (Vassiliev and Matsen 2003), there is mounting evidence that the theoretical treatment is quantitatively inaccurate (Maurer et al. 1998). Most concerning is a recent demonstration that the treatment only produces sensible results due to an approximation of the scattering function, $S(\boldsymbol{k})$; when the full expression in Eq. (2.282) is used, the short-wavelength fluctuations cause a catastrophic divergence (Kudlay and Stepanow 2003).

Fortunately, new and more rigorous approaches have started to emerge. They go to the heart of the problem and attempt to improve upon the saddle-point approximation. The first by Stepanow (1995) attempts this by generating an expansion, where fluctuation corrections are represented by a series of diagrammatic graphs, much like in quantum electrodynamics (QED). However, this approach has not yet been pushed to the point of producing experimentally verifiable predictions. More promising are field-theoretic simulations (Fredrickson et al. 2002; Müller and Schmid 2004).

To help with the full evaluation of Z, we take advantage of the usual quadratic form of $U[\Phi_A, 1-\Phi_A]$ and employ a Hubbard–Stratonovich transformation (Müller and Schmid 2004)

$$\exp\left(n\chi N \int d\boldsymbol{r}\, \Phi_A^2\right)$$

$$\propto \int \mathcal{D}W_- \exp\left(-\frac{n}{\mathcal{V}} \int d\boldsymbol{r}\, \left[\frac{W_-^2}{\chi N} + 2W_-\Phi_A\right]\right) \qquad (2.313)$$

to reduce the functional integrations in Eq. (2.198) from three to two. This is done by inserting the transformation into Eq. (2.198), which then allows the integration over $\Phi_A(r)$, producing the Dirac delta functional $\delta[2W_- - W_A + W_B]$. This identifies

$$W_-(\boldsymbol{r}) = \tfrac{1}{2}(W_A(\boldsymbol{r}) - W_B(\boldsymbol{r})) \qquad (2.314)$$

and, in turn, allows the integration to be performed over $W_A(\mathbf{r})$. With the substitution of $W_B(\mathbf{r})$ for a new function $W_+(\mathbf{r})$ defined by

$$W_+(\mathbf{r}) = \tfrac{1}{2}(W_A(\mathbf{r}) + W_B(\mathbf{r})) \qquad (2.315)$$

the partition function for binary polymer blends in Eq. (2.198) becomes

$$Z \propto \int \mathcal{D}W_- \, \mathcal{D}W_+ \, \exp\left(-\frac{F[W_-, W_+]}{k_B T}\right) \qquad (2.316)$$

where now

$$\frac{F}{nk_B T} = \phi \ln\left(\frac{\phi \mathcal{V}}{\mathcal{Q}_A[W_+ + W_-]}\right) + (1-\phi)\ln\left(\frac{(1-\phi)\mathcal{V}}{\mathcal{Q}_B[W_+ - W_-]}\right)$$
$$+ \frac{1}{\mathcal{V}} \int d\mathbf{r} \left[\frac{W_-^2}{\chi N} - W_+\right] \qquad (2.317)$$

Similarly, the ensemble average of, for example, the A-segment concentration becomes

$$\langle \hat{\phi}_A(\mathbf{r}) \rangle = \frac{1}{Z} \int \mathcal{D}W_- \, \mathcal{D}W_+ \, \phi_A(\mathbf{r}) \, \exp\left(-\frac{F[W_-, W_+]}{k_B T}\right) \qquad (2.318)$$

where $\phi_A(\mathbf{r})$ is the concentration from n_A non-interacting A-type polymers subjected to the complex field, $W_A(\mathbf{r}) = W_+(\mathbf{r}) + W_-(\mathbf{r})$. In this case, $\phi_A(\mathbf{r})$ is a complex quantity with no physical significance. The same procedure can be adapted equally well to the partition functions of other systems such as the diblock copolymer melts considered in Section 2.9.1.

Since the integrand of Z in Eq. (2.316) involves a Boltzmann factor, $\exp(-F/k_B T)$, standard simulation techniques of statistical mechanics can be applied, where the coordinates are now the fluctuating fields $W_-(\mathbf{r})$ and $W_+(\mathbf{r})$. Ganesan and Fredrickson (2001) have used Langevin dynamics, while Düchs et al. (2003) have formulated a Monte Carlo algorithm for generating a sequence of configurations with the appropriate Boltzmann weights. The ensemble average $\langle \hat{\phi}_A(\mathbf{r}) \rangle$ can then be approximated by averaging $\phi_A(\mathbf{r})$ over the configurations generated by the simulation. Even though $\phi_A(\mathbf{r})$ is generally complex, the imaginary part will average to zero. The need to consider two separate fluctuating fields is still computationally demanding, but in practice the functional integral over the field $W_+(\mathbf{r})$ can be performed with the saddle-point approximation, which implies that, as $W_-(\mathbf{r})$ fluctuates, $W_+(\mathbf{r})$ is continuously adjusted so that $\phi_A(\mathbf{r}) + \phi_B(\mathbf{r}) = 1$ (Düchs et al. 2003). Even with the problem reduced to a single fluctuating field, it is still an immense challenge to perform simulations in full three-dimensional space. However, with the rapid improvement in computational performance and the development of improved algorithms, it is just a matter of time before these field-theoretic simulations become viable.

2.12
Appendix: The Calculus of Functionals

SCFT is a continuum field theory that relies heavily on the calculus of functionals. For those not particularly familiar with this more obscure variant of calculus, this section reviews the essential aspects required for a comprehensive understanding of SCFT. In short, the term *functional* refers to a function of a function. A specific example of a functional is

$$\mathcal{F}[f] \equiv \int_0^1 dx \exp(f(x)) \tag{2.319}$$

This definition provides a well-defined rule for producing a scalar number from the input function $f(x)$. For example, $\mathcal{F}[f]$ returns the value e for the input $f(x) = 1$, and it returns $(e - 1)$ for $f(x) = x$. Note that we follow a convention where the arguments of a functional are enclosed in square brackets, rather than the round brackets generally used for ordinary functions.

The calculus of functionals is, in fact, very much like multivariable calculus, where the value of $f(x)$ for each x acts as a separate variable. Take the case where $0 \leq x \leq 1$, and imagine that the x coordinate is divided into a discrete array of equally spaced points, $x_m \equiv m/M$ for $m = 0, 1, 2, \ldots, M$. Provided that M is large, the function $f(x)$ is well represented by the set of values $\{f_0, f_1, \ldots, f_M\}$, where $f_m \equiv f(x_m)$. In turn, the functional in Eq. (2.319) is well approximated by the multivariable function

$$\mathcal{F}(\{f_m\}) \equiv \frac{\exp(f_0)}{2M} + \frac{1}{M} \sum_{m=1}^{M-1} \exp(f_m) + \frac{\exp(f_M)}{2M} \tag{2.320}$$

The correspondence between calculus of functionals and multivariable calculus is easily understood in terms of this approximation. Functional derivatives are related to partial derivatives by the correspondence

$$\frac{\mathcal{D}}{\mathcal{D}f(x)}\mathcal{F}[f] \iff \frac{\partial}{\partial f_n}\mathcal{F}(\{f_m\}) \tag{2.321}$$

and functional integrals are related to ordinary multidimensional integrals by

$$\int \mathcal{D}f \, \mathcal{F}[f] \iff \int \mathrm{d}f_0 \, \mathrm{d}f_1 \cdots \mathrm{d}f_M \, \mathcal{F}(\{f_m\}) \tag{2.322}$$

While Eq. (2.321) offers a useful intuitive definition of a functional derivative, the more rigorous definition in terms of limits is

$$\frac{\mathcal{D}\mathcal{F}[f]}{\mathcal{D}f(y)} \equiv \lim_{\epsilon \to 0} \frac{\mathcal{F}[f + \epsilon \delta] - \mathcal{F}[f]}{\epsilon} \tag{2.323}$$

where $\mathcal{F}[f + \epsilon \delta]$ represents the functional evaluated for the input function $f(x) + \epsilon \delta(x - y)$. There is one subtle point in that the Dirac delta function is itself defined in terms of a limit involving a set of finite functions such as

$$\delta(x) = \lim_{\sigma \to 0} \frac{\exp(-x^2/\sigma)}{\sqrt{\pi \sigma}} \tag{2.324}$$

The definition in Eq. (2.323) assumes that the limit $\epsilon \to 0$ is taken before the limit $\sigma \to 0$, so that $\epsilon \delta(x - y)$ can be treated as infinitesimally small. For the specific example in Eq. (2.319),

$$\mathcal{F}[f + \epsilon \delta] \equiv \int_0^1 \mathrm{d}x \, \exp(f(x) + \epsilon \delta(x - y))$$

$$\approx \int_0^1 \mathrm{d}x \, \exp(f(x))[1 + \epsilon \delta(x - y)]$$

$$= \mathcal{F}[f] + \epsilon \exp(f(y)) \tag{2.325}$$

where terms of order $\mathcal{O}(\epsilon^2)$ have been dropped. Inserting this into the definition of a functional derivative (Eq. 2.323) and renaming y as x, it immediately follows that

$$\frac{\mathcal{D}\mathcal{F}[f]}{\mathcal{D}f(x)} = \exp(f(x)) \tag{2.326}$$

Let us now consider a general class of functionals, which are defined as an integral

$$\mathcal{I}[f] \equiv \int_a^b \mathrm{d}x \, \mathcal{L}(x, f(x), f'(x)) \tag{2.327}$$

with an arbitrary integrand $\mathcal{L}(x, f(x), f'(x))$, involving x, the function $f(x)$, and its first derivative $f'(x)$. To calculate its functional derivative, we start with

$$\mathcal{I}[f+\epsilon\delta] \equiv \int_a^b \mathrm{d}x\, \mathcal{L}(x, f(x)+\epsilon\delta(x-y), f'(x)+\epsilon\delta'(x-y))$$

$$\approx \mathcal{I}[f] + \epsilon \int_a^b \mathrm{d}x\, \delta(x-y)\frac{\partial}{\partial f}\mathcal{L}(x, f(x), f'(x))$$

$$+ \epsilon \int_a^b \mathrm{d}x\, \delta'(x-y)\frac{\partial}{\partial f'}\mathcal{L}(x, f(x), f'(x))$$

$$= \mathcal{I}[f] + \epsilon \frac{\partial}{\partial f}\mathcal{L}(y, f(y), f'(y))$$

$$- \epsilon \frac{\mathrm{d}}{\mathrm{d}y}\left(\frac{\partial}{\partial f'}\mathcal{L}(y, f(y), f'(y))\right) \tag{2.328}$$

In the last step, the integration involving $\delta'(x-y)$ is converted to one involving $\delta(x-y)$ by performing an integration by parts, and then both integrals are evaluated using the sifting property of the Dirac delta function,

$$\int \mathrm{d}x\, \delta(x-y)g(x) = g(y) \tag{2.329}$$

where $g(x)$ is an arbitrary function. Inserting Eq. (2.328) into the definition of a functional derivative, we have

$$\frac{\mathcal{D}\mathcal{I}[f]}{\mathcal{D}f(x)} = \frac{\partial}{\partial f}\mathcal{L}(x, f(x), f'(x)) - \frac{\mathrm{d}}{\mathrm{d}x}\left(\frac{\partial}{\partial f'}\mathcal{L}(x, f(x), f'(x))\right) \tag{2.330}$$

This more general result is indeed consistent with our prior example. If we choose $\mathcal{L}(x, f, f') = \exp(f)$, then

$$\frac{\partial}{\partial f}\mathcal{L}(x, f, f') = \exp(f) \quad \text{and} \quad \frac{\partial}{\partial f'}\mathcal{L}(x, f, f') = 0$$

Substituting these into Eq. (2.330) reduces it to the specific case of Eq. (2.326).

Many problems in physics can be expressed in terms of either a minimization or maximization of an integral of the form in Eq. (2.327). One such example involves the Lagrangian formalism of classical mechanics (Goldstein 1980), where the trajectory of an object, $r_p(t)$, is determined by minimizing the so-called action S, which is an integral over time involving its position $r_p(t)$ and velocity $r'_p(t)$. The condition for an extremum of the integral $\mathcal{I}[f]$ is simply

$$\frac{\mathrm{d}}{\mathrm{d}x}\left(\frac{\partial}{\partial f'}\mathcal{L}(x, f(x), f'(x))\right) - \frac{\partial}{\partial f}\mathcal{L}(x, f(x), f'(x)) = 0 \tag{2.331}$$

which is a differential equation referred to as the Euler–Lagrange equation. This result is used numerous times throughout this chapter.

Of course, calculus involves integration as well as differentiation. Here we demonstrate some of the intricacies of functional integration by developing a useful expression for the Dirac delta functional $\delta[f]$. This is a generalization of the ordinary delta functional defined in Eq. (2.324), although we now use the alternative integral representation

$$\delta(x) = \frac{1}{2\pi} \int_{-\infty}^{\infty} dk \, \exp(ikx) \qquad (2.332)$$

which is derived using Fourier transforms (Eqs. 2.80 and 2.81) combined with the sifting property from Eq. (2.329). It is in fact the sifting property that is the essential characteristic by which the delta function is defined. The functional version of this sifting property is

$$\int \mathcal{D}f \, \delta[f - g] \mathcal{F}[f] = \mathcal{F}[g] \qquad (2.333)$$

where $\mathcal{F}[f]$ is an arbitrary functional. The generalization of Eq. (2.332) is arrived at by first constructing the approximate discrete version

$$\delta[f] \approx \delta(f_0)\delta(f_1)\cdots\delta(f_M)$$

$$= \frac{1}{(2\pi)^{(M+1)}} \int dk_0 \, dk_1 \cdots dk_M \, \exp\left(i \sum_{m=0}^{M} k_m f_m\right) \qquad (2.334)$$

Then taking the limit $M \to \infty$ gives

$$\delta[f] \propto \int \mathcal{D}k \, \exp\left(i \int dx \, k(x) f(x)\right) \qquad (2.335)$$

where now we have a functional integral over $k(x)$. Notice that there is a slight difficulty in taking the limit; as M increases, the proportionality constant approaches zero, although this is compensated for by the increasing number of integrations. In practice, this is not a problem, because functional integrals are always evaluated for finite M, but it does prevent us from specifying a proportionality constant in Eq. (2.335). Nevertheless, the constant is unimportant in the applications to SCFT, and so Eq. (2.335) is sufficient for our purposes.

References

Almdal, K., Rosedale, J. H., Bates, F. S., Wignall, G. D., and Fredrickson, G. H., 1990, *Phys. Rev. Lett.* **65**, 1112.

Ball, R. C., Marko, J. F., Milner, S. T., and Witten, T. A., 1991, *Macromolecules* **24**, 693.

Bates, F. S. and Fredrickson, G. H., 1999, *Phys. Today* **52**, 32.

Bates, F. S., Rosedale, J. H., Fredrickson, G. H., and Glinka, C. J., 1988, *Phys. Rev. Lett.* **61**, 2229.

Bates, F. S., Schulz, M. F., Khandpur, A. K., Förster, S., Rosedale, J. H., Almdal, K., and Mortensen, K., 1994, *Faraday Discuss.* **98**, 7.

Brazovskii, S. A., 1975, *Sov. Phys. JETP* **41**, 85.

Carrier, G. F., Krook, M., and Pearson, C. E., 1966, *Functions of a Complex Variable*, Section 6-3. McGraw-Hill, New York.

de Gennes, P.-G., 1979, *Scaling Concepts in Polymer Physics*. Cornell University Press, Ithaca, NY.

Düchs, D., Ganesan, V., Fredrickson, G. H., and Schmid, F., 2003, *Macromolecules* **36**, 9237.

Edwards, S. F., 1965, *Proc. Phys. Soc. London* **85**, 613.

Fetters, L. J., Lohse, D. J., Richter, D., Witten, T. A., and Zirkel, A., 1994, *Macromolecules* **27**, 4639.

Feynman, R. P. and Hibbs, A. R., 1965, *Quantum Mechanics and Path Integrals*. McGraw-Hill, New York.

Fredrickson, G. H. and Helfand, E., 1987, *J. Chem. Phys.* **87**, 697.

Fredrickson, G. H., Ganesan, V., and Drolet, F., 2002, *Macromolecules* **35**, 16.

Ganesan, V. and Fredrickson, G. H., 2001, *Europhys. Lett.* **55**, 814.

Gido, S. P. and Thomas, E. L., 1994, *Macromolecules* **27**, 6137.

Goldstein, H., 1980, *Classical Mechanics*, 2nd edn. Addison-Wesley, Reading, MA.

Hajduk, D. A., Harper, P. E., Gruner, S. M., Honeker, C. C., Kim, G., Thomas, E. L., and Fetters, L. J., 1994, *Macromolecules* **27**, 4063.

Hajduk, D. A., Harper, P. E., Gruner, S. M., Honeker, C. C., Kim, G., Thomas, E. L., and Fetters, L. J., 1995, *Macromolecules* **28**, 2570.

Hajduk, D. A., Takenouchi, H., Hillmyer, M. A., Bates, F. S., Vigild, M. E., and Almdal, K., 1997, *Macromolecules* **30**, 3788.

Hamley, I. W. and Podneks, V. E., 1997, *Macromolecules* **30**, 3701.

Hamley, I. W., Koppi, K. A., Rosedale, J. H., Schulz, M. F., Bates, F. S., Almdal, K., and Mortensen, K., 1993, *Macromolecules* **26**, 5959.

Hashimoto, T., Yamasaki, K., Koizumi, S., and Hasegawa, H., 1993, *Macromolecules* **29**, 2895.

Hecht, E. and Zajac, A., 1974, *Optics*. Addison-Wesley, Reading, MA.

Helfand, E., 1975, *J. Chem. Phys.* **62**, 999.

Henry, N. F. M. and Lonsdale, K. (eds.), 1969, *International Tables for X-Ray Crystallography*. Kynoch Press, Birmingham.
Hong, K. M. and Noolandi, J., 1981, *Macromolecules* **14**, 727.
Kudlay, A. and Stepanow, S., 2003, *J. Chem. Phys.* **118**, 4272.
Leibler, L., 1980, *Macromolecules* **13**, 1602.
Leibler, L., Ajdari, A., Mourran, A., Coulon, G., and Chatenay, D., 1994, in *OUMS Conference on Ordering in Macromolecular Systems*, eds. A. Teramoto, M. Kobayashi, and T. Norisuje, p. 301. Springer-Verlag, Berlin.
Likhtman, A. E. and Semenov, A. N., 1994, *Macromolecules* **27**, 3103.
Likhtman, A. E. and Semenov, A. N., 1997, *Macromolecules* **30**, 7273.
Likhtman, A. E. and Semenov, A. N., 2000, *Europhys. Lett.* **51**, 307.
Matsen, M. W., 1995a, *Phys. Rev. Lett.* **74**, 4225.
Matsen, M. W., 1995b, *J. Chem. Phys.* **103**, 3268.
Matsen, M. W., 1996, *J. Chem. Phys.* **104**, 7758.
Matsen, M. W., 1997, *J. Chem. Phys.* **107**, 8110.
Matsen, M. W., 1998, *J. Chem. Phys.* **108**, 785.
Matsen, M. W., 1999, *J. Chem. Phys.* **110**, 4658.
Matsen, M. W., 2002a, *J. Phys.: Condens. Matter* **14**, R21.
Matsen, M. W., 2002b, *J. Chem. Phys.* **117**, 2351.
Matsen, M. W., 2003a, *Macromolecules* **36**, 9647.
Matsen, M. W., 2003b, *Phys. Rev. E* **67**, 023801.
Matsen, M. W., 2004, *J. Chem. Phys.* **121**, 1938.
Matsen, M. W. and Bates, F. S., 1996a, *Macromolecules* **29**, 1091.
Matsen, M. W. and Bates, F. S., 1996b, *Macromolecules* **29**, 7641.
Matsen, M. W. and Bates, F. S., 1997a, *J. Polym. Sci.: Part B: Polym. Phys.* **35**, 945.
Matsen, M. W. and Bates, F. S., 1997b, *J. Chem. Phys.* **106**, 2436.
Matsen, M. W. and Gardiner, J. M., 2001, *J. Chem. Phys.* **115**, 2794.
Matsen, M. W. and Schick, M., 1994a, *Phys. Rev. Lett.* **72**, 2660.
Matsen, M. W. and Schick, M., 1994b, *Macromolecules* **27**, 6761.
Matsen, M. W. and Schick, M., 1994c, *Macromolecules* **27**, 7157.
Matsen, M. W. and Whitmore, M. D., 1996, *J. Chem. Phys.* **105**, 9698.
Maurer, W. W., Bates, F. S., Lodge, T. P., Almdal, K., Mortensen, K., and Fredrickson, G. H., 1998, *J. Chem. Phys.* **108**, 2989.
Merzbacher, E., 1970, *Quantum Mechanics*, Chap. 8. Wiley, New York.
Milner, S. T., Witten, T. A., and Cates, M. E., 1988, *Macromolecules* **21**, 2610.
Morse, D. C. and Fredrickson, G. H., 1994, *Phys. Rev. Lett.* **73**, 3235.
Müller, M. and Schmid, F., 2005, *Adv. Polym. Sci.* **185**, 1.
Naughton, J. R. and Matsen, M. W., 2002, *Macromolecules* **35**, 5688.
Netz, R. R. and Schick, M., 1996, *Phys. Rev. Lett.* **77**, 302.
Netz, R. R. and Schick, M., 1998, *Macromolecules* **31**, 172201.

Olmsted, P. D. and Milner, S. T., 1998, *Macromolecules* **31**, 4011.
Papadakis, C. M., Mortensen, K., and Posselt, D., 1998, *Eur. Phys. J. B* **4**, 325.
Rasmussen, K. O. and Kalosakas, G., 2002, *J. Polym. Sci.: Part B: Polym. Phys.* **40**, 1777.
Reif, F., 1965, *Fundamentals of Statistical and Thermal Physics*. McGraw-Hill, New York.
Reiter, G. and Khanna, R., 2000, *Phys. Rev. Lett.* **85**, 5599.
Sakamoto, N., Hashimoto, T., Han, C. D., Kim, D., and Vaidya, N. Y., 1997, *Macromolecules* **30**, 1621.
Schmid, F., 1998, *J. Phys.: Condens. Matter* **10**, 8105.
Schulz, M. F., Bates, F. S., Almdal, K., and Mortensen, K., 1994, *Phys. Rev. Lett.* **73**, 86.
Semenov, A. N., 1985, *Sov. Phys. JETP* **61**, 733.
Semenov, A. N., 1993, *Macromolecules* **26**, 6617; see ref. 21 therein.
Sferrazza, M., Xiao, C., Jones, R. A. L., Bucknall, D. G., Webster, J., and Penfold, J., 1997, *Phys. Rev. Lett.* **78**, 1997.
Shefelbine, T. A., Vigild, M. E., Matsen, M. W., Hajduk, D. A., Hillmyer, M. A., and Bates, F. S., 1999, *J. Am. Chem. Soc.* **121**, 8457.
Shi, A.-C., Noolandi, J., and Desai, R. C., 1996, *Macromolecules* **29**, 6487.
Shull, K. R., Mayes, A. M., and Russell, T. P., 1993, *Macromolecules* **26**, 3929.
Stepanow, S., 1995, *Macromolecules* **28**, 8233.
Takahashi, K. and Yunoki, Y., 1967, *J. Phys. Soc. Japan* **22**, 219.
Thomas, E. L., Kinning, D. J., Alward, D. B., and Henkee, C. S., 1987, *Nature* **334**, 598.
Thompson, R. B., Rasmussen, K. O., and Lookman, T., 2004, *J. Chem. Phys.* **120**, 31.
Vassiliev, O. N. and Matsen, M. W., 2003, *J. Chem. Phys.* **118**, 7700.
Vavasour, J. D. and Whitmore, M. D., 1992, *Macromolecules* **25**, 2041.
Wang, J. F., Wang, Z.-G., and Yang, Y. L., 2005, *Macromolecules* **38**, 1979.
Wang, Z.-G., 1995, *Macromolecules* **28**, 570.
Whitmore, M. D. and Vavasour, J. D., 1995, *Acta Polym.* **46**, 341.
Wickham, R. A. and Shi, A.-C., 2001, *Macromolecules* **34**, 6487.
Yeung, C., Desai, R. C., Shi, A.-C., and Noolandi, J., 1994, *Phys. Rev. Lett.* **72**, 1834.
Yeung, C., Shi, A.-C., Noolandi, J., and Desai, R. C., 1996, *Macromol. Theory Simul.* **5**, 291.
Zhu, L., Huang, P., Cheng, S. Z. D., Ge, Q., Quirk, R. P., Thomas, E. L., Lotz, B., Wittmann, J.-C., Hsiao, B. S., Yeh, F., and Liu, L., 2001, *Phys. Rev. Lett.* **86**, 6030.

3
Comparison of Self-Consistent Field Theory and Monte Carlo Simulations

Marcus Müller

Abstract

We quantitatively compare the properties of polymer interfaces obtained from self-consistent field theory (SCFT) and Monte Carlo simulations. Interfaces in binary polymer melts and surfaces of a dense polymer liquid in coexistence with its vapor are considered.

In the first case, the interface is much wider than the size of a monomeric unit, resulting in a rather universal behavior of polymer blends. The molecular characteristics on the segmental scale influence the properties on large length scales only via a small number of coarse-grained parameters: the incompatibility per chain χN, the chain extension R_e, and the invariant polymerization index $\mathcal{N}$. We discuss how to identify these parameters for a particle-based simulation model and utilize them to compare simulations and field-theoretic calculations quantitatively. The SCFT neglects fluctuation effects. Short-range fluctuations, like the liquid-like packing of monomeric units, can be accounted for by a suitable identification of the coarse-grained parameters. Long-range correlations, like composition fluctuations in the vicinity of the critical point of demixing or capillary waves, cannot be described by the SCFT. Notwithstanding their practical relevance for finite chain lengths, these long-range fluctuations formally become less important for large invariant polymerization indices, $\mathcal{N} \to \infty$.

When the width of the polymer surface is narrow, the structural details of the monomeric units matter and cannot be parameterized by a small number of coarse-grained parameters. Local intramolecular correlations can be captured by a detailed chain model, and the corresponding single-chain problem can be solved by a partial enumeration over a large ensemble of single-chain conformations. Liquid-like packing effects on the segmental scale can be incorporated via a density-functional ansatz for the interaction free energy.

Soft Matter, Vol. 1: Polymer Melts and Mixtures. Edited by G. Gompper and M. Schick
Copyright © 2006 WILEY-VCH Verlag GmbH & Co. KGaA, Weinheim
ISBN: 3-527-30500-9

Quantitative comparison with computer simulations demonstrates that the SCFT accurately describes many interface properties, and it is a powerful and versatile tool to explore the properties of spatially inhomogeneous multicomponent polymeric systems on large length scales.

3.1
Introduction

Computational methods to predict the properties of inhomogeneous systems that contain polymers have attracted longstanding interest. On the one hand, these systems find ample applications. Polymeric materials in daily life are generally multicomponent systems in which chemically different polymers are "alloyed" so as to design a material that combines the favorable characteristics of the individual components (Paul and Newman 1978; Cahn et al. 1993; Garbassi et al. 2000). One commercially important example is rubber-toughened polystyrene, where the brittle polystyrene is mixed with the rubbery polybutadiene, thereby significantly improving the mechanical properties of the material. Such melt blending of polymers is a promising route for tailoring materials to specific application properties. Unlike metallic alloys, however, chemically different polymers often do not mix on microscopic length scales. Rather, a complicated morphology of droplets of one component dispersed into the other component forms on a mesoscopic length scale of a few micrometers, and the blend can be conceived as an assembly of interfaces. While the detailed structure on this mesoscopic length scale strongly depends on the way the material is processed, the local properties of interfaces are certainly crucial for understanding the material properties. Miscibility on a microscopic length scale is desirable for a high tensile strength of the material. For instance, the interfacial width sets the length scale on which entanglements between polymers of the different components form. Experiments by Creton et al. (1991) suggest that the mechanical strength significantly increases if the interfacial width exceeds a threshold value due to entanglements between polymers of different species across the interfaces. Alternatively, the interface tension is important for breaking up droplets under shear (Taylor 1932; Milner 1997): the lower the interface tension, the finer the two components are dispersed.

On the other hand, inhomogeneous polymer systems often are an ideal test-bed for investigating problems of the statistical mechanics of bulk phase separation, interface properties, and wetting. By virtue of the extended size of the polymers, phenomena of interest occur on length scales that are larger than the size of a monomeric unit. The large extension implies that many properties are universal, and coarse-grained models, which do not describe the details of the chemical structure of the monomeric units but only the relevant properties on large length scales, can be employed successfully. Additionally, the larger length scales facilitate experimental techniques, and a wealth of structural properties is accessible. Due to the molecular extension and fractal shape, one macromolecule interacts with many neighbors. This effect leads to a strong reduction of fluctuation effects, and inhomogeneous polymer systems can be rather well described by mean-field theories (see Chapter 2 by Matsen).

In this chapter we will focus on the quantitative comparison between Monte Carlo (MC) simulations and numerical self-consistent field (SCF) calculations. This mean-field theory is highly successful in predicting the behavior of inhomogeneous polymer systems, and it is often applied to analyze experiments. Two systems will be discussed: interfaces between two immiscible polymers in a melt, and the surface of a dense polymer liquid in contact with its vapor. In the case of a polymer blend, the typical length scale of the inhomogeneous system (e.g. the width of the interface between two coexisting phases) is much larger than the size of a monomeric unit, and the monomeric interactions that give rise to phase separation (i.e. repulsion between unlike species) are much smaller than the thermal energy scale, $k_B T$ (T being temperature and k_B Boltzmann's constant). This is in marked contrast to the energy scale of liquid–vapor interfaces, which is on the order of $k_B T$ per monomeric unit, and the interfacial width is comparable to the size of a monomeric unit. Thus, the latter behavior is much less universal, and a quantitative description has to incorporate more details of the local fluid structure and polymer architecture.

This chapter is arranged as follows. In the next section, coarse-grained models for polymer blends will be introduced, and their scope and limitations will be discussed. Then, we will describe the self-consistent field theory for a specific computational model and discuss its predictions. In order to compare computer simulations and mean-field calculations, coarse-grained parameters (e.g. the Flory–Huggins parameter) have to be identified, and different ways to extract the Flory–Huggins parameter from simulations and experiments of spatially homogeneous systems will be discussed. Then, we will present a detailed quantitative comparison between simulations and theory for interfaces. In the following section, similar questions will be addressed for liquid–vapor interfaces. Various ways to extend the self-consistent field theory to incorporate more details on short length scales will be discussed and compared to computer simulations. The chapter closes with an outlook on future directions.

3.2
Polymer Blends

3.2.1
Coarse-Grained Polymer Models

Polymers are formed by chemically linking monomeric units to form a long macromolecule. The most prominent feature of this class of molecules is their larger extension. Different types of polymer architectures have been discussed in previous chapters. Here, we restrict ourselves to the simplest one – homopolymers, which consist of a linear string of monomeric units all of the same type. The architecture of polymers is characterized by widely spread time and length scales, and different techniques have been devised to analyze the behavior of complex macromolecular systems. The particular choice of the level of detail, or abstraction, depends on the question one wants to address. For instance, one may be interested in the structure and dynamics on the segment level in a binary blend (Chung et al. 1994; Ngai and Plazek 1995; Kumar et al. 1996; Lodge and McLeish 2000; Kamath et al. 2003) or, at the other extreme, the morphology of a phase-separated state with a length scale of several micrometers. Just as different experimental techniques are suited for studying different phenomena, so theoretical approaches vary in their ability to describe different aspects of macromolecular systems.

In the following, we employ coarse-grained models that emphasize the universal properties of polymers on the length scale of the size of a single molecule. These models do not capture the structure on the atomistic scale, but lump a small number of chemical repeat units into a monomer of the coarse-grained model. These monomers interact via coarse-grained, simplified interactions. Electrostatic and torsional potentials typically are neglected in these models. The reduced number of degrees of freedom and the softer interactions on a coarse scale lead to a significant computational speed-up. Hence, larger system sizes and longer time scales can be studied that are inaccessible in atomistic simulations (Müller 2001; Shelly et al. 2001; Müller et al. 2003a; Nielsen et al. 2004).

The concept of coarse-graining relies on assuming that the structural details on the small length scale of a single chemical monomeric unit enter the model only via a small number of coarse-grained parameters like the incompatibility of the two molecular species or the size of the polymer coil. Which coarse-grained parameters or which interactions on the coarse-grained level have to be retained depends on the phenomena one wants to investigate. These parameters are input into the coarse-grained description and can be identified by comparison to experiments or simulations for a specific system. They are difficult to predict from the chemical structure of the monomeric units.

The use of coarse-grained models to describe polymeric systems has a long-standing tradition (Binder 1995; Kremer and Müller-Plathe 2000; Baschnagel et al. 2000; Müller-Plathe 2002). Conceptually, the coarse-graining procedure consists of grouping a small number of monomeric units of the original chain into an effective segment and integrating out all internal degrees of freedom associated with the monomeric units inside an effective segment (e.g. bond stretching, bending and torsional potentials, etc.). From the probability distribution of original chain conformations one can construct the probability distribution of distances along the chain of effective, coarse-grained segments. From the latter distribution one can back out the interactions of the effective segments. Typically, the interactions on the coarse-grained scale are softer and the mapping, in principle, generates higher-order interactions, i.e. even if the monomeric units originally interacted only via pairwise interactions, the coarse-graining procedure might introduce three-body (or many-body) interactions among the effective segments. In practice, these higher-order interactions are often omitted and approximated by pairwise interactions. The use of these effective potential on the coarse-grained scale then involves subtle problems related to the transferability and thermodynamic consistency (Louis 2002).

In polymer solutions and melts, the elimination of the degrees of freedom is justified due to the self-similar structure on a large range of length scales from the statistical segment length a to the polymer's end-to-end distance R_e. If the monomeric units repel each other (excluded-volume chain), the molecule adopts the conformation of a self-avoiding walk. If the molecule consists of N monomeric units, the mean square end-to-end distance obeys a power law $\langle R_e^2 \rangle \sim N^{2\nu}$ with $\nu = 0.588\ldots$ The power law marks the self-similarity of the conformations; there is no intrinsic length scale. Similarly, freely jointed chains without excluded-volume interactions adopt Gaussian, self-similar behavior on large length scales, which is characterized by an exponent $\nu = 1/2$. The way that this limiting behavior is approached and the invariance of the large-scale conformations with respect to grouping segments together are explicitly illustrated in Chapter 2 by Matsen. In semidilute solutions and melts, excluded-volume interactions are screened on large length scales and, to leading order, polymer coils also adopt self-similar, Gaussian conformations (Doi and Edwards 1986; des Cloizeaux and Jannink 1990).

The coarse-graining procedure of dilute and semidilute solutions can formally be performed exactly within the framework of the renormalization group (Freed 1987; des Cloizeaux and Jannink 1990; Schäfer 1999). It was de Gennes (1972) who first realized that the behavior of polymers in the limit of long chain length, $N \to \infty$, corresponds to the critical behavior of the n-component vector model in the limit $n \to 0$. Similar to the universality at a second-order phase transition, the behavior of long chains (i.e. at the

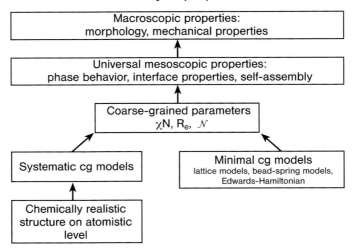

Fig. 3.1 Illustration of properties on different length scales for a binary polymer blend. Coarse-grained models rely on the assumption that properties on large length scales depend on microscopic, chemical details only via a small number of coarse-grained parameters. The set of these parameters depends on the problem, and the figure reflects a choice for the static equilibrium properties of dense binary homopolymer blends.

critical point) does not depend on the details of the microscopic interactions, as long as the relevant interactions – connectivity along the backbone of the polymer and excluded volume of the segments – are incorporated into the coarse-grained model. The renormalization group calculations are not limited to isolated chains (i.e. dilute solutions) but are also able to describe the crossover to the semidilute regime, where the polymer coils strongly overlap (i.e. $\rho_0 R_e^3/N \gg 1$, but the monomer number density ρ_0 is small).

To predict the properties of polymer liquids within the framework of coarse-grained models, one has to identify which interactions on the coarse-grained scale are relevant for the phenomena under investigation, and relate the properties of the coarse-grained model to experimental systems. Roughly, two (not mutually exclusive) strategies can be distinguished (see Fig. 3.1):

1. More often than not, coarse-graining is a concept rather than a practical procedure. Minimal coarse-grained models incorporate only the relevant interactions to bring about the phenomena one is interested in. The interaction parameters can be identified by comparing the coarse-grained model to experiments or simulations (see Section 3.2.4), and then one can use the coarse-grained model to make further predictions. The main advantage of minimal coarse-grained models is that their predictions apply to a whole class of systems (e.g. binary blends with negligible volume change

upon mixing). However, the uncertainty as to which interactions are relevant, and which interactions can be omitted for the sake of computational efficiency, limits their potential to make quantitative predictions.

Notwithstanding the limitations of coarse-grained polymer models, they offer direct, important insights into phenomena on the length scale from a few nanometers to a micrometer. They have been employed to address a large variety of questions: liquid–vapor (Smit et al. 1995; Mackie et al. 1995; Wilding et al. 1996; Escobedo and dePablo 1996; Panagiotopoulos et al. 1998) and liquid–liquid phase equilibria (Sariban and Binder 1988; Deutsch and Binder 1992a; Kumar 1994; Müller and Binder 1995; Grest et al. 1996; Escobedo and dePablo 1999), polymers at surfaces and interfaces (Bitsanis and Hadziiouanou 1990; Yethiraj 1994; Müller and Binder 1998; Müller and MacDowell 2000; Yethiraj 2002), single-chain dynamics (Kremer and Grest 1990; Paul et al. 1991a) and kinetics of phase separation (Sariban and Binder 1991; Reister et al. 2001), and the glass transition of macromolecules in the bulk and thin films (Baschnagel and Binder 1995; Binder et al. 2003).

2. Recently, much effort has been directed towards coarse-grained polymer models that not only capture the generic features of polymers on the coarse-grained scale, but also retain information about the underlying chemical structure. These systematic coarse-graining approaches aim at designing models that bridge the length and time scales from atomistic to macroscopic (Tries et al. 1997; Kremer and Müller-Plathe 2000; Müller-Plathe 2002). To this end, one chooses a set of structural and thermodynamic quantities of the underlying atomistic systems (e.g. extracted from an atomistic simulation or measured in experiments), and constructs interactions between the coarse-grained degrees of freedom so as to reproduce those quantities. Typical choices (Baschnagel et al. 2000; Müller-Plathe 2002; Shelly et al. 2001) include geometrical characteristics of the molecules, the distribution of distances between entities, and thermodynamic properties.

An example of this type of coarse-graining is the modeling of polycarbonate at a nickel surface (Site et al. 2002; Abrams et al. 2003). Car–Parinello type of quantum-mechanical density-functional calculations reveal a strong and orientation-dependent adsorption of the chain ends onto the metal surface. Unfortunately, these *ab initio* calculations are only feasible for a fragment of a chemical repeat unit of polycarbonate. Nevertheless, those calculations identified a relevant interaction for this material, and this interaction has to be accounted for in a coarse-grained model. Then, this coarse-grained model can be employed to investigate a melt of short polycarbonate polymers confined between two nickel surfaces. This strategy allows one to investigate the consequences of this specific interaction of the chain ends with the nickel surface on the atomistic scale for the molecular conformations on the length scale of the whole polymer and for the segmental density profiles of a melt in contact with the surface.

In principle, those models allow the reintroduction of atomic degrees of freedom and the study of structure on the smaller length scales they entail once the coarse-grained model has equilibrated on a large length scale. Although these coarse-grained models hold the promise of quantitatively predicting properties of polymer liquids, there remain serious caveats to date. As the interactions on the coarse-grained scale differ qualitatively from the atomistic interactions, they are in general not transferable (Müller-Plathe 2002; Louis 2002), i.e. the systematic coarse-graining procedure is specific to a particular state point described by temperature, pressure, etc. Moreover, a tiny inaccuracy in the free energy on the atomistic scale can give rise to dramatic changes on mesoscopic or macroscopic length scales. Note that the interactions on the atomistic scale (on the order 1 eV $\approx 40 k_B T$ for bonded and about $k_B T$ for non-bonded interactions) are much stronger than the typical interactions on the coarse-grained scale. For instance, the Flory–Huggins parameter that measures the repulsion of monomers in units of $k_B T$ typically takes values of 10^{-1}–10^{-4}. Recent large-scale simulation of a united-atom model of isotactic polypropylene/polyethylene blends (Heine et al. 2003) investigated the different contributions to the energy of mixing arising from bonded and non-bonded interactions. Each contribution is only on the order of $10^{-3} k_B T$ per united-atom site, and the availability of accurate interatomic potentials was considered a serious limitation of the simulation approach to predict the miscibility behavior. This problem becomes particularly pronounced in the vicinity of phase transitions where one encounters a singular dependence on system parameters. Thus, much of the quality of the coarse-graining depends on a careful choice of the set of quantities used for the mapping and the type of interactions in the coarse-grained model.

Of course, these two approaches – minimal coarse-grained models and systematic coarse-graining – are only the two extreme cases of a wide spectrum of coarse-graining methods (see Fig. 3.1). This spectrum spans from united-atom models, where one lumps the hydrogen atoms bonded to a carbon atom together into a coarse-grained segment, to the Edwards Hamiltonian, which represent polymers as Gaussian strings. It is the latter type of models we shall focus on in the following.

3.2.2
A Coarse-Grained Model for Binary Polymer Blends

In this section on polymer blends we restrict ourselves to a deceptively simple system: a dense polymer mixture of two components A and B. The excluded-volume interactions in a dense melt are screened, such that the chains adopt Gaussian conformations on scales larger than the (microscopic) mesh size, $\xi_{ev} \ll R_e$. On length scales larger than ξ_{ev}, density fluctuations are strongly suppressed and the system is effectively incompressible, i.e. when we coarse-grain the total density over the length scale ξ_{ev}, it hardly varies in space. Upon increasing the incompatibility, χ, between the two species, one encounters a critical point. At larger incompatibilities, the mixture will phase-separate into domains of A-rich and B-rich liquid. Due to the incompressibility, the system is characterized by a single order parameter, $\bar{\phi}_A$, the composition of the mixture.

Which interactions are necessary to describe this liquid–liquid demixing in binary polymer blends? This depends on the question one is interested in. If we are interested in the factors that determine the miscibility behavior of a specific pair of substances on the molecular scale, we would aim at describing the detailed fluid structure of the polymeric mixture. If we are interested in the universal equilibrium properties of dense multicomponent polymer systems and phenomena on large length scales, much insight could be gained from a highly coarse-grained description characterized by only three parameters: the end-to-end distance (R_e) of the molecules, the incompatibility per chain, χN, and the invariant polymerization index, $\mathcal{N}$.[1] The Flory–Huggins parameter, χ, describes the repulsion between unlike segments (Flory 1941; Huggins 1941) and N is the number of segments per molecule. The end-to-end vector (R_e) sets the characteristic length scale of spatial inhomogeneities, e.g. the width of interfaces between coexisting phases. Since this length scale is much larger than the size of a chemical repeat unit along the backbone of the polymer chain, one can use a coarse-grained chain model (e.g. the Edwards Hamiltonian), which only captures the long-wavelength behavior of the polymer conformations in a melt. Note that R_e and χN denote properties on the scale of a whole molecule. They identify the length and energy scale of the model. In the same spirit, one can measure the polymer number density Φ in units of the chain volume R_e^3, and define a third dimensionless parameter $\mathcal{N} \equiv (\Phi R_e^3)^2 = (\rho_0 R_e^3/N)^2$, where ρ_0 denotes the number density of segments. In the dense melt, the chain conformations are Gaussian

---

[1] We emphasize that the choice of coarse-grained parameters depends on the problem. For instance, if we were interested in the kinetics of phase separation (or the morphology on long length scales), we would also require some information about the dynamics which could be parameterized by the viscosity.

on large length scales, $R_e^2 \sim N$, and hence $\mathcal{N} \sim N$. Therefore, we refer to this quantity as the "invariant polymerization index". Phenomenologically, it describes the number of other chains inside the volume of a reference chain. Of course, one can also describe the properties of the blend by the characteristics on the scale of effective segments: the number of segments N, the statistical segment length $a = \sqrt{R_e^2/N}$, the Flory–Huggins parameter χ, and the number density ρ_0 of segments. However, since the grouping of the chemical monomeric units into effective segments is somewhat arbitrary, universal results must not change upon representing the molecules by a different number of effective segments N'. Such a reparameterization leaves the combinations χN, R_e, and $\mathcal{N}$ invariant.

The parameters R_e, χN, and $\mathcal{N}$ encode the chemical structure of the polymers on microscopic length scales. The assumption of the coarse-grained modeling is that the properties on the mesoscopic length scale depend on the microscopic structure only via the parameters R_e, χN, and $\mathcal{N}$. Even in simple cases, the Flory–Huggins parameter χ typically results from subtle differences of dispersion forces between the different chemical constituents. The dispersion forces between all segments are strongly attractive when the liquid coexists with its vapor at vanishingly small pressure. The differences between the attractive interactions make up the Flory–Huggins parameter and these differences can be orders of magnitude smaller. Likewise, the extension of a polymer in a homogeneous melt results from a delicate screening of excluded-volume interactions along the chain by surrounding molecules, and R_e depends on density and temperature. Accurately predicting those coarse-grained parameters, R_e and χN, from a microscopic model requires extremely accurate atomistic force fields. When these coarse-grained parameters are determined independently (e.g. by comparison to experiments) and used as an input, however, coarse-grained models are successful in making quantitative predictions.

One role of coarse-grained models is to predict the behavior on large length scales starting from a coarse-grained yet particle-based model. This allows one to test phenomenological approaches (e.g. Ginzburg–Landau theories or effective interface Hamiltonians) and provide input parameters (e.g. interface tension and rigidity, interface potentials). Another role of coarse-grained models is to investigate the qualitative dependence of the coarse-grained parameters (like, e.g., the Flory–Huggins parameter) on the gross features of the molecular architecture or structure of the polymeric liquid.

A simple, yet efficient, realization of a coarse-grained polymer model is the bond fluctuation model by Carmesin and Kremer (1988). In this model, effective segments are represented by unit cells on a simple cubic lattice. The linear dimensions of the lattice are denoted by L_x, L_y, and L_z, and it comprises $\mathcal{V} = L_x L_y L_z$ lattice sites. Each monomer blocks the eight corners of the unit cell from double occupancy. This can be described by a pairwise potential

$$V_{\text{ev}}(r) = \begin{cases} c_{\text{ev}} & \text{for } |r| < \sqrt{3} \\ 0 & \text{otherwise} \end{cases} \quad (3.1)$$

where r denotes the distance between effective segments measured in units of the lattice spacing. This interaction represents the excluded-volume interactions of the effective segments in the limit $c_{\text{ev}}/k_\text{B}T \to \infty$. To model a melt or concentrated solution, one occupies half the lattice sites, i.e. the number density of effective segments is $\rho_0 = 1/16$. At this density, the chains obey Gaussian statistics on large length scales, i.e. the single-chain structure factor is well describable by the Debye function (cf. Eq. 3.62). The gradual crossover from Gaussian statistics to self-avoiding-walk statistics on short length scales occurs around $\xi_{\text{ev}} \approx 7$ lattice spacings (Paul et al. 1991b). The choice of the density, $\rho_0 = 1/16$, is a compromise. If the density is high, the excluded-volume screening length ξ_{ev} will decrease and the chains will obey Gaussian statistics over a wide range of length scales. However, the relaxation of the chain conformations on the lattice will slow down considerably, because the acceptance rate of many MC moves strongly decreases with density.

Since the size of a segment is larger than the size of a vacancy, the fluid of segments exhibits packing effects. Those packing effects are less pronounced than in off-lattice models (cf. Section 3.2.4), but stronger than what one would obtain from an explicit coarse-graining of a chemically realistic model, which would result in much softer potentials.

Monomeric units along the backbone of a chain are connected via bond vectors that can take the lengths 2, $\sqrt{5}$, $\sqrt{6}$, 3, and $\sqrt{10}$ in units of the lattice spacing. Each polymer contains N segments. This bonding of neighboring segments along the chain molecules represents the chain connectivity. The 108 possible bond vectors allows for 87 distinct bond angles. The large number of bond vectors results in a rather faithful approximation of continuum space and offers the opportunity to augment the model by additional interactions – for instance, bond length and angle potential or different monomer shapes (Müller 1999).

The blend comprises two types of homopolymers – denoted A and B. There are n_A polymers of type A and n_B polymers of type B. For simplicity, we assume that each polymer contains the same number of segments, $N \equiv N_\text{A} \equiv N_\text{B}$. The composition of the blend is $\bar{\phi}_\text{A} \equiv n_\text{A}/(n_\text{A} + n_\text{B})$. Both types of segments have identical shape and bonds to their neighbors, but unlike segments repel each other while like segments attract each other via a short-range potential. For computational convenience we choose a square-well potential, which comprises the $Z_c = 54$ nearest-neighbor sites with a distance smaller than $\sqrt{6}$ in units of the lattice spacing. This square-well potential also applies to segments that are neighbors along the backbone of the polymer. If two segments of the same type are closer than $\sqrt{6}$, the energy will be reduced by an amount ϵ, while the contact between unlike segments increases the energy by ϵ. Thus

$$V_{\text{AB}}(r) = -V_{\text{AA}}(r) = -V_{\text{BB}}(r) = \begin{cases} \epsilon & \text{for } |r| \leq \sqrt{6} \\ 0 & \text{otherwise} \end{cases} \quad (3.2)$$

The interaction strength should be compared to the thermal energy scale, $k_{\text{B}}T$, and the relevant parameter is $\epsilon/k_{\text{B}}T$.

One advantage of coarse-grained models is that one can choose the interactions to cause only the phenomena of interest (in this case, demixing), while avoiding additional complications. For instance, if we were to model the interactions as (strongly) attractive with (slightly) different strengths between unlike species,

$$-V_{ij}(r) \sim \mathcal{O}(k_{\text{B}}T) \quad \text{and} \quad 2V_{\text{AB}} - V_{\text{AA}} - V_{\text{BB}} \sim \mathcal{O}(k_{\text{B}}T/N) \quad (3.3)$$

(which is a more faithful modeling of interactions in view of the experimental situation), the presence of vacancies would allow for a liquid–vapor phase separation between a polymer melt and a dilute phase in our ternary system consisting of A and B monomers and vacancies. Then, the fluid structure (packing and density) would strongly depend on temperature, i.e. there would be a coupling between composition and density fluctuations (van Konynenburg and Scott 1980), which, in general, results in a complex interplay between liquid–liquid and liquid–vapor phase coexistences.

3.2.3
Self-Consistent Field (SCF) Theory and Fluctuation Effects

Literal Mean-Field Approximation
In order to compare the results of the MC simulations to the predictions of the SCF theory, we derive the SCF theory for a microscopic model of a dense binary polymer blend. To be specific, we use the bond fluctuation model, but the derivation is more general and basically relies on the fact that we only consider pairwise interactions. Our starting point is the partition function of the interacting multi-chain system:

$$Z \propto \frac{1}{n_{\text{A}}!n_{\text{B}}!} \int \prod_{\alpha=1}^{n_{\text{A}}} \tilde{\mathcal{D}}_{\text{A}} \boldsymbol{R}_\alpha \prod_{\beta=1}^{n_{\text{B}}} \tilde{\mathcal{D}}_{\text{B}} \boldsymbol{R}_\beta$$

$$\times \exp\left(-\frac{U_{\text{ex}}[\hat{\phi}_{\text{A}},\hat{\phi}_{\text{B}}]}{k_{\text{B}}T}\right) \exp\left(-\frac{U[\hat{\phi}_{\text{A}},\hat{\phi}_{\text{B}}]}{k_{\text{B}}T}\right) \quad (3.4)$$

The factorials take account of the indistinguishability of the A and B polymers. The integration $\tilde{\mathcal{D}}_A \boldsymbol{R}_\alpha$ sums over all conformations of the A polymer α within the microscopic model with the appropriate statistical weight due to intramolecular interactions or external fields (see Eq. 3.5 below). Here $\boldsymbol{R}_\alpha$ denotes the coordinates of all N segments, $\boldsymbol{r}_{\alpha,1}, \ldots, \boldsymbol{r}_{\alpha,N}$, of A polymer α; U_{ex} is the excluded-volume constraint; while U is the square-well repulsion or attraction between effective segments of different or identical types, respectively.

Specifically, for the bond fluctuation model with its set of bond vectors, $\{\boldsymbol{b}\}_{\text{BFM}}$, we can uniquely characterize the conformation of chain α by the position of the first monomer, $\boldsymbol{r}_{\alpha,1}$, and a sequence of $N-1$ bond vectors, $\{\boldsymbol{b}_{\alpha,i} \equiv \boldsymbol{r}_{\alpha,i+1} - \boldsymbol{r}_{\alpha,i} \text{ for } i = 1, \ldots, N-1\}$, and explicitly write

$$\tilde{\mathcal{D}}_A \boldsymbol{R}_\alpha = \sum_{\boldsymbol{r}_{\alpha,1} \in \mathcal{V}} \left(\prod_{i=1}^{N-1} \sum_{\boldsymbol{b}_{\alpha,i} \in \{\boldsymbol{b}\}_{\text{BFM}}} \right) \tag{3.5}$$

The first sum runs over all positions of the first monomer on the lattice of volume $\mathcal{V}$, while the other sums run over all allowed bond vectors of the bond fluctuation model. This completes the definition of the partition function of the bond fluctuation model. Similarly, we can write down a partition function for other coarse-grained polymer models like, for instance, a bead–spring model (cf. Section 3.3.2).

Similar to the treatment of the Gaussian chain model in Chapter 2 by Matsen, we define the microscopic normalized density of A segments, which depends on the positions of all segments, $\{\boldsymbol{R}_\alpha\}$, by

$$\hat{\phi}_A(\boldsymbol{r}) = \frac{1}{\rho_0} \sum_{\alpha=1}^{n_A} \sum_{i=1}^{N} \delta(\boldsymbol{r} - \boldsymbol{r}_{\alpha,i}) \tag{3.6}$$

and an analogous expression holds for the density of B segments. The normalization is chosen such that the spatial average, $\int d\boldsymbol{r}\, \hat{\phi}_A(\boldsymbol{r})$, equals the average composition, $0 \leq \bar{\phi}_A = n_A N/\rho_0 \mathcal{V} \leq 1$. The energy of excluded-volume interaction, U_{ex}, then takes the form

$$U_{\text{ev}}[\hat{\phi}_A, \hat{\phi}_B] = \tfrac{1}{2}\rho_0^2 \int d\boldsymbol{r}\, d\boldsymbol{r}'\, (\hat{\phi}_A(\boldsymbol{r}) + \hat{\phi}_B(\boldsymbol{r})) V_{\text{ev}}(\boldsymbol{r} - \boldsymbol{r}')(\hat{\phi}_A(\boldsymbol{r}') + \hat{\phi}_B(\boldsymbol{r}'))$$

$$- \tfrac{1}{2}(n_A + n_B) N V_{\text{ev}}(0) \tag{3.7}$$

where V_{ev} is the excluded-volume interaction defined in Eq. (3.1). The last term subtracts the self-interactions that are included in the first term. A similar expression holds for the interactions that drive the phase separation

$$U[\hat{\phi}_A, \hat{\phi}_B] = \rho_0^2 \int d\boldsymbol{r}\, d\boldsymbol{r}'\, [\tfrac{1}{2}\hat{\phi}_A(\boldsymbol{r}) V_{AA}(\boldsymbol{r}-\boldsymbol{r}')\hat{\phi}_A(\boldsymbol{r}')$$

$$+ \hat{\phi}_A(\boldsymbol{r}) V_{AB}(\boldsymbol{r}-\boldsymbol{r}')\hat{\phi}_B(\boldsymbol{r}') + \tfrac{1}{2}\hat{\phi}_B(\boldsymbol{r}) V_{BB}(\boldsymbol{r}-\boldsymbol{r}')\hat{\phi}_B(\boldsymbol{r}')]$$

$$- \tfrac{1}{2} n_A N V_{AA}(0) - \tfrac{1}{2} n_B N V_{BB}(0) \tag{3.8}$$

where V_{AA}, V_{AA}, and V_{AA} are the square-well potentials in Eq. (3.2). In the last two expressions, self-interactions have been explicitly subtracted. In the following, they turn out to be irrelevant, because they contribute only a constant to the energy independent of the chain conformations and only linearly dependent on composition. Therefore, they are omitted in the following.

Choosing linear combinations of the densities, $\hat{\phi}_A$ and $\hat{\phi}_B$, we can eliminate the cross-term in the interactions. By virtue of the symmetry of the interactions, $V_{AA}(r) = V_{BB}(r)$, these linear combinations are simply the total segment density, $\hat{\varrho} \equiv \hat{\phi}_A + \hat{\phi}_B$, and the difference of densities, $\hat{\phi} \equiv \hat{\phi}_A - \hat{\phi}_B$. Then, the interaction energy is quadratic in each linear combination, $\hat{\varrho}$ and $\hat{\phi}$:

$$\frac{U_{\text{ev}} + U}{k_B T} = \frac{\rho_0}{2N} \int d\boldsymbol{r}\, d\boldsymbol{r}'\, [\hat{\varrho}(\boldsymbol{r}) N V_+(\boldsymbol{r}-\boldsymbol{r}')\hat{\varrho}(\boldsymbol{r}') - \hat{\phi}(\boldsymbol{r}) V_-(\boldsymbol{r}-\boldsymbol{r}')\hat{\phi}(\boldsymbol{r}')]$$
(3.9)

with

$$\frac{k_B T}{\rho_0} V_+(r) = V_{\text{ev}}(r) + \frac{1}{2}\left(\frac{V_{AA}(r) + V_{BB}(r)}{2} + V_{AB}(r)\right) \tag{3.10}$$

$$\frac{k_B T}{\rho_0} V_-(r) = \frac{N}{2}\left(V_{AB}(r) - \frac{V_{AA}(r) + V_{BB}(r)}{2}\right) \tag{3.11}$$

Specifically, for the bond fluctuation model, both interactions, $V_+(r) \sim V_{\text{ev}}(r)$ and $V_-(r) \sim V_{AB}(r)$, are non-negative and short-range. We have defined the interactions such that their integrated strength is of order unity, i.e. $\int d\boldsymbol{r}\, V_\pm(\boldsymbol{r}) \sim \mathcal{O}(1)$. The explicit chain-length dependence shows that the interactions that couple to the total density are an order N stronger than the interactions that involve the density difference. Note also that the two interactions contribute to the energy with different signs. Physically, this is reflected in the qualitatively different behavior of $\hat{\varrho}$ and $\hat{\phi}$. A dense polymer melt is nearly incompressible, therefore $\hat{\varrho}$ will hardly vary in space and exhibit only minor thermal fluctuations. This is in marked contrast to the density difference, $\hat{\phi}$. It distinguishes the A-rich and B-rich phase and it exhibits strong thermal fluctuations in the vicinity of the demixing transition.

It is common practice in SCF calculations not to treat the effect of the interactions $V_+(\boldsymbol{r})$ explicitly but rather to model their effect on large length scales phenomenologically. Physically, those interactions restrict fluctuations of the total density, $\hat{\varrho}$, on length scales larger than the excluded-volume screening length, ξ_{ev}, and one can describe their main effect by enforcing incompressibility, $\hat{\varrho} = 1$, or using a Gaussian penalty for deviations from the reference density (Helfand 1975):

$$\frac{U_{\mathrm{ev}} + U}{k_{\mathrm{B}}T} = \frac{\rho_0}{2N} \int \mathrm{d}\boldsymbol{r}\, \mathrm{d}\boldsymbol{r}' \bigg([\hat{\varrho}(\boldsymbol{r}) - 1] \frac{\rho_0 N}{k_{\mathrm{B}}T\kappa} \delta(\boldsymbol{r} - \boldsymbol{r}')[\hat{\varrho}(\boldsymbol{r}') - 1]$$

$$- \hat{\phi}(\boldsymbol{r}) V_-(\boldsymbol{r} - \boldsymbol{r}') \hat{\phi}(\boldsymbol{r}') \bigg) \qquad (3.12)$$

where κ denotes the compressibility of the fluid. Since linear terms in the total density, $\hat{\varrho}$, are immaterial for the phase behavior or interface properties, we can relate this Gaussian compressibility to Eq. (3.9) by identifying $V_+(\boldsymbol{r} - \boldsymbol{r}') = \rho_0 \delta(\boldsymbol{r} - \boldsymbol{r}')/(k_{\mathrm{B}}T\kappa)$, i.e. the Gaussian compressibility simply corresponds to a local, repulsive interaction, and strict incompressibility can be recovered in the limit $\kappa \to 0$. No fluid-like structure can be described by such a treatment and the relation to the microscopic model is rather indirect. However, such a phenomenological approach is well justified on length scales that exceed ξ_{ev}, where packing effects have decayed. In fact, as will be demonstrated in Section 3.2.6, such a phenomenological approach is rather necessary.

With these definitions, the partition function of the bond fluctuation model resembles the expression of the Gaussian chain model in the previous chapter. In principle, we can now literally follow the standard procedure to derive the mean-field approximation to this partition function (see Chapter 2 by Matsen). In the following, however, we use an equivalent formulation that only introduces auxiliary fields to decouple the interaction between different molecules, but does not require auxiliary fields (and a concomitant saddle-point approximation) for the microscopic densities.

As a first step, we use the Hubbard–Stratonovich formula

$$\exp(\tfrac{1}{2} x \alpha x) = \frac{1}{\sqrt{2\pi\alpha}} \int_{-\infty}^{+\infty} \mathrm{d}y \, \exp\left[-\left(\frac{y^2}{2\alpha} + xy\right)\right] \qquad (3.13)$$

at each point in space. The field W_- is introduced by identifying $\alpha = \rho_0 V_-/N$, $x = \hat{\phi}$, and $y = \rho_0 W_-/N$. The field W_+ is also inserted by the Hubbard–Stratonovich formula, but as $\alpha = -\rho_0 V_+ < 0$ we choose $x = \hat{\varrho}$ and $y = \rho_0 \mathrm{i} W_+/N$ with $\mathrm{i} = \sqrt{-1}$ to make the integrals well-behaved. Note that the field W_+, which is conjugated to the total density $\hat{\varrho}$, gives rise to an imaginary contribution. This Hubbard–Stratonovich transformation leads to the exact rewriting of the partition function:

$$Z \propto \frac{1}{n_A! n_B!} \int \prod_{\alpha=1}^{n_A} \tilde{\mathcal{D}}_A \mathbf{R}_\alpha \prod_{\beta=1}^{n_B} \tilde{\mathcal{D}}_B \mathbf{R}_\beta \, \mathcal{D}W_+ \, \mathcal{D}W_-$$

$$\times \exp\left[-\int d\mathbf{r} \, d\mathbf{r}' \frac{\rho_0}{2N^2} W_+(\mathbf{r}) V_+^{-1}(\mathbf{r}-\mathbf{r}') W_+(\mathbf{r}') \right.$$

$$\left. -i \int d\mathbf{r} \frac{\rho_0}{N} W_+(\mathbf{r}) \hat{\varrho}(\mathbf{r}) \right]$$

$$\times \exp\left[-\int d\mathbf{r} \, d\mathbf{r}' \frac{\rho_0}{2N} W_-(\mathbf{r}) V_-^{-1}(\mathbf{r}-\mathbf{r}') W_-(\mathbf{r}') \right.$$

$$\left. -\int d\mathbf{r} \frac{\rho_0}{N} W_-(\mathbf{r}) \hat{\phi}(\mathbf{r}) \right]$$

$$\propto \frac{1}{n_A! n_B!} \int \mathcal{D}W_+ \, \mathcal{D}W_- \, e^{\mathcal{S}[W_+, W_-]}$$

$$\times \int \prod_{\alpha=1}^{n_A} \tilde{\mathcal{D}}_A \mathbf{R}_\alpha \prod_{\beta=1}^{n_B} \tilde{\mathcal{D}}_B \mathbf{R}_\beta \, \exp\left[-\frac{\rho_0}{N} \int d\mathbf{r} \, (iW_+ \hat{\varrho} + W_- \hat{\phi}) \right] \quad (3.14)$$

with

$$\mathcal{S}[W_+, W_-] = -\frac{\rho_0}{2N} \int d\mathbf{r} \, d\mathbf{r}' \left[W_+(\mathbf{r}) \frac{V_+^{-1}(\mathbf{r}-\mathbf{r}')}{N} W_+(\mathbf{r}') \right. \quad (3.15)$$

$$\left. + W_-(\mathbf{r}) V_-^{-1}(\mathbf{r}-\mathbf{r}') W_-(\mathbf{r}') \right]$$

Here $V_+^{-1}(\mathbf{r})$ denotes the functional inverse of $V_+(\mathbf{r})$, which is defined by the relation

$$\int d\mathbf{r} \, V_+^{-1}(\mathbf{r}''-\mathbf{r}) V_+(\mathbf{r}-\mathbf{r}') = \delta(\mathbf{r}''-\mathbf{r}') \quad (3.16)$$

and a similar definition holds for V_-^{-1}. If the pairwise interactions were zero-ranged,

$$V_-(\mathbf{r}-\mathbf{r}') = \tfrac{1}{2}\chi N \delta(\mathbf{r}-\mathbf{r}')$$

or

$$V_+(\mathbf{r}-\mathbf{r}') = [\rho_0/(k_B T \kappa)] \delta(\mathbf{r}-\mathbf{r}')$$

then the functional inverses would simply be

$$V_-^{-1}(\mathbf{r} - \mathbf{r}') = 2\delta(\mathbf{r} - \mathbf{r}')/\chi N$$

or

$$V_+^{-1}(\mathbf{r} - \mathbf{r}') = k_B T \kappa \delta(\mathbf{r} - \mathbf{r}')/\rho_0$$

respectively. With the use of the explicit form of the microscopic density, the argument of the last exponential in Eq. (3.14) takes the form

$$\rho_0 \int d\mathbf{r} \left(iW_+ \hat{\varrho} + W_- \hat{\phi} \right) = \sum_{\alpha=1}^{n_A} \sum_{i=1}^{N} W_A(\mathbf{r}_{\alpha,i})$$

$$+ \sum_{\beta=1}^{n_B} \sum_{i=1}^{N} W_B(\mathbf{r}_{\beta,i}) \quad (3.17)$$

with $W_A = iW_+ + W_-$ and $W_B = iW_+ - W_-$. Thus, the chains do not mutually interact, but we have reformulated the partition function in terms of independent chains subjected to the external, fluctuating fields, W_+ and W_-. Let us define the partition function of a single A polymer in the external field $W_A(\mathbf{r})$ as

$$\mathcal{Q}_A[W_A] = \int \tilde{\mathcal{D}}_A \mathbf{R} \exp\left[-\frac{1}{N} \sum_{i=1}^{N} W_A(\mathbf{r}_i) \right] \quad (3.18)$$

and a similar expression defines $\mathcal{Q}_B[W_B]$. Then, we obtain for the partition function of the multi-chain system:

$$Z \propto \int \mathcal{D}W_+ \mathcal{D}W_- \, e^{S[W_+,W_-]} \frac{(\mathcal{Q}_A[iW_+ + W_-])^{n_A}}{n_A!} \frac{(\mathcal{Q}_B[iW_+ + W_-])^{n_B}}{n_B!}$$

$$\propto \int \mathcal{D}W_+ \mathcal{D}W_- \, \exp\left[-\frac{\mathcal{F}[W_+, W_-]}{k_B T} \right] \quad (3.19)$$

where the free-energy functional $\mathcal{F}$ takes the form

$$\frac{\mathcal{F}[W_+, W_-]}{k_B T (V/R_e^3)\sqrt{\bar{\mathcal{N}}}} = \bar{\phi}_A \ln\left(\frac{\bar{\phi}_A \rho_0 V}{eN \mathcal{Q}_A[iW_+ + W_-]} \right)$$

$$+ \bar{\phi}_B \ln\left(\frac{\bar{\phi}_B \rho_0 V}{eN \mathcal{Q}_B[iW_+ + W_-]} \right)$$

$$+ \frac{1}{N}\frac{1}{2V} \int d\boldsymbol{r}\, d\boldsymbol{r}'\, W_+(\boldsymbol{r}) V_+^{-1}(\boldsymbol{r}-\boldsymbol{r}') W_+(\boldsymbol{r}')$$

$$+ \frac{1}{2V} \int d\boldsymbol{r}\, d\boldsymbol{r}'\, W_-(\boldsymbol{r}) V_-^{-1}(\boldsymbol{r}-\boldsymbol{r}') W_-(\boldsymbol{r}') \qquad (3.20)$$

with $\bar\phi_A \equiv n_A N/\rho_0 \mathcal{V}$ and $\bar\phi_B \equiv n_B N/\rho_0 \mathcal{V}$ being the average compositions. Up to this stage we have reformulated the original problem in terms of a problem of independent chains in the fields W_+ and W_- without invoking any approximation. The initial difficulty due to the interactions between different molecules is now shifted to the functional integration over the fluctuating fields, W_+ and W_-. Unfortunately, we cannot perform the functional integrals over W_+ and W_-. To make further progress, we evaluate the functional integrals by a saddle-point approximation, i.e. instead of integrating over all fields, we estimate the integral by the most probable value of the integrand. It is at this stage that we neglect fluctuations of the external fields and correlations between the chains. To illustrate the consequences of this approximation, in the Appendix (Section 3.5.1) we discuss fluctuations in the bulk phases starting from Eq. (3.20).

We split the saddle-point integration into two steps. We start by approximating the functional integral over W_+ by the most probable value of the integrand:

$$\begin{aligned} \mathcal{Z} &\propto \int \mathcal{D}W_+\, \mathcal{D}W_-\, \exp\left[-\frac{\mathcal{F}[W_+, W_-]}{k_B T}\right] \\ &\approx \int \mathcal{D}W_-\, \exp\left[-\frac{\mathcal{G}_{\mathrm{EP}}[W_-]}{k_B T}\right] \end{aligned} \qquad (3.21)$$

with

$$\mathcal{G}_{\mathrm{EP}}[W_-] = \min_{W_+(\boldsymbol{r})} \mathcal{F}[W_+, W_-]$$

The real field W_+ gives rise to imaginary contributions to W_A and W_B, which, in turn, correspond to highly oscillating behavior of the integrand. To evaluate those highly oscillating contributions, the standard procedure is to extend the auxiliary field W_+ into the complex plane and perform the integration parallel to the real axis. The imaginary shift perpendicular to the real axis can be chosen such that $\mathcal{D}\mathcal{F}/\mathcal{D}W_+(\boldsymbol{r}) = 0$ at $W_+ = w_+$, where w_+ is purely imaginary. Consequently, the integrand has a stationary phase at w_+ along the shifted path of integration, and this region yields the dominant contribution to the integral. From the condition of stationary phase, we obtain

$$\frac{1}{k_\text{B}T}\frac{\mathcal{D}\mathcal{F}}{\mathcal{D}W_+(\boldsymbol{r})} = -\frac{i\rho_0}{N^2}\int \mathrm{d}\boldsymbol{r}'\, V_+^{-1}(\boldsymbol{r}-\boldsymbol{r}')W_+(\boldsymbol{r}')$$

$$-n_\text{A}\frac{\mathcal{D}\ln \mathcal{Q}_\text{A}[W_\text{A}]}{\mathcal{D}W_\text{A}(\boldsymbol{r})} - n_\text{B}\frac{\mathcal{D}\ln \mathcal{Q}_\text{B}[W_\text{B}]}{\mathcal{D}W_\text{B}(\boldsymbol{r})} \stackrel{!}{=} 0 \quad (3.22)$$

and the field that satisfies the saddle-point equation is denoted by a lower-case letter, w_+. The last terms are proportional to the densities, $\phi_\text{A}^*[W_\text{A}](\boldsymbol{r})$ and $\phi_\text{B}^*[W_\text{B}](\boldsymbol{r})$, that are created by a single A polymer or B polymer in the external field W_A or W_B, respectively. To make this explicit, we use the definition of the single-chain partition function (Eq. 3.18) and calculate the functional derivative

$$\frac{\mathcal{D}\ln \mathcal{Q}_\text{A}[W_\text{A}]}{\mathcal{D}W_\text{A}(\boldsymbol{r})} = \frac{1}{\mathcal{Q}_\text{A}}\frac{\mathcal{D}}{\mathcal{D}W_\text{A}(\boldsymbol{r})}\int \tilde{\mathcal{D}}_\text{A}\boldsymbol{R}_\alpha$$

$$\times \exp\left[-\int \mathrm{d}\boldsymbol{r}'\, W_\text{A}(\boldsymbol{r}')\frac{1}{N}\sum_{i=1}^{N}\delta(\boldsymbol{r}'-\boldsymbol{r}_{\alpha,i})\right]$$

$$= -\frac{1}{\mathcal{Q}_\text{A}}\int \tilde{\mathcal{D}}_\text{A}\boldsymbol{R}_\alpha \frac{1}{N}\sum_{i=1}^{N}\delta(\boldsymbol{r}-\boldsymbol{r}_{\alpha,i})$$

$$\times \exp\left[-\int \mathrm{d}\boldsymbol{r}'\, W_\text{A}(\boldsymbol{r}')\frac{1}{N}\sum_{i=1}^{N}\delta(\boldsymbol{r}'-\boldsymbol{r}_{\alpha,i})\right]$$

$$\equiv -\left\langle \frac{1}{N}\sum_{i=1}^{N}\delta(\boldsymbol{r}-\boldsymbol{r}_{\alpha,i})\right\rangle_{\text{single chain}} \quad (3.23)$$

$$\equiv -\frac{\rho_0}{n_\text{A}N}\phi_\text{A}^*[W_\text{A}](\boldsymbol{r}) \quad (3.24)$$

Inserting this expression into the saddle-point equation for w_+, we obtain

$$\int \mathrm{d}\boldsymbol{r}'\, \frac{V_+^{-1}(\boldsymbol{r}-\boldsymbol{r}')}{N}iw_+(\boldsymbol{r}') = (\phi_\text{A}^*(\boldsymbol{r}) + \phi_\text{B}^*(\boldsymbol{r}))$$

$$iw_+(\boldsymbol{r}'') = N\int \mathrm{d}\boldsymbol{r}\, V_+(\boldsymbol{r}''-\boldsymbol{r})\,(\phi_\text{A}^*(\boldsymbol{r}) + \phi_\text{B}^*(\boldsymbol{r})) \quad (3.25)$$

where, in the last step, we have multiplied the expression by $V_+(\boldsymbol{r}''-\boldsymbol{r})$ and integrated over $\boldsymbol{r}$ using Eq. (3.16).

Substituting back the saddle-point value, w_+, into the partition function, Eq. (3.19), we obtain

$$\frac{\mathcal{G}_{\text{EP}}[W_-]}{k_{\text{B}}T} = n_{\text{A}} \ln\left(\frac{n_{\text{A}}}{e\mathcal{Q}_{\text{A}}[iw_+ + W_-]}\right) + n_{\text{B}} \ln\left(\frac{n_{\text{B}}}{e\mathcal{Q}_{\text{B}}[iw_+ + W_-]}\right)$$

$$+ \frac{\rho_0}{2N} \int d\boldsymbol{r}\, d\boldsymbol{r}'\, W_- V_-^{-1} W_-$$

$$- \frac{\rho_0}{2} \int d\boldsymbol{r}\, d\boldsymbol{r}'\, (\phi_{\text{A}}^* + \phi_{\text{B}}^*) V_+ (\phi_{\text{A}}^* + \phi_{\text{B}}^*) \qquad (3.26)$$

Since iw_+ is real, one has to evaluate the single-chain partition functions, $\mathcal{Q}_{\text{A}}$ and $\mathcal{Q}_{\text{B}}$, and single-chain densities, ϕ_{A}^* and ϕ_{B}^*, of chains subjected to real, but fluctuating, fields, $W_{\text{A}} = iw_+[W_-] + W_-$ and $W_{\text{B}} = iw_+[W_-] - W_-$. At this stage, we have eliminated the fluctuation of the field that is conjugated to the total density, but retained the fluctuation of the field that couples to the composition of the blend. We refer to this scheme as "external potential theory" (Maurits and Fraaije 1997; Reister et al. 2001; Müller and Schmid 2005). The fluctuations can be sampled using real Langevin dynamics simulations (Reister et al. 2001) or Monte Carlo simulations (Düchs et al. 2003). In practice, it turns out that retaining the fluctuations in W_- is sufficient to capture most of the long-wavelength composition fluctuations.

To proceed in deriving the self-consistent field theory for the binary polymer blend, we also approximate the functional integral over the field W_- by its saddle-point value. The condition for the extremum becomes

$$\frac{1}{k_{\text{B}}T} \frac{\mathcal{DF}}{\mathcal{D}W_-(\boldsymbol{r})} = +\frac{\rho_0}{N} \int d\boldsymbol{r}'\, V_-^{-1}(\boldsymbol{r} - \boldsymbol{r}') W_-(\boldsymbol{r}')$$

$$- n_{\text{A}} \frac{\mathcal{D} \ln \mathcal{Q}_{\text{A}}[W_{\text{A}}]}{\mathcal{D}W_{\text{A}}(\boldsymbol{r})} + n_{\text{B}} \frac{\mathcal{D} \ln \mathcal{Q}_{\text{B}}[W_{\text{B}}]}{\mathcal{D}W_{\text{B}}(\boldsymbol{r})} \stackrel{!}{=} 0 \qquad (3.27)$$

$$w_-(\boldsymbol{r}'') = -\int d\boldsymbol{r}\, V_-(\boldsymbol{r}'' - \boldsymbol{r})\, (\phi_{\text{A}}^*(\boldsymbol{r}) - \phi_{\text{B}}^*(\boldsymbol{r})) \qquad (3.28)$$

The two Eqs. (3.25) and (3.28) relate the real saddle-point values, iw_+ and w_-, to $\phi_{\text{A}}^*[iw_+ + w_-](\boldsymbol{r})$ and $\phi_{\text{B}}^*[iw_+ - w_-](\boldsymbol{r})$, which are themselves functionals of the fields (cf. Eq. 3.24). Solving the interacting multi-chain problem within mean-field theory amounts to satisfying the Eqs. (3.24), (3.25), and (3.28) self-consistently.

Substituting the saddle-point values, w_+ and w_-, back into the free-energy functional, we obtain the mean-field estimate for the free energy:

$$\frac{F}{k_\mathrm{B} T} = \frac{\mathcal{F}[w_+, w_-]}{k_\mathrm{B} T} \tag{3.29}$$

$$= n_\mathrm{A} \ln\left(\frac{n_\mathrm{A}}{e\mathcal{Q}_\mathrm{A}[iw_+ + w_-]}\right) + n_\mathrm{B} \ln\left(\frac{n_\mathrm{B}}{e\mathcal{Q}_\mathrm{B}[iw_+ + w_-]}\right)$$

$$- \frac{\rho_0}{2N} \int \mathrm{d}\mathbf{r}\,\mathrm{d}\mathbf{r}'\, (\phi_\mathrm{A}^*(\mathbf{r}) + \phi_\mathrm{B}^*(\mathbf{r}))\, NV_+(\mathbf{r}-\mathbf{r}')\, (\phi_\mathrm{A}^*(\mathbf{r}') + \phi_\mathrm{B}^*(\mathbf{r}'))$$

$$+ \frac{\rho_0}{2N} \int \mathrm{d}\mathbf{r}\,\mathrm{d}\mathbf{r}'\, (\phi_\mathrm{A}^*(\mathbf{r}) - \phi_\mathrm{B}^*(\mathbf{r}))\, V_-(\mathbf{r}-\mathbf{r}')\, (\phi_\mathrm{A}^*(\mathbf{r}') - \phi_\mathrm{B}^*(\mathbf{r}'))$$

$$\tag{3.30}$$

To calculate the thermal average of the composition, $\langle \hat{\phi} \rangle$, we go back to the exact expression for the partition function in Eq. (3.14):

$$\langle \hat{\phi}(\mathbf{r}) \rangle \equiv \frac{1}{\mathcal{Z}} \int \mathcal{D}W_+ \, \mathcal{D}W_-\, e^{\mathcal{S}[W_+, W_-]}$$

$$\times \int \prod_{\alpha=1}^{n_\mathrm{A}} \tilde{\mathcal{D}}_\mathrm{A} \mathbf{R}_\alpha \prod_{\beta=1}^{n_\mathrm{B}} \tilde{\mathcal{D}}_\mathrm{B} \mathbf{R}_\beta\, \hat{\phi}(\mathbf{r}) \exp\left[-\frac{\rho_0}{N} \int \mathrm{d}\mathbf{r}\, (iW_+ \hat{\varrho} + W_- \hat{\phi})\right]$$

$$= \frac{1}{\mathcal{Z}} \int \mathcal{D}W_+ \, \mathcal{D}W_-\, e^{\mathcal{S}[W_+, W_-]}$$

$$\times \int \prod_{\alpha=1}^{n_\mathrm{A}} \tilde{\mathcal{D}}_\mathrm{A} \mathbf{R}_\alpha \prod_{\beta=1}^{n_\mathrm{B}} \tilde{\mathcal{D}}_\mathrm{B} \mathbf{R}_\beta \left(-\frac{N}{\rho_0} \frac{\mathcal{D}}{\mathcal{D}W_-(\mathbf{r})}\right)$$

$$\times \exp\left[-\frac{\rho_0}{N} \int \mathrm{d}\mathbf{r}\, (iW_+ \hat{\varrho} + W_- \hat{\phi})\right]$$

$$= \frac{1}{\mathcal{Z}} \int \mathcal{D}W_+ \, \mathcal{D}W_-\, e^{\mathcal{S}[W_+, W_-]} \left(-\frac{N}{\rho_0} \frac{\mathcal{D}}{\mathcal{D}W_-(\mathbf{r})}\right) \mathcal{Q}_\mathrm{A}^{n_\mathrm{A}} \mathcal{Q}_\mathrm{B}^{n_\mathrm{B}} \tag{3.31}$$

$$= \frac{\int \mathcal{D}W_+ \, \mathcal{D}W_-\, e^{\mathcal{S}[W_+, W_-]} \mathcal{Q}_\mathrm{A}^{n_\mathrm{A}} \mathcal{Q}_\mathrm{B}^{n_\mathrm{B}} (\phi_\mathrm{A}^*(\mathbf{r}) - \phi_\mathrm{B}^*(\mathbf{r}))}{\int \mathcal{D}W_+ \, \mathcal{D}W_-\, e^{\mathcal{S}[W_+, W_-]} \mathcal{Q}_\mathrm{A}^{n_\mathrm{A}} \mathcal{Q}_\mathrm{B}^{n_\mathrm{B}}}$$

$$= \langle \phi_\mathrm{A}^*(\mathbf{r}) - \phi_\mathrm{B}^*(\mathbf{r}) \rangle_W \tag{3.32}$$

where the last average $\langle \cdots \rangle_W$ is performed over all fields, W_+ and W_-, with the weight $e^{\mathcal{S}[W_+, W_-]} \mathcal{Q}_\mathrm{A}^{n_\mathrm{A}} \mathcal{Q}_\mathrm{B}^{n_\mathrm{B}}$. Within the mean-field approximation, this expression simplifies to

$$\langle \hat{\phi}(\mathbf{r}) \rangle = \phi_\mathrm{A}^*[iw_+ + w_-] - \phi_\mathrm{B}^*[iw_+ - w_-] = \langle \hat{\phi}_\mathrm{A} \rangle - \langle \hat{\phi}_\mathrm{B} \rangle \tag{3.33}$$

Similarly, we arrive at

$$\langle \hat{\varrho}(\mathbf{r}) \rangle = \phi_A^*[iw_+ + w_-] + \phi_B^*[iw_+ - w_-] = \langle \hat{\phi}_A \rangle + \langle \hat{\phi}_B \rangle \qquad (3.34)$$

These equations identify the density that a single chain in the external field creates as the thermodynamic average of the microscopic density. One can also start from Eq. (3.31) and integrate by parts to obtain a relation between the average of the microscopic density and the average of the field W_-:

$$\langle \hat{\phi}(\mathbf{r}) \rangle = -\frac{1}{\mathcal{Z}} \int \mathcal{D}W_+ \, \mathcal{D}W_- \, \mathcal{Q}_A^{n_A} \mathcal{Q}_B^{n_B} \left(-\frac{N}{\rho_0} \frac{\mathcal{D}}{\mathcal{D}W_-(\mathbf{r})} \right) e^{\mathcal{S}[W_+, W_-]}$$

$$= -\frac{1}{\mathcal{Z}} \int \mathcal{D}W_+ \, \mathcal{D}W_- \, \mathcal{Q}_A^{n_A} \mathcal{Q}_B^{n_B} \int d\mathbf{r}' \, V_-^{-1}(\mathbf{r}' - \mathbf{r}) W_-(\mathbf{r}') e^{\mathcal{S}[W_+, W_-]}$$

$$= -\int d\mathbf{r}' \, V_-^{-1}(\mathbf{r}' - \mathbf{r}) \langle W_-(\mathbf{r}') \rangle_W \qquad (3.35)$$

After the saddle-point approximation, $\langle W_-(\mathbf{r}) \rangle_W \approx w_-(\mathbf{r})$, we obtain Eq. (3.28). By the same token, we can also relate correlations of the fields to correlations of the microscopic densities:

$$\langle \hat{\phi}(\mathbf{r}) \hat{\phi}(\mathbf{r}') \rangle = \frac{1}{\mathcal{Z}} \int \mathcal{D}W_+ \, \mathcal{D}W_- \, \mathcal{Q}_A^{n_A} \mathcal{Q}_B^{n_B} \left(-\frac{N}{\rho_0} \frac{\mathcal{D}}{\mathcal{D}W_-(\mathbf{r})} \right)$$

$$\times \left(-\frac{N}{\rho_0} \frac{\mathcal{D}}{\mathcal{D}W_-(\mathbf{r}')} \right) e^{\mathcal{S}[W_+, W_-]}$$

$$= \int d\bar{\mathbf{r}} \, d\bar{\mathbf{r}}' \, V_-^{-1}(\bar{\mathbf{r}} - \mathbf{r}) V_-^{-1}(\bar{\mathbf{r}}' - \mathbf{r}') \langle W_-(\bar{\mathbf{r}}) W_-(\bar{\mathbf{r}}') \rangle_W$$

$$- \frac{N}{\rho_0} V_-^{-1}(\mathbf{r} - \mathbf{r}') \qquad (3.36)$$

In the first section of the Appendix we will use this expression to derive the random-phase approximation for the collective scattering function (cf. also Eq. 3.60). In contrast to Eq. (3.32), correlations of the microscopic densities, $\langle \hat{\phi}(\mathbf{r}) \hat{\phi}(\mathbf{r}') \rangle_W$, differ from the analogous expression $\langle [\phi_A^*(\mathbf{r}) - \phi_B^*(\mathbf{r})][\phi_A^*(\mathbf{r}') - \phi_B^*(\mathbf{r}')] \rangle$ utilizing the average densities of a single chain in the external field.

Solution for a Homogeneous Phase

It is instructive to discuss the homogeneous bulk solution of the self-consistent field equations, where densities and fields do not vary in space. The single-chain partition function, Eq. (3.18), in a spatially constant field is given by

$$\mathcal{Q}_A[w_A] = \int \tilde{\mathcal{D}}_A \boldsymbol{R}_\alpha \exp[-w_A] = \mathcal{V}\mathcal{Z}_0 \exp[-w_A] \qquad (3.37)$$

where $\mathcal{V}$ is the volume and, to be specific, the configurational single-chain partition function of a non-interacting chain takes the value $\mathcal{Z}_0 = 108^{N-1}$ in the bond fluctuation model with its 108 bond vectors. This allows us to compute

$$\phi_A^* = \frac{n_A N}{\rho_0 \mathcal{V}} = \bar{\phi}_A \quad \text{and} \quad \phi_B^* = \frac{n_B N}{\rho_0 \mathcal{V}} = \bar{\phi}_B \qquad (3.38)$$

where $\bar{\phi}_A$ and $\bar{\phi}_B$ denote the normalized, average densities. Inserting these expressions into the saddle-point equations (3.25) and (3.28), we find

$$iw_+ = N(\bar{\phi}_A + \bar{\phi}_B) \int d\boldsymbol{r}\, V_+(\boldsymbol{r}) \equiv \frac{N\rho_0(\bar{\phi}_A + \bar{\phi}_B)}{k_B T \kappa} \qquad (3.39)$$

$$w_- = -(\bar{\phi}_A - \bar{\phi}_B) \int d\boldsymbol{r}\, V_-(\boldsymbol{r}) \equiv -\frac{\chi N}{2}(\bar{\phi}_A - \bar{\phi}_B) \qquad (3.40)$$

where we have defined the compressibility, κ, according to

$$\frac{1}{\kappa} = \int d\boldsymbol{r}\, \frac{k_B T}{\rho_0} V_+(\boldsymbol{r})$$

$$= \int d\boldsymbol{r} \left[V_{\text{ev}}(r) + \frac{1}{2}\left(\frac{V_{AA}(r) + V_{BB}(r)}{2} + V_{AB}(r)\right) \right] \qquad (3.41)$$

and the segmental repulsion, χ, is given by

$$\chi = \frac{2}{N}\int d\boldsymbol{r}\, V_-(\boldsymbol{r}) = \frac{\rho_0}{k_B T}\int d\boldsymbol{r}\left(V_{AB}(r) - \frac{V_{AA}(r) + V_{BB}(r)}{2}\right) \qquad (3.42)$$

Substituting the results for the homogeneous system into the expression for the free energy (Eq. 3.30), one gets

$$\frac{F(\bar{\phi}_A, \bar{\phi}_B)}{k_B T} = n_A \ln\left(\frac{n_A}{e\mathcal{V}\mathcal{Z}_0}\right) + n_B \ln\left(\frac{n_B}{e\mathcal{V}\mathcal{Z}_0}\right)$$

$$+ (n_A + n_B)iw_+ + (n_A - n_B)w_-$$

$$- \frac{\mathcal{V}}{2k_B T \kappa}\rho_0^2(\bar{\phi}_A + \bar{\phi}_B)^2 + \frac{\chi \rho_0 \mathcal{V}}{4}(\bar{\phi}_A - \bar{\phi}_B)^2$$

$$= \frac{\rho_0(\bar{\phi}_A + \bar{\phi}_B)\mathcal{V}}{N}\ln\left(\frac{\rho_0}{eN\mathcal{Z}_0}\right) + \frac{\mathcal{V}}{2k_B T \kappa}\rho_0^2(\bar{\phi}_A + \bar{\phi}_B)^2$$

$$+ \frac{\bar{\phi}_A \rho_0 \mathcal{V}}{N} \ln(\bar{\phi}_A) + \frac{\bar{\phi}_B \rho_0 \mathcal{V}}{N} \ln(\bar{\phi}_B)$$

$$- \frac{\chi \rho_0 \mathcal{V}}{4} (\bar{\phi}_A - \bar{\phi}_B)^2 \qquad (3.43)$$

The first term in Eq. (3.43) corresponds to the free energy of an ideal gas of polymers containing $n_A + n_B$ polymer coils. The second term represents the energy due to the average pairwise interactions. Neither term depends on the composition of the mixture, $\bar{\phi}_A$, and therefore they do not influence the phase behavior of the mixture. The last two contributions constitute the excess free energy of mixing, ΔF_{FH}, commonly referred to as the Flory–Huggins free energy of mixing (Flory 1941; Huggins 1941):

$$\frac{\Delta F_{FH}}{k_B T (\mathcal{V}/R_e^3)\sqrt{\bar{N}}} = \bar{\phi}_A \ln(\bar{\phi}_A) + \bar{\phi}_B \ln(\bar{\phi}_B) - \tfrac{1}{4}\chi N (\bar{\phi}_A - \bar{\phi}_B)^2 + \tfrac{1}{4}\chi N \qquad (3.44)$$

Within the literal mean-field approximation, the segmental incompatibility, χ, defined in Eq. (3.42), is the Flory–Huggins parameter.

The first contribution to the Flory–Huggins free energy of mixing, ΔF_{FH}, stems from the combinatorial entropy of mixing; there is no further entropic contribution that might arise from the dependence of the number of possible chain conformations on the composition of the mixture. The second term represents the excess energy of mixing, and it is solely determined by the pairwise, enthalpic interactions.

One of the most important features of the Flory–Huggins free energy is that it provides a very simple, analytical expression for the free energy of mixing and that it is able to rationalize the ubiquitous observation that long polymers tend not to mix. The combinatorial entropy of mixing, which strongly favors miscibility in mixtures of small molecules, is only proportional to the number of molecules, while the energy of mixing is proportional to the number of segments, i.e. a factor of N larger. Hence, a minuscule repulsion between unlike segments – such as, for instance, the difference between protonated and deuterated chemical repeat units (Bates et al. 1985) – is sufficient to cause phase separation in long macromolecules.

From the free energy of the homogeneous system (cf. Eq. 3.43) we calculate the chemical potential of an A molecule:

$$\frac{\mu_A}{k_B T} = \frac{\partial F/k_B T}{\partial n_A}$$

$$= \ln\left(\frac{\rho_0}{N \mathcal{Z}_0}\right) + \frac{N \rho_0 (\bar{\phi}_A + \bar{\phi}_B)}{k_B T \kappa} + \ln \bar{\phi}_A - \frac{\chi N}{2}(\bar{\phi}_A - \bar{\phi}_B) \qquad (3.45)$$

and the analogous expression holds for the chemical potential of a B molecule. Using these expressions, we obtain the grand potential, $\Omega(\mu_A, \mu_B)$, via a Legendre transformation:

$$-\frac{\Omega}{k_B T \mathcal{V}} \equiv -\frac{F - n_A \mu_A - n_B \mu_B}{k_B T \mathcal{V}}$$

$$= \frac{n_A}{\mathcal{V}} + \frac{n_B}{\mathcal{V}} + \frac{\rho_0^2 (\bar{\phi}_A + \bar{\phi}_B)^2}{2 k_B T \kappa} - \frac{\rho_0 \chi (\bar{\phi}_A - \bar{\phi}_B)^2}{2} \quad (3.46)$$

Since $\Omega = -P\mathcal{V}$ (P being the pressure), we recognize that the equation of state is a second-order virial expression as expected for strictly pairwise interactions in mean-field approximation.

Predictions of the SCF Theory: Analytical Expressions in the Weak and Strong Segregation Limits

From the excess free energy of mixing, ΔF_{FH}, we can estimate the phase diagram and the fluctuations around the mean composition in the spatially homogeneous bulk phase. In the following, we assume the mixture to be incompressible, $\bar{\phi}_A + \bar{\phi}_B = 1$. The free energy of mixing in a volume R_e^3 is plotted in Fig. 3.2 for $\chi N = 3$ and two, small degrees of polymerization, $\mathcal{N} = 10$ and $\mathcal{N} = 100$, respectively. The two curves simply differ by a factor of 10 in the free-energy scale.

The first derivative of the Flory–Huggins free energy yields a relation between the exchange potential, $\Delta \mu = \mu_A - \mu_B$, and the composition, $\bar{\phi}_A$, of the mixture:

$$\frac{\Delta \mu}{k_B T} = \frac{R_e^3}{k_B T \mathcal{V} \sqrt{\mathcal{N}}} \frac{d \Delta F_{FH}}{d \bar{\phi}_A} = \ln(\bar{\phi}_A) - \ln(1 - \bar{\phi}_A) + \chi N (1 - 2\bar{\phi}_A) \quad (3.47)$$

This relation can be directly tested in semi-grand-canonical MC simulations, where $\Delta \mu$ is the control variable and one observes the composition and its fluctuations.

Fluctuations of the composition are described by the second derivative of the free energy of mixing:

$$\frac{d^2 \Delta F_{FH}}{d \bar{\phi}_A^2} = \frac{k_B T \mathcal{V} \sqrt{\mathcal{N}}}{R_e^3} \left(\frac{1}{\bar{\phi}_A} + \frac{1}{1 - \bar{\phi}_A} - 2\chi N \right) \quad (3.48)$$

$$= \frac{k_B T}{\langle (\bar{\phi}_A - \langle \bar{\phi}_A \rangle)^2 \rangle} \quad (3.49)$$

Note that the scale of the free energy in a volume of order R_e^3 is $k_B T \sqrt{\mathcal{N}}$. Thus fluctuations of the composition on the length scale R_e are only on the order

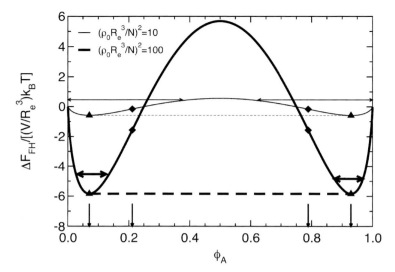

Fig. 3.2 Free energy as a function of composition, $\bar{\phi}_A$, of a binary blend at $\chi N = 3$ for two values of the invariant polymerization index, $\mathcal{N} = 10$ and 100. The compositions of the two coexisting phases and the stability limits (mean-field spinodals) are marked by triangles and diamonds, respectively. Those compositions are independent of $\mathcal{N}$. The amplitude of the free-energy variation is proportional to $\sqrt{\mathcal{N}}$. The composition fluctuation that corresponds to $1 k_B T$ away from the free-energy minima (indicated by horizontal arrows) decreases with $\mathcal{N}$.

$1/\sqrt{\mathcal{N}}$. The scale of the composition fluctuations is marked by horizontal arrows in Fig. 3.2, and one observes that they decrease upon increasing the invariant polymerization index, $\mathcal{N}$. These fluctuations can also be measured in simulations and experiments by the collective structure factor, $S(\bm{k})$, of composition fluctuations:

$$S(\bm{k}) = \frac{1}{4N(n_A + n_B)} \left\langle \left| \sum_{\alpha,i} e^{i\bm{k}\cdot\bm{r}_{\alpha,i}} - \sum_{\beta,j} e^{i\bm{k}\cdot\bm{r}_{\beta,j}} \right|^2 \right\rangle \quad (3.50)$$

$$= \frac{\rho_0}{V} \left\langle \left| \int d\bm{r}\, \hat{\phi}_A(\bm{r}) e^{i\bm{k}\cdot\bm{r}} \right|^2 \right\rangle \quad (3.51)$$

In the limit $\bm{k} \to 0$ the structure factor is related to fluctuations of the average composition, $\bar{\phi}_A$, via

$$\frac{S(k \to 0)}{N} = \frac{\rho_0 V}{N} \langle (\bar{\phi}_A - \langle \bar{\phi}_A \rangle)^2 \rangle = \left(\frac{1}{\bar{\phi}_A} + \frac{1}{1 - \bar{\phi}_A} - 2\chi N \right)^{-1} \quad (3.52)$$

At the (mean-field) spinodals, the homogeneous phase becomes unstable and phase separation occurs spontaneously. This is marked by a divergence of composition fluctuations, which corresponds to turning points of the Flory–Huggins free energy as a function of composition. For the onset of the instability, $\chi_s(\bar\phi_A)N$, the equation above yields

$$\chi_s(\bar\phi_A)N = \frac{1}{2}\left(\frac{1}{\bar\phi_A} + \frac{1}{1-\bar\phi_A}\right) \quad (3.53)$$

The locations of the spinodals are indicated by diamonds in Fig. 3.2.

The binodals can be obtained by a double-tangent construction. By virtue of the symmetry, $A \leftrightarrow B$, the tangent is horizontal in Fig. 3.2, i.e. the coexistence value of the exchange chemical potential is $\Delta\mu_{\rm coex} = 0$. Then, Eq. (3.47) yields for the binodal, $\phi_{A,\rm coex}$:

$$\chi N = \frac{1}{2\bar\phi_{A,\rm coex} - 1} \ln\left(\frac{\bar\phi_{A,\rm coex}}{1 - \bar\phi_{A,\rm coex}}\right) \quad (3.54)$$

At the critical point,

$$\bar\phi_{A,c} = 1/2 \quad \text{and} \quad \chi_c N = 2 \quad (3.55)$$

the difference in the composition between the two phases vanishes. Using the deviation from the critical composition, $\delta\bar\phi_A = \bar\phi_{A,\rm coex} - 1/2$, we can expand Eq. (3.54) as

$$\chi N = \frac{1}{2\delta\bar\phi_A} \ln\left(\frac{1 + 2\delta\bar\phi_A}{1 - 2\delta\bar\phi_A}\right) = \frac{1}{\delta\bar\phi_A}\left(\frac{2\delta\bar\phi_A}{1} + \frac{(2\delta\bar\phi_A)^3}{3} + \cdots\right) \quad (3.56)$$

and obtain for the shape of the binodal in the vicinity of the critical point:

$$2\bar\phi_{A,\rm coex} - 1 = \sqrt{3}(\tfrac{1}{2}\chi N - 1)^{1/2} = At^{\beta_{\rm MF}} \quad (3.57)$$

where the prefactor, $A = \sqrt{3}$, is called the "critical amplitude", and $\beta_{\rm MF} = 1/2$ is the value of the critical exponent within mean-field theory. Such power laws are characteristic of the universal behavior in the vicinity of a critical point. In the ultimate vicinity of the critical point, the spatial extent and amplitude of composition fluctuations diverge and the behavior of a polymer mixture resembles the behavior of a mixture of small molecules (de Gennes 1977).

If we increase the incompatibility, χN, along the critical isotherm, $\bar\phi_{A,c} = 1/2$, we obtain from Eq. (3.52) for $\chi N < 2$

$$4\langle(\bar\phi_A - \tfrac{1}{2})^2\rangle = \frac{1}{\sqrt{\mathcal{N}}(V/R_e^3)}(1 - \tfrac{1}{2}\chi N)^{-1} \quad (3.58)$$

and

$$4\langle(\bar{\phi}_A - \bar{\phi}_{A,\text{coex}})^2\rangle = \frac{1}{\sqrt{\mathcal{N}}(\mathcal{V}/R_e^3)}\frac{1}{\sqrt{2}}(\tfrac{1}{2}\chi N - 1)^{-1} \qquad (3.59)$$

for fluctuations of the average composition around the coexistence value for $\chi N > 2$. In both cases, composition fluctuations diverge with a critical exponent $\gamma_{\text{MF}} = 1$, and the amplitudes above and below the critical point differ by a factor of $\sqrt{2}$. The fluctuations of a thermodynamically intensive variable, like the average composition, decrease like $1/\mathcal{V}$. In the case of a polymer blend, however, the fluctuations are additionally suppressed by the factor $1/\sqrt{\mathcal{N}}$.

In order to study the spatial extent of composition fluctuations (the correlation length, ξ), we have to quantify the free-energy cost of a composition fluctuation as a function of its length scale. Within the random-phase approximation (RPA) for Gaussian chains (see the first section of the Appendix), one can extend Eq. (3.52) to non-zero wavevectors:

$$\frac{1}{S(k)} = \frac{1}{\bar{\phi}_A S_0(k)} + \frac{1}{(1-\bar{\phi}_A)S_0(k)} - 2\chi \qquad (3.60)$$

where $S_0(k)$ denotes the single-chain structure factor:

$$S_0(k) = \frac{1}{N}\left\langle \left|\sum_{i=1}^{N}\exp[i\boldsymbol{k}\cdot\boldsymbol{r}_i]\right|^2 \right\rangle \qquad (3.61)$$

For a Gaussian chain with end-to-end distance R_e, the single-chain structure factor is given by $S_0(k) = Ng(k)$ with the Debye function

$$g(\boldsymbol{k}) = \frac{2}{(k^2R_e^2/6)^2}[\exp(-k^2R_e^2/6) - 1 + k^2R_e^2/6] \qquad (3.62)$$

In the limit of long wavelength (Guinier regime), the Debye function can be approximated by $g(k) = 1 - (kR_e)^2/18 + \cdots$ and we obtain

$$\frac{S(k)}{N} = \frac{1}{2\chi_s(\bar{\phi}_A)N/[1 - (kR_e)^2/18 + \cdots] - 2\chi N}$$

$$\approx \frac{1/[2(\chi_s - \chi)N]}{1 + [\chi_s N/(\chi_s - \chi)N](kR_e)^2/18} \qquad (3.63)$$

This can be compared to the Ornstein–Zernicke form of the structure factor, $S(k) \sim 1/[1 + (\xi k)^2]$, and we read off the correlation length:

$$\frac{\xi}{R_e} = \frac{1}{\sqrt{18(1 - \chi N/\chi_s N)}} = \frac{1}{\sqrt{18[1 - 2\chi N\phi(1-\phi)]}}. \qquad (3.64)$$

The correlation length, ξ, characterizes the spatial extent of composition fluctuations. Its length scale is set by the size of the molecule, R_e, and at the critical point it diverges with an exponent $\nu_{\mathrm{MF}} = 1/2$. This length scale also characterizes the width of an interface between coexisting phases slightly below the critical point (in the weak segregation limit).

In the ultimate vicinity of the critical point, SCF theory will fail, because the saddle-point approximation neglects fluctuations. The SCF theory provides us, however, with a prediction for its breakdown – the Ginzburg criterion (Ginzburg 1960): the neglect of composition fluctuations will become important if the composition fluctuations in a volume of size ξ^d ($d = 3$ being the spatial dimension) are comparable to the composition difference between the two coexisting phases. Comparing the two quantities

$$\langle (\bar{\phi}_A - \bar{\phi}_{A,\mathrm{coex}})^2 \rangle \big|_\xi \sim \frac{1}{\sqrt{\mathcal{N}}(\xi/R_e)^d} \left(\frac{\chi N}{2} - 1\right)^{-1}$$

$$\sim \frac{1}{\sqrt{\mathcal{N}}} \left(\frac{\chi N}{2} - 1\right)^{(d-2)/2} \quad (3.65)$$

and

$$(2\phi_{A,\mathrm{coex}} - 1)^2 \sim \left(\frac{\chi N}{2} - 1\right) \quad (3.66)$$

we obtain the condition (de Gennes 1977; Joanny 1978; Binder 1984)

$$t^{4-d} \gg \frac{1}{\mathcal{N}} \quad \text{with} \quad t = \left|\frac{\chi N}{2} - 1\right| \quad \text{(Ginzburg criterion)} \quad (3.67)$$

for the SCF theory not to be affected by long-wavelength, critical composition fluctuations. In spatial dimensions d ($d > d_u = 4$) higher than the upper critical dimension, long-range composition fluctuations will change the mean-field critical behavior. For $d = 3$, however, mean-field theory fails and 3D Ising critical behavior is observed. In contrast to the behavior of small molecules, the large invariant polymerization index strongly suppresses long-wavelength composition fluctuations and the saddle-point approximation for those fluctuations remains valid up to the ultimate vicinity of the critical point. Physically, large $\mathcal{N}$ corresponds to a molecule interacting with many neighbors [$\mathcal{O}(\sqrt{\mathcal{N}})$].

At stronger segregation, $2 \ll \chi N \ll N$, the polymeric nature becomes more prominent, because the width of a planar interface becomes smaller than the extension of the molecules and the fractal, self-similar properties of the molecules become important. In Fig. 3.3 we present a configuration snapshot from an MC simulation and a schematic picture of the interface.

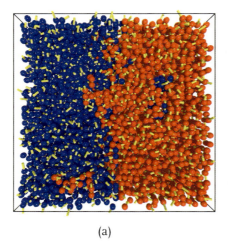 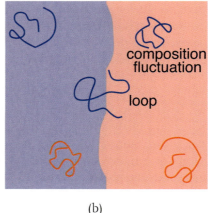

(a) (b)

Fig. 3.3 (a) Snapshot of an interface between two coexisting phases in a binary polymer blend in the bond fluctuation model (invariant polymerization index $\mathcal{N}=91$, incompatibility $\chi N \approx 17$, linear box dimension $L \approx 7.5 R_e$, or number of effective segments $N=32$, interaction $\epsilon/k_B T = 0.1$, monomer number density $\rho_0 = 1/16$). (b) Cartoon of the configuration illustrating loops of a chain into the domain of opposite type, fluctuations of the local interface position (capillary waves), and composition fluctuations in the bulk and the shrinking of the chains in the minority phase.

The interface properties in a blend of long chains are determined by loops of polymers across the interface into the opposite phase. Those loops carry a free energy of the order of thermal fluctuations, $k_B T$. Each segment of the loop contributes an amount $\chi k_B T$ to this free energy. Thus, the typical number of segments in one loop is $1/\chi$. In the strong segregation limit, $2 \ll \chi N \ll N$, the phases are well segregated, $\bar{\phi}_{A,\text{coex}} \approx 0$ or 1 for $\chi N \gg 2$, but each loop contains many segments, because $\chi \ll 1$. Then, we can assume that the statistics of a long loop is similar to the Gaussian statistics of the whole polymer, and the spatial extension of a typical loop is $R_e \sqrt{(1/\chi)/N}$. This provides an estimate for the width of a planar interface within the SCF theory:

$$\frac{w_{\text{SSL}}}{R_e} = \frac{1}{\sqrt{6\chi N}} \qquad (3.68)$$

The numerical coefficient corresponds to the asymptotic limit $\chi N \to \infty$ (but $\chi \ll 1$) of the SCF theory (Helfand and Tagami 1972; Helfand 1975). Then, a planar interface is characterized by a profile of the form

$$\phi_A(z) = \frac{1}{2}\left[1 + \tanh\left(\frac{z}{w_{\text{SSL}}}\right)\right] \qquad (3.69)$$

where z denotes the coordinate normal to the interface. Each monomer in the interfacial region contributes an amount χ to the interface tension, and the interface tension scales as $\chi \rho_0 R_e / \sqrt{\chi N}$. The result of the SCF theory (including a prefactor of order unity) is (Helfand and Tagami 1972; Semenov 1996):

$$\frac{\gamma_{\text{SSL}} R_e^2}{k_B T} = \sqrt{\mathcal{N}} \sqrt{\frac{\chi N}{6}} \qquad (3.70)$$

In the strong segregation limit (SSL), the width of a planar interface is much thinner than the extension R_e of the molecules, and loops determine the interface properties. In the weak segregation limit (WSL), the interface width is set by the bulk correlation length of composition fluctuations, $w_{WSL} = 2\xi$. The SCF theory describes both limits, WSL and SSL, and the crossover between them.

Comparison of the Literal Mean-Field Approximation for the Homogeneous Phase to Computer Simulations

To assess the consequences of the saddle-point approximation, let us compare the predictions of the SCF theory for the homogeneous phase with experimental observations and computer simulations of the bond fluctuation model. The direct comparison with computer simulations allows us to quantify the agreement or discrepancies for a wide variety of quantities within a well-defined model:

1. *Chain extension in a one-component system, $\bar{\phi}_A = 1$.* One of the most basic quantities is the extension of a polymer in a melt that consists only of A polymers. Within the *literal* SCF theory for a homogeneous system, chains are subjected to a constant field, $w_A = \int d\mathbf{r} \, (NV_+ - V_-)$. Thus, the energy of a chain conformation does not depend on the thermodynamic state specified by density $n_A/\mathcal{V}$ and temperature T, and the probability distribution $\tilde{\mathcal{D}}_A \mathbf{R}_\alpha$ describes the conformational statistics. For the simple distribution of the bond fluctuation model (Eq. 3.5), the bond vectors along the chain are completely uncorrelated:

$$\langle \mathbf{b}_i \mathbf{b}_j \rangle_{\text{SCF}} = b^2 \delta_{ij} \qquad (3.71)$$

where $b^2 = 7.389$ is the mean square bond length of the set of allowed bond vectors in the bond fluctuation model. From $R_e^2 = (\sum_{i=1}^N \mathbf{b}_i)^2$ one obtains

$$\langle R_e^2 \rangle_{\text{SCF}} = b^2(N-1) \qquad (3.72)$$

If we measure the end-to-end distance in the simulations, we obtain a value that decreases with increasing density or decreasing temperature.

For a melt-like density of $\rho_0 = 1/16$ and high temperature, one measures $\langle R_e^2 \rangle_{\text{MC}} \approx a^2(N-1)$ with a statistical segment length $a^2 \approx 10.3$ in units of the lattice spacing. Hence, the literal SCF theory underestimates the chain extension R_e by 18%, for the bond fluctuation model at $\rho_0 = 1/16$, and it does not predict its density dependence. This does not come as a surprise, because locally the excluded-volume interactions V_{ev} lead to a swelling of the chain conformations, and only at distances larger than the excluded-volume screening length ξ_{ev} does the chain approximately obey Gaussian statistics.[2] This swelling of the chain at distances smaller than ξ_{ev} and the concomitant dependence of the chain extension on temperature and density go along with fluctuations of the total density on these short length scales. Those fluctuations are neglected in the saddle-point approximation, and hence the literal SCF approximation cannot accurately predict the chain extension R_e.

2. *Critical temperature in the long-chain-length limit, $N \to \infty$.* Within the literal SCF theory, the Flory–Huggins parameter is given by the expression (3.42). Specifically, for the bond fluctuation model, the interactions comprise the nearest $Z_c = 54$ lattice sites, and thus we obtain $\chi = 2Z_c\rho_0\epsilon/k_B T$. This value is completely independent of the chain architecture and purely enthalpic. Using $\chi_c N = 2$, we obtain the prediction of the literal SCF theory for the critical temperature:

$$\frac{k_B T_c^{\text{SCF}}}{\epsilon} = Z_c \rho_0 N = 3.375 N \qquad (3.73)$$

Indeed, both experiments (Gehlsen et al. 1992) as well as MC simulations (Deutsch and Binder 1992b) confirm the linear scaling of the critical temperature with chain length N. For flexible, linear chains in the bond fluctuation model, one observes in the MC simulations a behavior of the form (Deutsch and Binder 1992b; Müller and Binder 1995)

$$\frac{k_B T_c^{\text{MC}}}{\epsilon} \approx 2.1 N \quad \text{for} \quad N \to \infty \qquad (3.74)$$

For the bond fluctuation model, the literal SCF theory overpredicts the critical temperature by 60% even in the limit of infinitely long chains, $N \to \infty$. Additionally, the prefactor depends on the chain architecture – it is larger for stiff chains (Müller 1995), which have a larger chain extension,

---

[2] Intriguingly, correlation effects are most pronounced and long-range for the bond–bond correlation function (Wittmer et al. 2004). In a dense melt, it decays *algebraically*, $\langle \boldsymbol{b}_i \boldsymbol{b}_j \rangle \to \{\sqrt{(6/\pi^3)}/[4\rho_0 \langle a^2 \rangle^{3/2}]\} |i-j|^{-3/2}$, for large distances $|i-j|$ independent of the details of the excluded-volume potential V_{ev}.

and it is smaller for branched, star or ring polymers (Müller 1999), which possess a less open chain structure.

3. *Scaling of the critical temperature in ultrathin films (quasi-2D systems)*. In the derivation of the literal SCF theory, we have not made any reference to the spatial dimension. Therefore, the theory also predicts a linear scaling of the critical temperature with chain length in a polymer blend that is confined to ultrathin films, where the chains adopt quasi-two-dimensional conformations. This is not observed in simulations, where one finds (Cavallo et al. 2003)

$$\frac{k_B T_c^{MC}}{\epsilon} \sim N^x \qquad (3.75)$$

The exponent estimated from MC simulations is $x \approx 0.65$ and $x = 5/8$ is the result of the scaling theory by Semenov and Johner (2003). Qualitatively, the pronounced increase of miscibility upon confinement can be rationalized by the packing of the chain conformations in two dimensions. In marked contrast to three dimensions, two-dimensional polymer coils do not interdigitate strongly and interactions between different polymers only occur at the fractal contact line between the segregated coils.

4. *Binodals in the vicinity of the critical point.* As in any mean-field theory, the SCF theory predicts mean-field critical behavior that is characterized by the parabolic shape of the binodals close to the critical point. In the ultimate vicinity of the critical point, however, the correlation length of composition fluctuations becomes much larger than the molecular extension, and the behavior of a binary polymer blend falls into the same universality class as an incompressible binary mixture of small molecules that is characterized by a single scalar order parameter (de Gennes 1977). Thus, simulations (Sariban and Binder 1988) as well as experiments (Bates et al. 1990; Schwahn et al. 1994) show 3D Ising critical behavior in the ultimate vicinity of the critical point, which, for instance, manifests itself in much flatter binodals

$$2\bar{\phi}_{A,\text{coex}} - 1 \sim t^{\beta_{3DI}} \quad \text{with} \quad \beta_{3DI} = 0.324 \text{ and } t = \frac{T_c - T}{T_c} \qquad (3.76)$$

5. *Chain extension in a mixture.* The same intermolecular forces that determine the miscibility behavior alter the conformation of the extended flexible macromolecules. Monte Carlo simulations (Sariban and Binder 1988; Müller 1998; Heine et al. 2003) for rather short chain lengths reveal a contraction of the polymer coils in the minority phase. Experiments in highly incompatible poly(methyl methacrylate) (PMMA) and poly(vinyl acetate)

(PVAc) blends of rather low molecular weight indicate a relative contraction of isolated PMMA chain extensions by 13–15% (Peterson et al. 1990). These observations illustrate a possible coupling between the single-chain conformations and the thermodynamic state (i.e. temperature and composition of the mixture).

These examples show that, even in the simple situation of a spatially homogeneous system, the saddle-point approximations and the neglect of correlations in the framework of the *literal* SCF theory impart quite significant quantitative inaccuracies. This does not mean, however, that the SCF theory is not useful. Going beyond the saddle-point approximations, one can estimate the importance of fluctuation effects. For a recent discussion, see Wang (2002) and the Appendix (Section 3.5.1).

In order to proceed, it proves useful to divide fluctuation effects into two classes (Müller and Binder 1995): On the one hand, there are long-range fluctuation effects that decrease in importance as we increase the invariant polymerization index, $\mathcal{N}$. This suppression of fluctuations in the limit $\mathcal{N} \to \infty$ is associated with the fact that the scale of the free energy is $k_B T \sqrt{\mathcal{N}}$ (cf. Eq. 3.20). The issues 4 and 5 above belong to this class, as we shall explicitly demonstrate in Section 3.2.4. Typically, these fluctuation effects are related to long length scales comparable to or larger than the polymer extension R_e.

On the other hand, there are correlation effects that do not vanish in the limit of very long chains (cf. issues 1–3 above). They are related to correlations due to the liquid-like packing of the effective segments, the interplay of excluded-volume constraints and back-folding, the screening of excluded-volume interactions or correlations between intermolecular packing and chain architecture. Those short-range fluctuation effects cannot be described by the SCF theory. The magnitude of deviations due to short-range fluctuation effects is not universal and depends on the specific microscopic model.

There are powerful theoretical approaches – the polymer reference interaction site (P-RISM) theory by Schweizer and Curro (1997) and the lattice cluster theory by Freed and Dudowicz (1998) – that can successfully calculate these fluctuation effects on short and intermediate length scales from microscopic model systems taking into account the liquid-like packing of segments and properly distinguishing inter- and intramolecular correlations. These approaches are mostly restricted to spatially homogeneous systems. They successfully relate the microscopic structure of model systems to the Flory–Huggins parameter χ, and also provide an estimate for the dependence of the chain extensions on the thermodynamic state. This theory yields important insights into the correlation between molecular architecture and blend thermodynamics, and thereby greatly enlarges the predictive power of the theory. We shall come back to this issue in Section 3.3 on liquid–vapor interfaces.

In the following we take a less ambitious and more pragmatic approach. We recall that the coarse-grained properties of polymer blends depend only on a small number of parameters: the incompatibility χN, the length scale R_e, and the invariant polymerization index $\mathcal{N}$. These effective parameters encode the microscopic structure *including effects due to fluctuations* on short and intermediate length scales. Thus, it is tempting to identify these coarse-grained parameters by comparing the predictions of the SCF theory to simulations or experiments. This is common experimental practice. Unlike the direct comparison with computer simulations, where the Hamiltonian is exactly known and one can test the theory in all details, the interactions between the components of the blend are not known (with sufficient accuracy) and it is impracticable to predict the Flory–Huggins parameter for a specific pair of chemically distinct polymers via Eq. (3.42). Thus, χN and R_e are used as adjustable parameters or extracted from experiments. Using the so-determined coarse-grained parameters, the SCF theory indeed has proven remarkably successful in predicting many experimental observations. Even if it was not possible to relate the Flory–Huggins parameter to the microscopic structure, once identified by comparing a specific property to the SCF predictions, it would be useful, for instance, for predicting the properties of interfaces between coexisting phases, the kinetics of phase separation, the behavior at surfaces or the self-assembly in copolymers.

3.2.4
Identifying the Effective Flory–Huggins Parameter, $\tilde{\chi}$

Mapping Observations and the Predictions of the SCF Theory
The identification of the chain extension R_e is straightforward. It is immediately accessible in computer simulations and it can be measured by scattering experiments. The invariant polymerization index $\mathcal{N} = (\rho_0 R_e^3/N)^2$ can also be calculated. Note that the predictions of the SCF theory do not depend on $\mathcal{N}$; its value influences fluctuation effects only.

A suitable identification of the Flory–Huggins parameter turns out to be more subtle. A wealth of properties of binary polymer mixtures can be extracted from computer simulations as well as from experiments. Comparing these results to the approximate mean-field expressions might give rise to quite distinct estimates for the Flory–Huggins parameter, because fluctuations neglected in the SCF theory affect estimates in different ways. In the following we argue that it is useful to compare results of experiments or simulations to the predictions of the SCF theory in the regime where long-wavelength fluctuations are not important. In this way, the Flory–Huggins parameter characterizes the microscopic structure of the blend.

A selection of possible identifications of the Flory–Huggins parameter is given below:

1. The relation between exchange potential $\Delta\mu$ and composition $\bar{\phi}_A$ of the blend (cf. Eq. 3.47) or composition fluctuations/scattering (cf. Eq. 3.60) in the one-phase region far away from the critical point can be used to identify χ.

2. Measuring the composition of the two coexisting phases at large incompatibility and comparing the result to the prediction of the Flory–Huggins theory (Eq. 3.54) for $\chi N > 2$, one can estimate the χ parameter.

3. The mean-field theory also relates the location of the critical point of demixing (cf. Eq. 3.55) and scattering in the vicinity of the critical point to the Flory–Huggins parameter.

4. The SCF theory makes explicit predictions for the width of the interface between coexisting phases, and simple analytical expressions are known in the weak and strong segregation limits (see Eqs. 3.69 or 3.68). Comparing experimental measurements or simulation results to those predictions also yields an estimate for χ.

All identifications basically treat the Flory–Huggins parameter χ as a phenomenological quantity, determined by matching the properties of the blend to the predictions of the SCF theory and thereby absorbing all unknown information concerning blend miscibility (Dudowicz et al. 2002). This is very much in the spirit of coarse-grained models, where the structure on the microscopic length scale determines the behavior on the large length scale only via a small number of parameters. The last two schemes are, however, affected by long-wavelength fluctuations, namely composition fluctuations near critical points, and capillary waves at interfaces. Let us discuss the first three identifications of the Flory–Huggins parameter, and postpone the discussion of item 4 above to Section 3.2.6 on interfaces.

Measuring the dependence of the composition on the exchange potential is straightforward in the semi-grand-canonical ensemble, where the exchange potential $\Delta\mu$ is a control parameter and one observes the composition and its fluctuations in the simulation (Müller and Binder 1995; Müller 1995). The dependence $\bar{\phi}_A(\Delta\mu)$ (cf. Eq. 3.47) and the composition fluctuations $\langle \bar{\phi}_A^2 \rangle - \langle \bar{\phi}_A \rangle^2$ (cf. Eq. 3.60) correspond to the first and second derivatives of the Flory–Huggins free energy of mixing with respect to the composition, and therefore constitute a direct test of the Flory–Huggins free energy (Flory 1941; Huggins 1941).

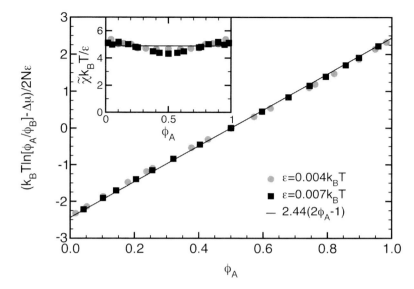

Fig. 3.4 Dependence of the composition on the exchange potential for a symmetric binary blend in the bond fluctuation model ($N = 64$, $\rho_0 = 1/16$, and two interaction strengths, $\epsilon/k_B T$ as indicated in the key). According to Eq. (3.47), the data will collapse onto a straight line and from the slope we can identify the effective Flory–Huggins parameter $\tilde\chi$. In the inset we show the composition dependence of

$$\frac{\tilde\chi}{\epsilon/k_B T} = \left[\frac{1}{N\phi_A} + \frac{1}{N(1-\phi_A)} - \frac{1}{S(k\to 0)}\right] \bigg/ \left(\frac{2\epsilon}{k_B T}\right)$$

according to Eq. (3.52). The thin horizontal line marks the value $\tilde\chi/(\epsilon/k_B T) = 4.88$ extracted from the main panel.

The simulation data for the bond fluctuation model are presented in Fig. 3.4, where we plot the difference between the exchange chemical potential and the ideal-gas contribution, $[\Delta\mu - k_B T \ln(\bar\phi_A/\bar\phi_B)]/2N\epsilon$, versus the composition $\bar\phi_A$. In accord with the Flory–Huggins estimate (Eq. 3.42), we observe a linear dependence on composition for all temperatures far above the demixing transition. This demonstrates that $\chi \sim \epsilon/k_B T$ in our model, and we can identify the Flory–Huggins parameter, $\tilde\chi = 2z_c\epsilon/k_B T$ with an *effective coordination number* $z_c = 2.44$. Note that this value is smaller than the prediction of the literal SCF approximation, $z_c = \rho_0 Z_c = 54/16$.

In the inset of Fig. 3.4 we present the estimate for the effective coordination number via the inverse of the collective structure factor (Eq. 3.51). Within the statistical uncertainties, we again find that $\chi \sim \epsilon/k_B T$ over a range of different compositions, with a slight upwards parabolic correction. This parabolic composition dependence becomes stronger as we approach the critical point (Sariban and Binder 1988; Deutsch and Binder 1992b). Far above the critical temperature, however, the ratio takes a value $\tilde\chi k_B T/\epsilon = 2z_c \approx 4.88$

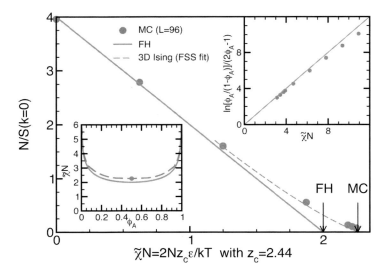

Fig. 3.5 Inverse maximum of the collective structure factor of composition fluctuations, $N/S(k\to 0)$, as a function of the incompatibility, $\tilde{\chi}N$. Symbols correspond to MC simulations, the dashed curve presents the results of a finite size scaling analysis of the simulation data in the vicinity of the critical point, and the straight solid line indicates the prediction of the Flory–Huggins theory. The critical incompatibilities predicted by the Flory–Huggins theory and obtained from MC simulations of the bond fluctuation model ($\mathcal{N} \approx 240$) are indicated by arrows. The left inset compares the phase diagram obtained from simulations with the prediction of the Flory–Huggins theory. The right inset replots the composition of the coexistence compositions such that the mean-field theory predicts them to fall onto a straight line.

consistent with the results of the main panel. Even though composition fluctuations are directly related to the experimental procedure, we prefer to extract the Flory–Huggins parameter from the excess of the exchange potential (Eq. 3.47) in the simulations, because the combinatorial contribution diverges only logarithmically for $\phi \to 0$ or 1. Also, in experiments, strongly asymmetric compositions might induce inaccuracies (Crist 1998) when one utilizes the structure factor (Eq. 3.52).

In Fig. 3.5 we use this identification of the Flory–Huggins parameter, $\tilde{\chi} = 2z_c\epsilon/k_BT$, to plot the inverse structure factor as a function of $\tilde{\chi}N$. Far above the critical temperature, $\chi N \ll \chi_c N = 2$, the data are consistently describable by the mean-field theory, i.e. for $\bar{\phi}_A = 1/2$ we find the simple relation $N/S(k=0) = 4 - 2\tilde{\chi}N$. This plot constitutes a common experimental procedure to measure the Flory–Huggins parameter in experiments. In the ultimate vicinity of the critical point, however, deviations from the SCF theory become pronounced. The inverse structure factor extracted from the MC simulations does not vanish at $\tilde{\chi}_c^{SCF}N = 2$, but the mean-field theory overestimates the true critical temperature by about 12% for the chain length $N = 64$ utilized in the simulation. Note that this value corresponds to an invariant polymerization index $\mathcal{N} = 240$. Similar to simulations (Sariban and Binder 1988) and experiments (Bates et al. 1990; Bates et al. 1992; Gehlsen and Bates 1994; Schwahn et al. 1994), in Fig. 3.5 we can see the crossover from the mean-field dependence to the Ising behavior. If we approach the critical point at the critical composition, $\phi_A = 1/2$, we will not observe a linear dependence of the inverse scattering factor with $\epsilon/k_B T$ all the way up to the critical point, but rather we find $1/S(k \to 0) \sim (\epsilon/k_B T - \epsilon/k_B T_c)^\gamma$, where $\gamma \approx 1.24$ denotes the critical exponent of the order-parameter fluctuations in the three-dimensional Ising model. By virtue of the universality at second-order phase transitions, a polymer mixture exhibits the same critical behavior as other mixtures characterized by a single scalar order parameter. The SCF theory, however, being a mean-field theory that assumes random mixing, cannot describe these critical, long-wavelength composition fluctuations.

By the same token, the SCF theory predicts a parabolic shape of the binodal in the vicinity of the critical point, while simulations and experiments show $|\bar{\phi}_{A,\text{coex}} - \bar{\phi}_{B,\text{coex}}| \sim (\epsilon/k_B T_c - \epsilon/k_B T)^\beta$, where the critical exponent of the order parameter takes the value of the 3D Ising universality class, $\beta \approx 0.32$. The binodals are presented in the insets of Fig. 3.5. While the SCF theory fails in the ultimate vicinity of the critical point, at larger $\tilde{\chi} N$ the binodals are well described by the SCF theory. Note that we have used rather short chains in the simulations, $\mathcal{N} = 240$, and we expect the region where long-wavelength composition fluctuations lead to a breakdown of the SCF theory to become even smaller for larger invariant polymerization index.

In the right inset of Fig. 3.5 we plot the binodals in the form $\ln[\bar{\phi}_A/(1-\bar{\phi}_A)]/(2\bar{\phi}_A - 1)$ versus $\tilde{\chi} N$. Within the mean-field theory, all data fall onto a line with unit slope. This is indeed what we observe in the simulations for an intermediate range of incompatibilities, and this diagram constitutes an alternative possibility to extract the Flory–Huggins parameter for a phase-separated blend. At low incompatibilities, in the ultimate vicinity of the critical point, we find deviations due to long-wavelength fluctuations as discussed above. At very large incompatibilities, the Flory–Huggins parameter becomes of the order unity for the short chain length considered in the simulations and then the repulsion between unlike segments is strong enough to change the fluid structure.

To summarize, fitting the results of simulations to the predictions of the SCF theory far away from the critical point (methods 1 and 2 above) yields consistent results, i.e. all observations can be described by a Flory–Huggins parameter that is proportional to the repulsion between unlike segments, $\epsilon/k_B T$. The proportionality constant is, however, smaller than the estimate of the literal SCF theory. In the ultimate vicinity of the critical point, the neglect of long-wavelength composition fluctuations by the SCF theory affects the estimate of the Flory–Huggins parameter (method 3). Identifying the Flory–Huggins parameter from properties in the ultimate vicinity of the critical point leads to spurious effects due to fitting a mean-field theory to data that exhibit critical fluctuations according to the 3D Ising universality class. Doing so, the estimate of the Flory–Huggins parameter not only accounts for short-range correlations but also incorporates some effects of long-range critical fluctuations. Since these critical fluctuations are only important in the ultimate vicinity of the critical point, which becomes smaller as the invariant polymerization index $\mathcal{N}$ becomes larger, this is not a useful identification of the Flory–Huggins parameter. If we estimated the Flory–Huggins parameter via method 3, the Flory–Huggins parameter would not give a useful description further away from the critical point or for longer chain length, because the effect of long-range composition fluctuations would become negligible but short-range correlations would still be important.

Relating the Flory–Huggins Parameter to the Local Structure
While the comparison between simulations and experiments far away from the critical point to the predictions of the SCF theory yields a consistent identification of the Flory–Huggins parameter, which parameterizes the repulsion on the segmental level, and which can be successfully employed to predict the properties in spatially inhomogeneous systems within the SCF theory, it does not yield any direct insight into what properties determine the degree of miscibility on the segmental scale and what are the local correlations that give rise to the deviations from the literal SCF approximation.

Much qualitative insight, however, can be gained from investigating the free-energy change upon mutating a B polymer into an A polymer (Müller 1995). To this end, we go back to the partition function (3.4) and calculate the exchange potential $\Delta\mu$ from a Widom "exchange" formula (Müller and Binder 1995; Buta and Freed 2002):

$$\frac{\Delta\mu}{k_B T} = -\ln \mathcal{Z}(n_A+1, n_B-1) + \ln \mathcal{Z}(n_A, n_B)$$

$$= -\ln \left[\frac{[1/(n_A+1)!(n_B-1)!] \int \prod_{\alpha=1}^{n_A+1} \tilde{\mathcal{D}}_A \boldsymbol{R}_\alpha \prod_{\beta=1}^{n_B-1} \tilde{\mathcal{D}}_B \boldsymbol{R}_\beta \, e^{-(U_{ex}+U)/k_B T}}{[1/n_A! n_B!] \int \prod_{\alpha=1}^{n_A} \tilde{\mathcal{D}}_A \boldsymbol{R}_\alpha \prod_{\beta=1}^{n_B} \tilde{\mathcal{D}}_B \boldsymbol{R}_\beta \, e^{-(U_{ex}+U)/k_B T}}\right]$$

$$= -\ln \left[\frac{(n_A-1)!(n_B+1)!}{n_A! n_B!}\right]$$

$$- \ln \left[\frac{\int \prod_{\alpha=1}^{n_A} \tilde{\mathcal{D}}_A \boldsymbol{R}_\alpha \prod_{\beta=1}^{n_B} \tilde{\mathcal{D}}_B \boldsymbol{R}_\beta \, e^{-(U_{ex}+U)/k_B T} e^{-\Delta E(\boldsymbol{R}_{n_B})/k_B T}}{\int \prod_{\alpha=1}^{n_A} \tilde{\mathcal{D}}_A \boldsymbol{R}_\alpha \prod_{\beta=1}^{n_B} \tilde{\mathcal{D}}_B \boldsymbol{R}_\beta \, e^{-(U_{ex}+U)/k_B T}}\right]$$

$$= \ln\left(\frac{n_A+1}{n_B}\right) - \ln \left\langle \exp\left[-\frac{\Delta E(\boldsymbol{R}_{n_B})}{k_B T}\right]\right\rangle_{\{n_A, n_B\}} \quad (3.77)$$

where $\Delta E(\boldsymbol{R}_{n_\mathrm{B}})$ denotes the energy change required to mutate the B polymer n_B into an A polymer. The thermal average $\langle\cdots\rangle_{\{n_\mathrm{A},n_\mathrm{B}\}}$ is performed in a system that contains n_A polymers of type A and n_B polymers of type B.

This exact, Widom-like expression relates the exchange potential to the work that is necessary to mutate a B polymer into an A polymer. To evaluate this work, we make three approximations. (1) We assume that the typical conformations do not depend on the composition. (2) We assume random mixing, i.e. neglect the correlation that it is more likely to find an A polymer in the vicinity of another A polymer than a B polymer. (3) We assume that the value of $\Delta E(\boldsymbol{R}_{n_\mathrm{B}})$ does not fluctuate strongly.

By virtue of assumption (1), the typical conformation of the n_Bth B polymer also corresponds to a typical configuration of an A polymer, and those typical configurations yield the dominant contribution to the average. If the typical conformations of both species differed, the Boltzmann factor, $\exp[-\Delta E(\boldsymbol{R}_{n_\mathrm{B}})/k_\mathrm{B}T]$, for a typical conformation of a B polymer would be vanishingly small, and only the rare cases where the conformation of the n_Bth B polymer corresponded to a typical configuration of an A polymer would yield a large Boltzmann factor, and those rare cases would dominate the average.

The energy change associated with mutating a B polymer into an A polymer can be expressed via the interaction potentials. To this end, it is important to distinguish between intramolecular interactions, which occur along the same chain, and intermolecular interactions between different molecules. If the chains are characterized by the same distribution function, $\tilde{\mathcal{D}}_\mathrm{A}\boldsymbol{R} = \tilde{\mathcal{D}}_\mathrm{B}\boldsymbol{R}$, there is no change in the intramolecular interactions. Even if the intramolecular distributions differed, there would be an energy contribution, but this contribution would not depend on composition [by virtue of assumption (1)] and could therefore be absorbed in the non-interacting single-chain partition function, $\mathcal{Z}_0$. Thus, intramolecular interactions do not contribute to the demixing behavior, and they are omitted in the following.

To calculate the intermolecular energy change, it is useful to define the intermolecular pair correlation function, $g_{IJ}^{\mathrm{inter}}(\boldsymbol{r})$ (with $I, J = \mathrm{A, B}$), which denotes the probability of finding a pair of segments of type I and J a distance $\boldsymbol{r}$ apart which belong to different molecules, e.g.

$$g_{\mathrm{AA}}^{\mathrm{inter}}(\boldsymbol{r}) = \frac{2V}{(n_\mathrm{A}N)^2}\left\langle \sum_{\alpha=1}^{n_\mathrm{A}}\sum_{i=1}^{N}\sum_{\alpha'=\alpha+1}^{n_\mathrm{A}}\sum_{j=1}^{N}\delta(\boldsymbol{r}-(\boldsymbol{r}_{\alpha,i}-\boldsymbol{r}_{\alpha',j}))\right\rangle \quad (3.78)$$

The correlation function is normalized such that $g_{IJ}^{\mathrm{inter}}(r\to\infty) = 1$. This allows us to calculate the average energy change:

$$\langle \Delta E \rangle = \rho_0 N \int d\mathbf{r} \big(\bar{\phi}_A g_{AB}^{\text{inter}}(\mathbf{r})[V_{AA}(\mathbf{r}) - V_{AB}(\mathbf{r})]$$

$$+ \bar{\phi}_B g_{BB}^{\text{inter}}(\mathbf{r})[V_{AB}(\mathbf{r}) - V_{BB}(\mathbf{r})] \big) \quad (3.79)$$

Using the random mixing assumption (2), we set $g^{\text{inter}}(\mathbf{r}) \equiv g_{AA}^{\text{inter}}(\mathbf{r}) \approx g_{BB}^{\text{inter}}(\mathbf{r}) \approx g_{AB}^{\text{inter}}(\mathbf{r})$. If the invariant polymerization index is large, one chain has many interactions, and fluctuations in ΔE are small [assumption (3)]. Thus we can make the approximation $\ln\langle \exp[-\Delta E] \rangle \approx -\langle \Delta E \rangle$. Using $\bar{\phi}_A + \bar{\phi}_B = 1$, we obtain

$$\ln \left\langle \exp\left(-\frac{\Delta E}{k_B T}\right) \right\rangle \approx \frac{\rho_0 N}{k_B T} \bar{\phi}_A \int d\mathbf{r}\, g^{\text{inter}}(\mathbf{r})(2V_{AB} - V_{AA} - V_{BB})$$

$$+ \text{constant} \quad (3.80)$$

where the constant does not depend on composition. Using this result in conjunction with Eqs. (3.47) and (3.77), we obtain for the Flory–Huggins parameter

$$\tilde{\chi} = \frac{\rho_0}{k_B T} \int d\mathbf{r}\, g^{\text{inter}}(\mathbf{r}) \left(V_{AB}(\mathbf{r}) - \frac{V_{AA}(\mathbf{r}) + V_{BB}(\mathbf{r})}{2} \right) \quad (3.81)$$

Note that this expression for the Flory–Huggins parameter resembles that of the literal SCF theory (cf. Eq. 3.42) except for the important factor $g^{\text{inter}}(\mathbf{r})$. Basically, this factor excludes intramolecular interactions, for they do not contribute to phase separation, and calculates the interaction energy using the true distribution of segment positions around a reference segment instead of a uniform density in a homogeneous state. If we identified the Flory–Huggins parameter via the excess energy of mixing

$$\frac{\Delta E_{\text{FH}}}{k_B T \rho_0 \mathcal{V}} = \tilde{\chi} \bar{\phi}_A \bar{\phi}_B \quad (3.82)$$

we would arrive at the same result.

The derivation also clarifies the assumptions under which the Flory–Huggins parameter can be identified via the simple estimate (3.81). The first and most crucial assumption of the typical chain conformations not to depend on composition is often not fulfilled for chemically realistic polymers, i.e. the chain structure does depend on the composition of the blend. Remember that an A segment corresponds to a small number of monomeric repeat units, and this group of monomeric repeat units might have a different number of (internal) arrangements in an A-rich environment than in a B-rich environment. In this case, there is an entropy change upon mixing, which is proportional to the number of segments (Müller 1995). This contribution outweighs the combinatorial entropy of mixing, which is proportional to the number of polymers, and therefore alters the temperature dependence of the phase behavior. Such an effect can also be parameterized in terms of a Flory–Huggins parameter, which then also acquires an entropic contribution. Experimentally, one often uses a form $\tilde{\chi} = A/k_\mathrm{B}T + B$. While the phase diagram as a function of $\tilde{\chi}N$ still looks as in the inset of Fig. 3.5, when plotted as a function of temperature one can also find lower critical solution points (LCSP), where phase separation occurs upon heating (Müller 1995; Dudowicz et al. 2002).

Predicting the intermolecular pair correlation function and relating it to the molecular structure is a key to understanding the miscibility of polymer blends. In Fig. 3.6 we present the intermolecular pair correlation function of symmetric polymer blends in the athermal limit and at the critical point of the mixture. In the athermal case, $\tilde{\chi}N = 0$, the distinction between the two species becomes irrelevant. The intermolecular pair correlation function mirrors two effects (Müller and Binder 1995). (1) Due to the extended monomer structure, the pair correlation function vanishes for distances smaller than the monomer site, $r < 2$. The presence of vacancies introduces local packing

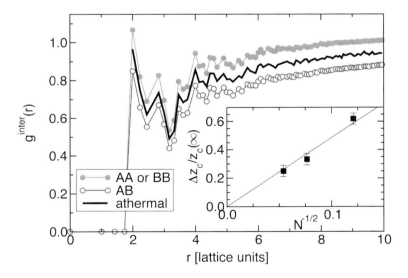

Fig. 3.6 Intermolecular pair correlation function in a binary polymer blend obtained from MC simulations of the bond fluctuation model ($\mathcal{N} \approx 267$). The pair correlations in the vicinity of the critical point and in the athermal ($\tilde{\chi}N = 0$) system are shown. The inset demonstrates how the difference between the AA and AB pair correlations decreases upon increasing the invariant polymerization index $\mathcal{N}$.
(Adapted from Müller 1999).

effects, which give rise to a highly structured function at short distances. One can identify several neighboring shells, which are characteristic of the fluid of segments. These packing effects are, of course, absent in simple lattice models where a segment occupies a single lattice site and are less pronounced in the bond fluctuation model than in continuum models like a Lennard-Jones bead–spring model. The length scale of these packing effects is set by the extension of a segment or the statistical segment length; the detailed shape depends strongly on the model and the degree of structure on local length scales. (2) Furthermore, the extended structure of the macromolecules manifests itself in a reduction of contacts with *other* chains on intermediate length scales (de Gennes 1979). The length of this polymeric correlation hole is set by the size of the molecules, R_e, and its universal shape is characteristic for the large-scale conformations of the molecule.

To a first approximation, one can assume that the fluid structure is determined by the packing of the hard cubes on the lattice, i.e. we use the fluid of unconnected monomeric units as a reference fluid. The influence of the connectivity of the monomers along a polymer or the thermal interactions on the total pair correlation function is neglected. Under this assumption, we can separate the fluid-like packing effects on the monomer scale and the polymeric correlation hole effect and approximate the intermolecular pair correlation function by (Müller and Binder 1995)

$$g^{\text{inter}}(r) = g_{N=1}(r)\left(1 - \frac{1}{\sqrt{\mathcal{N}}} f(r/R_e)\right) \quad (3.83)$$

where $g_{N=1}$ denotes the pair correlation function of the monomer fluid and the function f parameterizes the structure of the molecule on the scale R_e. The prefactor is determined by the requirement that the correlation hole contains one polymer:

$$\rho_0 \int d\mathbf{r} \left(\frac{g^{\text{inter}}(r)}{g_{N=1}(r)} - 1\right) = N \quad (3.84)$$

Indeed, this factorization works excellently for flexible molecules in the bond fluctuation model, as shown in Fig. 3.7. The ratio $g^{\text{inter}}(r)/g_{N=1}(r)$ is largely independent of packing effects and permits a distinction between microscopic segmental packing effects and universal polymeric correlation hole effects in the simulation – although the length scales are not clearly separated for short chains. Not surprisingly, the correlation hole becomes deeper and wider as we increase the molecular weight. This imparts a chain-length dependence on the Flory–Huggins parameter or the effective coordination number, respectively. The effective coordination number is related to the short-distance behavior of the intermolecular pair correlation function, $z_c \sim \rho_0 \sigma_e^3 g^{\text{inter}}(\sigma_e)$, where σ_e

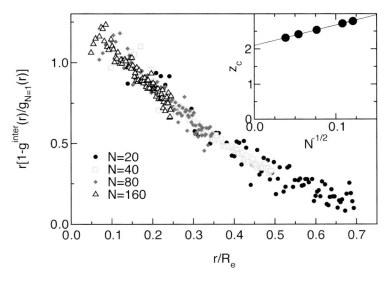

Fig. 3.7 Scaling of the correlation hole with chain extension. To eliminate packing effects, the intermolecular pair correlation function $g^{\text{inter}}(r)$ has been divided by the pair correlation function of a fluid of unconnected monomers. The inset shows the approach of the effective coordination number z_c towards the limit $\mathcal{N} \to \infty$. (Adapted from Müller and Binder 1995).

characterizes the range of the thermal interactions. More quantitatively, we calculate

$$z_c = \rho_0 \int_{r \leq \sqrt{6}} d\mathbf{r}\, g^{\text{inter}}(r) = z_c(\infty)\left(1 + \frac{\text{constant}}{\sqrt{\mathcal{N}}}\right) \qquad (3.85)$$

The scaling of the effective coordination numbers for athermal, flexible, linear chains is presented in the inset of Fig. 3.7. The effective coordination number approaches its limiting scaling behavior with a $1/\sqrt{\mathcal{N}}$ correction.

In Fig. 3.6 we also present the intermolecular pair correlation functions g_{AA} and g_{AB} for chain length $N = 80$ close to criticality. In accord with intuition, AA contacts are more likely than AB ones, and hence $g_{\text{AA}}^{\text{inter}} > g_{\text{BB}}^{\text{inter}}$. However, note that the sum of AA and AB correlations can be well approximated by the intermolecular pair correlation function $g_{\text{atherm}}^{\text{inter}}(\mathbf{r})$ in the athermal limit:

$$\tfrac{1}{2}[g_{\text{AA}}^{\text{inter}}(\mathbf{r}) + g_{\text{AB}}^{\text{inter}}(\mathbf{r})] \approx g_{\text{atherm}}^{\text{inter}}(\mathbf{r}) \qquad (3.86)$$

This relation shows that the weak difference in interactions between the monomers, $\tilde{\chi} \sim 1/N$, does not alter the structure of the monomer fluid. The energy of the system is mainly determined by composition fluctuations. The approximate decoupling between density fluctuations/packing effects and composition fluctuations in our model makes it possible to use the athermal value of the intermolecular pair correlation functions in Eq. (3.81). This identification corresponds to the high-temperature approximation in the framework of the P-RISM theory (Schweizer and Curro 1997).

The inset of Fig. 3.6 shows the integral of the difference of the correlation functions over the range of the square-well potential, i.e.

$$\Delta z_{\rm c} = \rho_0 \int_{r \leq \sqrt{6}} {\rm d}\boldsymbol{r}\, [g_{\rm AA}^{\rm inter}(\boldsymbol{r}) - g_{\rm AB}^{\rm inter}(\boldsymbol{r})]$$

at the critical point. The simulation results show that the difference between the AA and AB contacts decreases like $1/\sqrt{\mathcal{N}}$, when $\tilde{\chi}N$ is held constant. This exemplifies that the random mixing assumption (2) is justified in the limit $\mathcal{N} \to \infty$.

The vanishing of composition fluctuations for $\mathcal{N} \to \infty$ can be rationalized by estimating the scaling of non-random mixing effects with growing chain length. The strength of composition fluctuations in a volume V is of the order $N/\rho_0 V$. Expressing the composition fluctuations via the correlation functions, we obtain for a symmetric blend

$$1 \sim \frac{\rho_0 V}{N}(\langle \phi^2 \rangle - \langle \phi \rangle^2)$$

$$\sim \frac{\rho_0}{N} \int {\rm d}\boldsymbol{r} \left[\frac{g_{\rm AA}^{\rm inter}(\boldsymbol{r}) + g_{\rm BB}^{\rm inter}(\boldsymbol{r})}{2} - g_{\rm AB}^{\rm inter}(\boldsymbol{r}) \right]$$

$$\sim \sqrt{\mathcal{N}} \int {\rm d}\boldsymbol{x} \left[\frac{g_{\rm AA}^{\rm inter}(\boldsymbol{x}) + g_{\rm BB}^{\rm inter}(\boldsymbol{x})}{2} - g_{\rm AB}^{\rm inter}(\boldsymbol{x}) \right]$$

$$\text{with } \boldsymbol{x} = \frac{\boldsymbol{r}}{R_{\rm e}} \tag{3.87}$$

Therefore, we expect the difference in the AA and AB correlations to vanish like

$$g_{\rm AA}^{\rm inter}(\boldsymbol{r}) - g_{\rm AB}^{\rm inter}(\boldsymbol{r}) \sim 1/\sqrt{\mathcal{N}} \tag{3.88}$$

upon increasing the chain length. These mean-field arguments are in agreement with P-RISM calculations by Singh et al. (1995).

Non-random mixing effects lower the energy of mixing per segment by an amount of the order $\tilde{\chi}/\sqrt{\mathcal{N}}$. By the same token, there is a shift of the critical value of the Flory–Huggins parameter:

$$\tilde{\chi}_{\rm c} N = 2 \left(1 + \frac{\text{constant}}{\sqrt{\mathcal{N}}} \right) \tag{3.89}$$

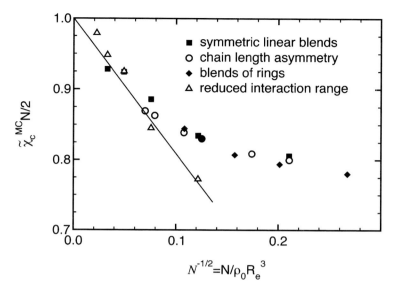

Fig. 3.8 Ratio between the critical incompatibility obtained from finite size scaling analysis of the MC data for the bond fluctuation model, $\tilde{\chi}_c = 2z_c\epsilon_c/k_BT$, and the mean-field prediction, $\chi_c = 2/N$ for binary mixtures as a function of the invariant polymerization index $\mathcal{N}$. The graph includes data for symmetric polymer blends (squares), asymmetric mixtures where the components differ in chain length (circles), blends of ring polymers that adopt less open structures than linear polymers (diamonds), and symmetric linear blends where the square-well repulsion extends only over the six nearest (instead of 54) lattice sites (triangles). The straight line indicates the common asymptotic behavior. (Adapted from Müller 1999).

i.e. non-random mixing effects give rise to a correction of relative order $1/\sqrt{\mathcal{N}}$. As those non-random mixing effects are associated with long-range composition fluctuations, they become smaller for large $\mathcal{N}$. This is illustrated by simulation data in Fig. 3.8, where we plot the ratio of the critical χ parameter, estimated by locating the critical temperature via finite size scaling from the MC simulations and utilizing Eq. (3.81) and the mean-field value $\tilde{\chi}_c^{\text{SCF}} = 2/N$. Indeed, one observes that for large $\mathcal{N}$ the critical temperatures tend to the value predicted by the SCF theory with a $1/\sqrt{\mathcal{N}}$ correction.

3.2.5
Single-Chain Conformations in the Bulk

The SCF theory cannot predict the extension of the chains in a dense polymer melt, because it results from a screening of excluded-volume interactions on a length scale $\xi_{\text{ev}} \ll R_e$. These correlations can be accounted for on large length scales by identifying R_e with the end-to-end distance in the melt rather than the extension of a non-interacting chain.

In addition to these short-range correlations, simulations and experiments also show that the chain conformations may change as a function of composition. One of the basic assumptions in relating the Flory–Huggins parameter $\tilde{\chi}$ to the local fluid structure via the energy of mixing is that the chain conformations do not depend on composition. Sophisticated theoretical approaches have been developed to study the dependence of chain conformations on the environment: self-consistent P-RISM theory (Grayce et al. 1994; Weinhold et al. 1999) and SCF theory for clusters of chains (Müller 1998). In the following we constrain ourselves to MC simulations and simple scaling arguments to study the qualitative behavior (Müller 1999).

MC simulations suggest that one possible mechanism of conformational changes in blends is associated with exchanging energetically unfavorable intermolecular contacts for attractive intramolecular contacts upon reducing the spatial extension of the molecule. Attributing this shrinking of the minority component to a balance between entropy loss due to deviations from the unperturbed conformations and reduction of the energy upon shrinking, we can estimate the magnitude of conformational changes (Müller 1998). Within the Gaussian chain model, a deviation from the unperturbed chain extension R_e gives rise to an entropic force of the form

$$\frac{\mathrm{d}S}{\mathrm{d}R} \sim \frac{(R - R_e)}{R_e^2} \tag{3.90}$$

This is opposed to an enthalpic force $\mathrm{d}E/\mathrm{d}R$. Here E denotes the single-chain energy, which comprises energetically favorable interactions Nz_{intra} among monomers of the same chain and Nz_c interactions with monomers of other polymers. The exchange of an intermolecular interaction with an intramolecular one lowers the single-chain energy by an amount of the order $\tilde{\chi}$. The number of intramolecular interactions per monomer z_{intra} is given by the density of monomers of the same chain inside of its volume $z_{\mathrm{intra}} \sim N/\rho_0 R_e^3 \sim 1/\sqrt{\mathcal{N}}$. Under the assumption that the reduction of the chain extension does not affect the total number of interactions, but merely exchanges intermolecular interactions into energetically favorable intramolecular ones, we estimate the chain-length dependence of the energy change as

$$\frac{\mathrm{d}E}{\mathrm{d}R} \sim -\frac{\tilde{\chi}N}{\rho_0 R_e^4/N} \tag{3.91}$$

Balancing the entropic force against the enthalpic one, we obtain

$$\frac{R_e - R}{R_e} \sim \frac{\tilde{\chi}N}{\rho_0 R_e^3/N} = \frac{\tilde{\chi}N}{\sqrt{\mathcal{N}}} \tag{3.92}$$

These scaling arguments are similar to those for a chain in a marginal solvent, and suggest that the perturbation of the chain conformations decreases upon increasing the chain length at $\tilde\chi N = \text{constant}$. In contrast to excluded-volume interactions, which change the large-scale conformations from Gaussian to self-avoiding-walk-like, the use of perturbation theory is justified because the composition-dependent interactions are only of the order $1/N$ per monomer.

The derivation of the scaling arguments relies on the number of intramolecular contacts and its dependence on the chain extension. Obviously, this estimate excludes contributions from the neighbors along the polymer, which give rise to $z_\text{intra} \sim N$. Although the neighbor contribution is important for the scaling of the number of intramolecular contacts with chain length, we assume them to be independent of the instantaneous shape/extension of the polymer.

Clearly, a detailed verification of this scaling behavior by MC simulations is warranted. Such a test is presented in Fig. 3.9(a). Using the scaling estimate above, the dependence of z_intra on the instantaneous extension R at fixed chain length is given by $\mathrm{d}z_\text{intra}/\mathrm{d}R \sim N/\rho_0 R_e^4 \sim 1/N$. The inset presents the average number of intramolecular contacts at fixed end-to-end distance for an athermal melt of chain length $N = 256$. At the mean end-to-end distance $\sqrt{\langle R^2 \rangle_0}$ we determine the slope $\mathrm{d}z/\mathrm{d}R$ as indicated by linear regression. The chain-length dependence of the derivatives of the number of inter- and intramolecular contacts with respect to the chain extension decreases like $1/N$ for large chain lengths. This confirms the scaling predictions. Moreover, the sum of intermolecular and intramolecular contacts depends much more weakly on the spatial extension R_e, and the dependence decreases faster than $1/N$. This indicates that the fluid structure of the monomers is mainly determined by packing and approximately decouples from the chain conformations. For long chains, the conformational changes merely result in an exchange of inter- and intramolecular contacts. From the MC data, we estimate $\mathrm{d}z/\mathrm{d}R = 0.77(7)/N$ for the bond fluctuation model at density $\rho_0 = 1/16$.

The scaling predictions can be made more quantitative in the framework of the Gaussian chain model. Let $P(\boldsymbol{R})$ denote the probability distribution of the end-to-end vector $\boldsymbol{R}$, which incorporates the dependence of the single-chain energy E on the chain extension:

$$E(\boldsymbol{R}) \approx E\left(\sqrt{\langle R_e^2\rangle_0}\right) + \frac{\mathrm{d}E}{\mathrm{d}R}\left(|\boldsymbol{R}| - \sqrt{\langle R_e^2\rangle_0}\right) \qquad (3.93)$$

where $\langle R_e^2\rangle_0 = a^2 N$ denotes the end-to-end distance in the athermal limit. Assuming Gaussian statistics for the unperturbed chain, we can write the probability distribution for a B polymer in an A-rich matrix in the form

$$P(\boldsymbol{R}) \sim \exp\left[-\frac{3\boldsymbol{R}^2}{2\langle R_e^2\rangle_0} - \frac{1}{k_\mathrm{B}T}\frac{\mathrm{d}E}{\mathrm{d}R}\left(|\boldsymbol{R}| - \sqrt{\langle R_e^2\rangle_0}\right)\right] \qquad (3.94)$$

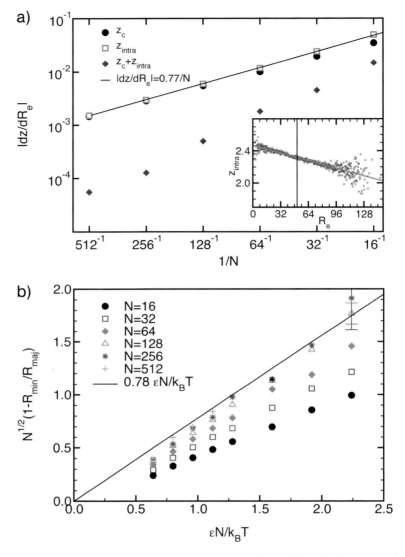

Fig. 3.9 (a) Dependence of the number of intra- and intermolecular contacts as well as their sum on the instantaneous end-to-end distance. Upon increasing the chain length, the dependence of the intra- and intermolecular contacts decreases like $1/N$. The total number of contacts depends only weakly on R_e. The inset shows the correlation between the instantaneous end-to-end distance R_e and the number of intramolecular contacts z_{intra} in the framework of the bond fluctuation model ($N = 256$, $\epsilon/k_B T = 0$, and $\rho_0 = 1/16$). The vertical line indicates the average value $\sqrt{\langle R_e \rangle}$, and the slope dz_{intra}/dR_e is indicated by a solid line. (b) Relative shrinkage of the molecular extension in the minority phase as a function of incompatibility. The symbols present the data for the bond fluctuation model using different chain lengths as indicated in the key, while the solid line corresponds to the prediction (3.95). (Adapted from Müller 1998).

The total energy change associated with the transfer of two intermolecular contacts of a B polymer $[\epsilon(2\langle\phi\rangle - 1)]$ into an intermolecular contact $[-\epsilon]$ and a contact between monomers not belonging to this B polymer $[-\epsilon(2\langle\phi\rangle - 1)^2]$ amounts to $\Delta E = 4\epsilon\langle\phi\rangle^2$. The number of intramolecular contacts per monomer increases by 2, and the number of intermolecular contacts decreases by the same amount. Therefore dE/dR equals $2\epsilon\langle\phi\rangle^2 \, dz/dR$. Using this estimate and assuming that the conformational changes are small, we calculate the mean square end-to-end distance

$$\langle R_e^2 \rangle \approx \langle R_e^2 \rangle_0 \left[1 - \sqrt{\frac{8}{27\pi}} \frac{1}{\sqrt{\mathcal{N}}} \left(\frac{\rho_0 R_e^4}{z_c N} \frac{dz_{\text{intra}}}{dR} \right) \tilde{\chi} N \langle\phi\rangle^2 \right] \tag{3.95}$$

We expect this asymptotic expression to hold only for very small values of $\tilde{\chi}N/\sqrt{\mathcal{N}}$, where the conformational changes can be treated perturbatively. This expression predicts a quadratic dependence of the chain extension on the composition of the mixture. The effect increases linearly with the $\tilde{\chi}$ parameter, but decreases at fixed $\tilde{\chi}N$ like $1/\sqrt{\mathcal{N}}$. Hence, the conformations of long macromolecules are only very weakly dependent on composition. This is also in qualitative agreement with field-theoretical calculations by Aksimentiev and Holyst (1998). A decreasing dependence of the single-chain conformations on the environment is also found in simulations and self-consistent P-RISM calculations for atomistic models (Heine et al. 2003).

In Fig. 3.9(b) we explore the scaling of the shrinking of the chains in the minority phase at the binodal. The simple estimate (3.95) predicts the relative shrinking $1 - R_{\min}/R_{\text{maj}}$ of the minority phase with respect to the unperturbed conformations of the majority component to increase linearly with $\epsilon N/\sqrt{\mathcal{N}}$. The straight line represents the prediction of Eq. (3.95). Surprisingly, the simulation data approach the asymptotic behavior very slowly. Only for chain length $N \geq 128$ do the simulations reach the scaling limit; for smaller chain lengths the estimate (3.95) overestimates the shrinking.

To summarize, the dependence of the chain conformations on composition results from the increase of the self-density on the length scale R_e, and the corresponding exchange of unfavorable intermolecular contacts of the minority component by favorable intramolecular contacts. This effect is related to long-range composition fluctuations and decreases at fixed R_e and $\tilde{\chi}N$, like $1/\sqrt{\mathcal{N}}$.

3.2.6
Interfaces

In the previous section, we have illustrated that short-range correlations, which are neglected by the saddle-point approximation, can be accounted for by using an effective Flory–Huggins parameter $\tilde{\chi}$. This effective $\tilde{\chi}$ parameter can be consistently extracted by comparing the predictions of the SCF theory for the homogeneous system with the results of simulations or experiments away from the critical point of the mixture. For model systems, like the bond fluctuation model of a binary polymer melt, this effective Flory–Huggins parameter can also be related to the local structure of the polymeric fluid.

In this section, we use the effective Flory–Huggins parameter $\tilde{\chi}$ to compare the predictions of the SCF theory concerning interfaces to simulation results of the bond fluctuation model. Properties of interfaces can broadly be divided into (1) excess properties (e.g. interface tension) and (2) profiles of quantities across the interface. The former can straightforwardly be predicted by the SCF theory; the latter quantities are strongly affected by long-range fluctuations of the local interface position, i.e. capillary waves (Buff et al. 1965; Helfrich 1973; Weeks 1977; Bedeaux and Weeks 1985).

Excess Properties of the Interface

In contrast to interface profiles, excess properties of the interface do not explicitly make reference to the position of the interface but rather describe properties per unit area. Often, these quantities can be obtained from integrals of profiles across the interface. For instance, the interface tension, i.e. the free-energy cost of the AB interface per unit area, is such a quantity. If the blend is compressible, the density at the center of the interface will be reduced in order to minimize the unfavorable contacts between different species (Müller et al. 1995). This effects can be quantified by the negative density excess at the interface. By the same token, a non-selective solvent will enrich at the interface (Helfand 1975) and this effect can be characterized by the solvent excess at the interface. Likewise, if the blend also contains diblock copolymers, those will adsorb at the interface so as to extend their blocks into the respective bulk phases (Leibler 1982).

In Fig. 3.10 we compare the interface free energy γ_I of a symmetric polymer blend extracted from simulations of the bond fluctuation model with the prediction of the SCF theory. The figure shows the ratio between the value of γ_I and the analytical estimate in the strong segregation limit (cf. Eq. 3.70). The results of the SCF theory approach the limit $\chi N \to \infty$ with a leading $1/\chi N$ correction(Semenov 1996)

$$\frac{\gamma_\mathrm{I} R_\mathrm{e}^2}{k_\mathrm{B} T} = \sqrt{\mathcal{N}} \sqrt{\frac{\tilde{\chi} N}{6}} \left(1 - \frac{4 \ln 2}{\tilde{\chi} N} + \frac{1.1842}{(\tilde{\chi} N)^{3/2}} \right) \qquad (3.96)$$

which is also shown in the figure.

Note that in such a plot simulation results for different chain lengths and incompatibilities collapse onto a single curve. This explicitly shows that the

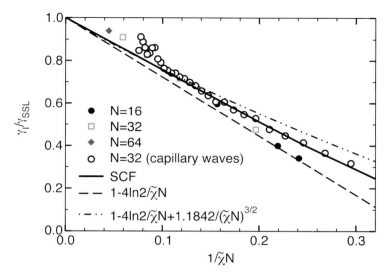

Fig. 3.10 Ratio between the interface tension $\gamma_{\rm I}$ and the simple expression for the strong segregation limit $\gamma_{\rm SSL}$ in Eq. (3.70) as a function of inverse incompatibility. Symbols correspond to MC results for the bond fluctuation model, the solid line shows the result of the SCF theory, and the dashed line presents first corrections to Eq. (3.70) calculated by (Semenov 1996). (Adapted from Schmid and Müller 1995).

parameters χN, $R_{\rm e}$, and $\mathcal{N}$ provide a useful description of the interface properties. The results of the SCF theory agree quantitatively with the simulation results. Thus, once the Flory–Huggins parameter has been identified from bulk properties, the SCF theory is quantitatively accurate.

Capillary Waves

If we wish to compare profiles across the AB interface, we encounter a basic difficulty. In the SCF theory the interface is perfectly flat; this is, indeed, the most probable configuration. Thus, all spatial variations occur only in the direction normal to the interface. In simulations (cf. Fig. 3.3a) or experiments, however, the local interface position fluctuates (Buff et al. 1965; Helfrich 1973; Weeks 1977; Bedeaux and Weeks 1985). Those thermal fluctuations cost vanishingly little energy in the long-wavelength limit. A snapshot of a coarse-grained local interface position $u(x, y)$ (where x and y denote the two coordinates parallel to the interface) is presented in Fig. 3.11(a) for a symmetric blend with chain length $N = 32$ in the bond fluctuation model (Werner et al. 1997). Such fluctuations of the local interface position will broaden the profiles measured in experiments or simulations, which are laterally averaged over a certain patch size.

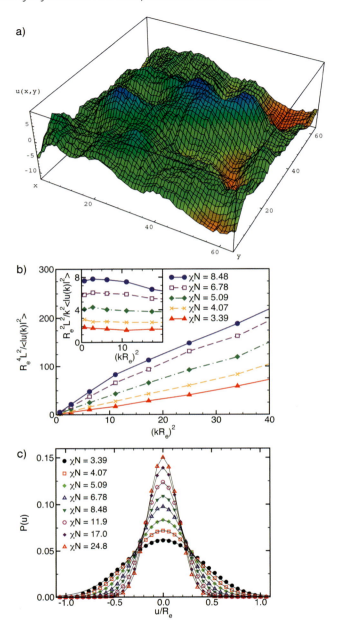

Fig. 3.11 (a) Snapshot of the local interface position obtained from MC simulations in the framework of the bond fluctuation model ($\mathcal{N} = 91$, $\tilde{\chi}N = 5.09$, $L = 3.77 R_e$). The local interface position is determined by analyzing the profiles on a lateral length scale $B = 0.47 R_e$. (Adapted from Werner et al. 1997). (b) Spectrum of interface fluctuations for different incompatibilities as indicated in the key. The inset shows the wavevector dependence of the interface tension. (c) Probability distribution of the local interface position for different incompatibilities. (Adapted from Werner et al. 1999).

In the simulations (Müller and Schick 1996a; Müller and Werner 1997; Werner et al. 1997) one can measure the local interface position $z = u(\boldsymbol{x})$ as a function of the two coordinates, $\boldsymbol{x} = (x, y)$, parallel to the (average) interface plane (Monge representation). At this stage, we assume that the configuration does not contain bubbles or overhangs. This can be achieved by smoothing the composition profile over a microscopic length scale. Rather than describing the configuration by a detailed, particle-based model that describes the location and conformations of the molecules in space, we just characterize the configuration by the location, $u(\boldsymbol{x})$, of the interface.

On large lateral length scales, the effect of the fluctuations of the local interface position is to increase the area of the interface, and the free-energy cost can be described by the capillary-wave Hamiltonian $\mathcal{H}[u(\boldsymbol{x})]$ (Buff et al. 1965; Helfrich 1973; Weeks 1977; Bedeaux and Weeks 1985):

$$\mathcal{H}[u(\boldsymbol{x})] = \gamma_\mathrm{I} \int \mathrm{d}^2\boldsymbol{x} \left[\sqrt{1 + (\nabla u)^2} - 1\right] \approx \tfrac{1}{2}\gamma_\mathrm{I} \int \mathrm{d}^2\boldsymbol{x}\, (\nabla u)^2 \qquad (3.97)$$

where γ_I denotes the interface tension, and the integration is extended over the projected area, $L \times L$. From this capillary-wave Hamiltonian, we can estimate the fluctuations of the local interface position. Utilizing the Fourier spectrum of interface fluctuations, $u(\boldsymbol{k})$,

$$u(\boldsymbol{x}) = \frac{1}{L^2} \sum_{\boldsymbol{k}} u(\boldsymbol{k}) \exp[\mathrm{i}\boldsymbol{k}\cdot\boldsymbol{x}] = \frac{1}{(2\pi)^2} \int \mathrm{d}^2\boldsymbol{k}\, u(\boldsymbol{k}) \exp[\mathrm{i}\boldsymbol{k}\cdot\boldsymbol{x}] \quad (3.98)$$

$$u(\boldsymbol{k}) = \int \mathrm{d}^2\boldsymbol{x}\, u(\boldsymbol{x}) \exp[-\mathrm{i}\boldsymbol{k}\cdot\boldsymbol{x}] \qquad (3.99)$$

one diagonalizes the capillary-wave Hamiltonian:

$$\mathcal{H}[u(\boldsymbol{k})] = \frac{\gamma_\mathrm{I}}{2} \frac{1}{(2\pi)^2} \int \mathrm{d}\boldsymbol{k}\, k^2 |u(\boldsymbol{k})|^2 \qquad (3.100)$$

The statistical mechanics of the interface position can be described by the partition function

$$\mathcal{Z}_\mathrm{int} = \int \mathcal{D}[u] \exp\left[-\frac{\mathcal{H}[u]}{k_\mathrm{B}T}\right] \qquad (3.101)$$

where the functional integral $\int \mathcal{D}[u]$ sums over all local positions of the interface. Since the Hamiltonian is the sum of independent, quadratic degrees of freedom, $u(\boldsymbol{k})$, the Fourier modes are uncorrelated and Gaussian distributed around zero, and their variance is given by the equipartition theorem:

$$\frac{\gamma_\mathrm{I}}{L^2} k^2 \langle |u(\boldsymbol{k})|^2 \rangle = k_\mathrm{B}T \qquad (3.102)$$

The spectrum of interface fluctuations is shown in Fig. 3.11(b) as a function of the wavevector k. When plotted as $R_e^4 L^2/\langle|u(\boldsymbol{k})|^2\rangle$ versus $(kR_e)^2$, the data for small wavevectors form a straight line, and one can extract the interface tension from the slope, $\gamma_I R_e^2/k_B T$. Indeed, this yields reliable results (Müller and Schick 1996a) that agree well with values determined independently from grand-canonical MC simulations and the predictions of the SCF theory (cf. Fig. 3.10).

Utilizing this result, we obtain for the fluctuations of the local interface position in a lateral patch of size $L \times L$:

$$\langle u^2(\boldsymbol{x})\rangle_L = \frac{1}{(2\pi)^4} \int d^2\boldsymbol{k}\, d^2\boldsymbol{k}'\, \langle u(\boldsymbol{k})u(\boldsymbol{k}')\rangle \exp[i\boldsymbol{k}\cdot\boldsymbol{x} + i\boldsymbol{k}'\cdot\boldsymbol{x}]$$

$$= \frac{1}{(2\pi)^2} \int d^2\boldsymbol{k}\, \frac{k_B T}{\gamma_I k^2}$$

$$= \frac{k_B T}{2\pi\gamma_I} \int_{k_{\min}}^{k_{\max}} dk\, \frac{1}{k}$$

$$= \frac{k_B T}{2\pi\gamma_I} \ln\left(\frac{k_{\max}}{k_{\min}}\right) \qquad (3.103)$$

Since the integral over the magnitude of the wavevector diverges logarithmically, we have introduced an upper and a lower cutoff, $k_{\min}$ and $k_{\max}$. The lower cutoff is set by the lateral system size, $k_{\min} = 2\pi/L$, or the lateral coherence length in a scattering experiment.[3] The upper cutoff is more subtle: it marks the lateral length scale, where the description of fluctuations via the capillary-wave Hamiltonian (Eq. 3.97) breaks down and the intrinsic structure of the interface becomes important (Semenov 1993), and we shall investigate this issue further below.

The distribution of the local interface position is presented in Fig. 3.11(c). Indeed, the fluctuations are Gaussian distributed and the variance increases as we decrease the incompatibility,

$$P_L(u) = \frac{1}{2\pi\langle u^2\rangle_L} \exp\left[-\frac{u^2}{2\langle u^2\rangle_L}\right] \qquad (3.104)$$

To describe the profile, $p_{Q,L}(z)$, of a quantity Q measured in simulations and experiments on a lateral length scale L and to compare it with the prediction of the SCF theory, one has to account for the fluctuation of the interface position. If the interface position $u(x, y)$ were fixed, the quantity Q

---

[3] An external field, e.g. gravity or in the vicinity of the surface, also gives rise to a lower cutoff.

would approximately be describable by

$$p_Q(x, y, z) = p_{Q,\text{intr}}(z - u(x, y)) \tag{3.105}$$

where $p_{Q,\text{intr}}(z)$ is the profile of quantity Q across a hypothetically flat interface and z denotes the coordinate normal to that interface. It is this quantity that can be calculated by the SCF theory, and we refer to it as the "intrinsic" profile. Averaging this expression over the distribution of the local interface positions, $P_L(u)$, we arrive at the convolution approximation for the apparent profile (Semenov 1993):

$$p_{Q,L}(z) = \int du\, P_L(u) p_{Q,\text{intr}}(z - u) \tag{3.106}$$

To estimate the qualitative effect, we assume that the "intrinsic" profile $p_{Q,\text{intr}}$ varies smoothly on the scale $\langle u^2 \rangle_L$. Then, one can expand $p_{Q,\text{intr}}(z-u)$ around $u = 0$, and obtain to lowest order (Lacasse et al. 1998):

$$p_{Q,L}(z) = p_{Q,\text{intr}}(z) + \tfrac{1}{2}\langle u^2 \rangle_L \frac{d^2 p_{Q,\text{intr}}(z)}{dz^2} + \mathcal{O}(\langle u^4 \rangle_L) \tag{3.107}$$

The intrinsic composition profile across the interface is describable by a hyperbolic tangent (tanh) form or an error function (erf):

$$p_{\phi,\text{intr}}(z) \approx \frac{1}{2} + \frac{2\bar{\phi}_{A,\text{coex}} - 1}{2} \tanh\left[\frac{z}{w_{\text{intr}}}\right] \tag{3.108}$$

$$\approx \frac{1}{2} + \frac{2\bar{\phi}_{A,\text{coex}} - 1}{2} \operatorname{erf}\left[\frac{\sqrt{\pi}\, z}{2 w_{\text{intr}}}\right] \tag{3.109}$$

Both profiles provide a good description of the interface profile, and we have chosen the numerical coefficients such that the slope of both profiles at the center of the interface (i.e. at $z = 0$) coincide. The inverse of this slope at the center of the interface defines the width:

$$w \equiv \frac{2\bar{\phi}_{A,\text{coex}} - 1}{2\, dp_\phi(z=0)/dz} \tag{3.110}$$

The SCF theory predicts a tanh profile for the "intrinsic" profile in the weak and strong segregation limits.[4] The erf profile will provide a good description of the apparent profile if the "intrinsic" width is smaller than $\langle u^2 \rangle_L$.

---

4) At intermediate segregation, however, the interface profile is characterized by two length scales (Müller et al. 2000a): the intrinsic width w_{intr}, which depends on χN and determines the slope of the profile at the center of the interface; and the bulk correlation length $\xi \sim R_e$, which controls the approach of the profile to the composition of the coexisting phases in the wings of the profile.

Using the erf profile we obtain for the broadening of the interface width (Werner et al. 1997)

$$w_L^2 = w_{\text{intr}}^2 + \tfrac{1}{2}\pi\langle u^2\rangle_L = w_{\text{intr}}^2 + \frac{k_B T}{4\gamma_I}\ln\left(\frac{L}{B_0}\right) \qquad (3.111)$$

where $B_0 = 2\pi/k_{\text{max}}$ defines the shortest length scale on which the description via the capillary-wave Hamiltonian still holds. Note that the prefactor in front of the logarithmic term depends on the details of the interface profile (namely the third derivative at the center of the interface). Thus, the dependence of the apparent width of the interface is less accurate than the spectrum of interface fluctuations (cf. Fig. 3.10b) for extracting the interface tension γ_I from simulations or experiments.

At this point, two remarks are in order.

(1) If we measured the interface width in the natural length scale R_e, we could rewrite Eq. (3.111) in the form

$$\frac{w_L^2}{R_e^2} = f_w^2(\chi N) + \frac{1}{4\sqrt{\mathcal{N}}f_\gamma(\chi N)}\ln\left(\frac{k_{\text{max}}}{k_{\text{min}}}\right) \qquad (3.112)$$

where f_w and f_γ are functions of the incompatibility χN that approach the limits $1/\sqrt{6\chi N}$ and $\sqrt{\chi N/6}$, respectively, for large χN. In this formal sense, the broadening by capillary waves is a long-wavelength fluctuation: the saddle-point approximation invoked in the SCF theory neglects this fluctuation, and capillary waves are also not included in the identification of the effective Flory–Huggins parameter $\tilde{\chi}$. If one extracted the effective Flory–Huggins parameter from the apparent width of the interface in simulations or experiments (method 4 in Section 3.2.4), the so-determined Flory–Huggins parameter would depend on the lateral size on which the interface is observed and it would not be consistent with the identification via the bulk phase behavior. Like other long-wavelength fluctuations (i.e. composition fluctuations in the vicinity of the critical point), the contribution of capillary waves to the width of the interface is only of the order $1/\sqrt{\mathcal{N}}$. Nevertheless, in simulations (Müller et al. 1995; Werner et al. 1997; Werner et al. 1999) as well as in experiments (Shull et al. 1993; Kerle et al. 1996; Sferrazza et al. 1997), the broadening due to interface fluctuations often constitutes a significant fraction of the apparent width.

(2) Eq. (3.111) also points to a rather fundamental problem (Werner et al. 1999): What is the significance of the "intrinsic" width that characterizes a hypothetical, perfectly flat interface? In simulations or experiments one can only measure the apparent width w_L of the interface as a function of the lateral length scale L. In Fig. 3.12(a) we plot the square of the apparent width w_L^2 versus $\ln L$. At large lateral length scales L, we find a linear behavior as expected

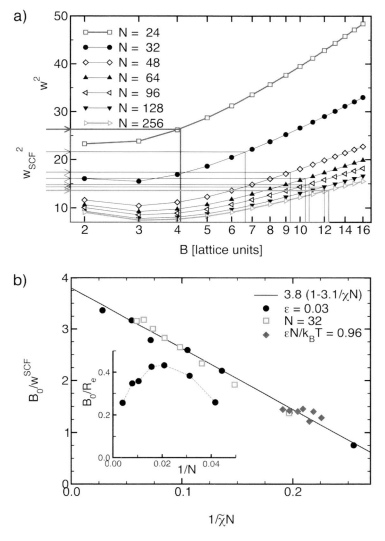

Fig. 3.12 (a) Square of the apparent width w of the interface obtained from analyzing MC simulations on a lateral size B for different chain lengths N as indicated in the figure (for $\epsilon/k_B T = 0.03$). The arrows on the left mark the prediction of the SCF theory. Reading off the lateral length scale B_0 at which the prediction of the SCF calculations agrees with the observation in the MC simulations, one identifies the short-length-scale cutoff.
(b) Short-length-scale cutoff B_0, measured in units of the intrinsic interface width w_{SCF}, as a function of the inverse chain length N. The graph includes results obtained by varying N at fixed $\epsilon/k_B T = 0.03$ (circles), increasing the incompatibility for a fixed chain length $N = 32$ (squares), and varying both N and ϵ such that the product $\epsilon N/k_B T = 0.96$ remains constant (diamonds). The inset presents the ratio B_0/R_e at fixed $\epsilon/k_B T = 0.03$. (Adapted from Werner et al. 1999).

from Eq. (3.111). From the slope, we can estimate the interface tension. From the intercept we only obtain the combination $w_{\text{intr}}^2 - (k_B T/4\gamma_I) \ln B_0$, but it is not obvious how to extract w_{intr} and B_0 separately. Physically, the length scale B_0 describes the crossover from long-wavelength capillary fluctuations to short-wavelength fluctuations that contribute to the "intrinsic" interfacial width (Semenov 1994). In the simulation data presented in Fig. 3.12(a), however, there is no abrupt change of behavior visible at small length scales; the simulation data only deviate very gradually from a straight line that describes the long-wavelength behavior as one analyzes smaller lateral length scales.

In view of the quantitative description of the behavior in the bulk and the excess properties of the interface by the SCF theory, polymer blends provide a unique opportunity to investigate the short-length-scale cutoff B_0. To this end, we *define* the cutoff B_0 as the lateral length scale where the apparent profiles of the simulation and the prediction of the SCF theory coincide. The cutoff has to be a quantity that characterizes the "intrinsic" structure of the interface. Possible candidates include the distance $\rho_0^{-1/3}$ between segments [a microscopic length that depends on neither chain length nor incompatibility], the intrinsic width w_{intr} of the interface as calculated by the SCF theory [a length that depends on incompatibility but becomes independent of chain length in the limit $\chi N \to \infty$], or the correlation length ξ of composition fluctuations in the bulk [a length that is proportional to R_e and becomes independent of χN in the strong segregation limit]. In Fig. 3.12(a) we illustrate our procedure to extract the short-wavelength cutoff B_0. As one can see, the length scale depends on chain length and incompatibility. In Fig. 3.12(b) we plot the resulting B_0, varying chain length and incompatibility. The different data sets collapse onto a single line and are consistently described by (Werner et al. 1999)

$$\frac{B_0}{w_{\text{intr}}} = 3.8 \left(1 - \frac{3.1}{\tilde{\chi} N}\right) \tag{3.113}$$

This observation is also in qualitative accord with a prediction of Semenov (1994), $B_0 = \pi w_{\text{intr}}$.

Profiles Across an Interface

To illustrate the level of quantitative accuracy of the SCF theory, we compare the predictions of the SCF theory with the simulation results for the bond fluctuation model (Müller and Werner 1997). We use a binary blend with chain length $N = 64$, where the A component is flexible while a bond-bending potential acts along the backbone of B polymers, i.e. the single-chain distribution defined in Eq. (3.5) takes for B polymers the form

$$\tilde{\mathcal{D}}_B \boldsymbol{R}_\beta = \sum_{\boldsymbol{r}_{\beta,1} \in \mathcal{V}} \left(\prod_{i=1}^{N-1} \sum_{\boldsymbol{b}_{\beta,i} \in \{\boldsymbol{b}\}_{\text{BFM}}} \right) \exp\left[f \sum_{i=1}^{N-2} \cos \Theta_{i,i+1} \right] \quad (3.114)$$

where $\Theta_{i,i+1}$ denotes the angle between two successive bonds $\boldsymbol{b}_{\beta,i}$ and $\boldsymbol{b}_{\beta,i+1}$. On increasing the parameter f from $f = 0$ (flexible chain) to $f = 2$ (semi-flexible chain), we increase the statistical segment length of B polymers by a factor 1.5. Such a moderate difference in statistical segment length is a quite common asymmetry in polymer blends. We use only small values of f such that the chain conformations on large length scales are still Gaussian and there is no nematic order in the fluid. Also the entropic contribution of the effective Flory–Huggins parameter is small (Müller 1995). For chain length $N = 32$ and stiffness parameter $f = 1$, we can measure the entropic contribution to the effective Flory–Huggins parameter via Eq. (3.47) and obtain the value $\tilde{\chi}_{\text{entropic}} = 0.0018(2)$, which, for the chain lengths and incompatibility considered, is small compared to the enthalpic contribution.

By tuning the stiffness parameter f, we can vary the ratio between the "intrinsic" width of the interface and the statistical segment length a. If the condition $a \ll w_{\text{intr}}$ is not fulfilled, i.e. in very strongly segregated blends or in the case of semi-flexible chains, the conformational statistics of the polymer on the length scale w_{intr} cannot be described by the Gaussian chain model, and one has to account for the local chain architecture. This allows us to illustrate the role of local chain architecture on interface properties. One could incorporate the effect of local stiffness by utilizing the worm-like chain model (Kratky and Porod 1949; Saito et al. 1967; Morse and Fredrickson 1995; Schmid and Müller 1995), which captures the crossover from the rod-like behavior on short length scales to the Gaussian behavior on large length scales. Alternatively, instead of mapping the chain conformations onto an analytically tractable chain model (like the Gaussian or worm-like chain model), one can directly use the chain conformation from the simulation of the bulk. To this end, one generates a large sample of single-chain conformations from a simulation of the bulk (Müller and Werner 1997) or from a modeling scheme like the rotational isomeric state model (Flory 1969; Mattice and Suter 1994). These configurations capture the conformation statistics on all length scales, not only the Gaussian behavior on large length scales. Then, one approximates the sum over all single-chain conformations by a partial enumeration over a large number (on the order of $n_{\text{sample}} = 10^7$) of explicit chain conformations $\{\boldsymbol{R}_n\}$ with $1 \leq n \leq n_{\text{sample}}$ (Szleifer et al. 1986; Szleifer and Carignano 1996):

$$\int \tilde{\mathcal{D}}_A \boldsymbol{R}_\alpha \cdots \approx \frac{\mathcal{Z}_0}{n_{\text{sample}}} \sum_{\{\boldsymbol{R}_n\}} \cdots \quad (3.115)$$

This allows us to calculate the single-chain partition function, $Q_A[W_A]$ (cf. Eq. 3.18) and the density created by a single chain in the external field (cf. Eq. 3.24). In practice, the partial enumeration over independent single-chain conformations can conveniently be performed on a parallel computer (Müller and Schick 1996b; Müller and Werner 1997; Müller and Schick 1998; Müller and Binder 1998). To this end, one distributes the chain conformations evenly among the processors, and each processor calculates the Boltzmann weight of the single-chain conformations in the external field and their contribution to the density.

In Fig. 3.13(a) we show the interface tension γ_I as a function of the stiffness asymmetry. As expected, γ_I increases as we increase the stiffness of the B polymers, and both the absolute value as well as the stiffness dependence are quantitatively described by the SCF calculations. For comparison, we also show the result for the Gaussian chain model in the strong segregation limit, which was obtained from Helfand and Sapse (1975):

$$\frac{\gamma_{SSL} R_e^2}{k_B T} = \sqrt{\mathcal{N}} \sqrt{\frac{\chi N}{6}} \frac{2[1 + a_B/a_A + (a_B/a_A)^2]}{3[1 + a_B/a_A]}$$

$$\frac{w_{intr,SSL}}{R_e} = \frac{1}{\sqrt{6\chi N}} \sqrt{\frac{1 + (a_B/a_A)^2}{2}} \qquad (3.116)$$

where $R_e = a_A\sqrt{N-1}$ refers to the stiffness-independent end-to-end distance of the flexible A polymers. This asymptotic prediction for the Gaussian chain model in the limit $\chi N \to \infty$ reproduce the increase of the interface tension with stiffness disparity, but the simple expression neglects $1/\chi N$ corrections (cf. Eq. 3.96) and therefore overestimates the interface tension at finite incompatibilities similar to what is observed in symmetric mixtures (cf. Fig. 3.10).

Fig. 3.13(b) presents the results for the interface width. Two effects are clearly visible.

(1) As we decrease the lateral width on which we observe the interface from the full lateral system size, $B = L \approx 3.8R_e$, to $B = 0.9R_e$, the apparent width of the interface decreases. By reducing the lateral system size, we partially eliminate the effect of capillary-wave broadening. However, the interface width on the lateral length scale $B = 0.9R_e$ is still about 20% larger than the prediction of the SCF theory, and we utilize this length scale to compare profiles of other quantities. For comparison, we also show the dependence of the interface width w_e as extracted from the excess energy of the interface (cf. Eq. 3.82):

$$\frac{\Delta E}{k_B T \rho_0} = \int d\mathbf{r}\, \tilde{\chi} \phi_A \phi_B \qquad (3.117)$$

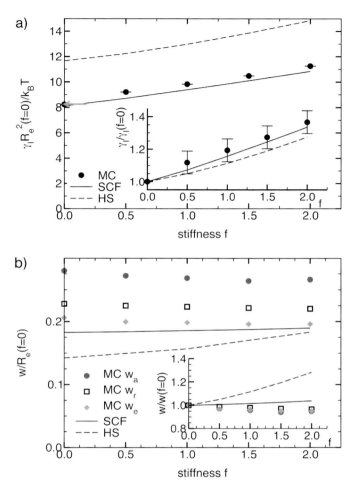

Fig. 3.13 (a) Interface tension γ_I as a function of stiffness f. Symbols correspond to MC simulations of the bond fluctuation model for chain length $N = 32$, which have been obtained by analyzing the spectrum of interface fluctuations. The arrow on the left-hand side (just visible by the "8") marks the interface tension of a symmetric blend obtained from grand-canonical MC simulations. The solid line corresponds to the prediction of the SCF theory using a partial enumeration scheme to incorporate the chain architecture on all length scales. The dashed line corresponds to the prediction of the Gaussian chain model in the strong segregation limit (see Eq. 3.116). The inset shows the relative increase of the interface tension due to stiffness disparity f.
(b) Estimates for the width of the interface as a function of stiffness on the lateral length scale $B = 3.8R_e$ (circles) and $B = 0.9R_e$ (squares). The diamonds correspond to an estimate of the intrinsic width from the excess energy of the interface (according to Eq. 3.118). The predictions of the SCF theory and the strong segregation limit for Gaussian chains are shown by solid and dashed lines, respectively. The inset shows the relative variation as a function of stiffness f. (Adapted from Müller and Werner 1997).

This quantity can be measured straightforwardly in the simulations. Utilizing a tanh profile for the "intrinsic" interface profile and assuming that the phases are strongly segregated, we arrive at the estimate (Müller and Schick 1996a; Müller and Werner 1997):

$$w_\mathrm{e} = \frac{2\Delta E}{\rho_0 \tilde{\chi} k_\mathrm{B} T \mathcal{A}} \qquad (3.118)$$

where $\mathcal{A}$ denotes the interface area. Note that the excess energy of the interface (like any other excess quantity) is related to the "intrinsic" interface profile. Thus, w_e provides us with an estimate of the "intrinsic" width, and, indeed, the values are very close to the prediction of the SCF theory.

(2) Upon increasing the stiffness, the interface width remains almost constant. This results from a cancellation of two effects. The simulation data are obtained at fixed repulsion between unlike segments, $\epsilon = 0.05 k_\mathrm{B} T$. As we increase the chain stiffness, the chain conformations become more open and the number of intermolecular contacts z_c increases. Thus, the effective Flory–Huggins parameter $\tilde{\chi}$ increases as we increase the stiffness disparity. The increase in $\tilde{\chi}$ tends to reduce the interface width. This effect is counterbalanced by an increase of the statistical segment length of the B polymers. The simulations as well as the SCF calculations using the detailed chain conformations quantitatively agree. Using the asymptotic expression for the Gaussian chain model (cf. Eq. 3.116), however, we predict an increase in the interface width by 21% (even if we account for the stiffness dependence of the effective Flory–Huggins parameter), in qualitative contrast to the observation.

The SCF theory not only predicts composition profiles, but also provides a detailed picture of all single-chain properties in a spatially inhomogeneous environment. In Fig. 3.14(a) we compare the orientation of the molecule's end-to-end vector at the interface. The alignment of the molecule at the interface can be quantified by the parameter $q_\mathrm{e}(z)$:

$$q_\mathrm{e}(z) = \frac{3\langle R_z^2 \rangle - \langle \boldsymbol{R}_\mathrm{e}^2 \rangle}{2\langle \boldsymbol{R}_\mathrm{e}^2 \rangle} \qquad (3.119)$$

where R_z denotes the component of the end-to-end vector normal to the interface. Negative values of the orientational parameter correspond to a parallel alignment of the end-to-end vector with respect to the interface plane. To obtain a profile, we associate the orientation of the end-to-end vector with the distance z of its midpoint from the local interface position. Note that the orientation effect on the length scale of the end-to-end vector is very pronounced. It is slightly larger for stiffer molecules than for flexible ones, and the orientation effects extends over $0.5 R_\mathrm{e} \approx 5.2 w_\mathrm{intr}$.

The analogous quantity for the orientation of the bond vectors, $\boldsymbol{b}$, is shown in Fig. 3.14(b). Note that the orientational parameter is about an order of

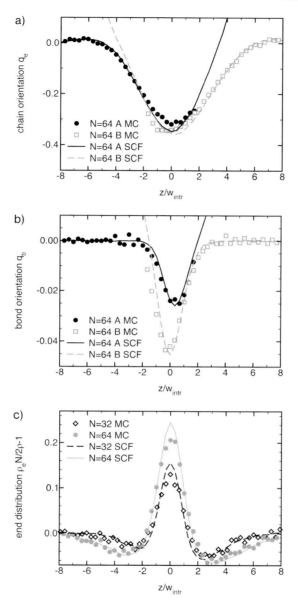

Fig. 3.14 Orientation of (a) the end-to-end vector and (b) the bond vectors as a function of the distance z from the center of the interface. In the simulations, we utilized chains comprising $N = 64$ effective segments. The chain stiffness parameter $f = 0$ was used for polymers of type B (enriched on the right-hand side of the interface). Lines correspond to the prediction of the SCF theory. (c) Relative enrichment of chain ends at the center of the interface for chain lengths $N = 32$ and 64. Symbols present the results of the MC simulations, while lines correspond to the prediction of the SCF theory. (Adapted from Müller and Werner 1997).

magnitude lower, i.e. orientation effects on short length scales are less pronounced, and the alignment is very localized at the center of the interface. On the short length scale, the effects of stiffness disparity are much more pronounced. The stiffer bonds exhibit a value of q_b that is twice as large as that for bonds along flexible molecules. Again, the SCF theory using the detailed chain conformations on all length scales is able to predict the orientation effects on all length scales. Within the Gaussian chain model there are no well-defined tangential vectors and the orientational parameter vanishes.

In Fig. 3.14(c) we present the relative enrichment of chain ends at the interface. The chain-end distribution is important for the interdiffusion and healing properties at interfaces between long polymers. They also play an important role for reactions at interfaces. In many experimental systems, chain ends have slightly different interactions than inner chain segments, which might result in a modification of the interface properties. Even if there are no specific interactions, chain-end effects give rise to rather large corrections to the interface tension. On the theoretical side, the behavior of chain ends is associated with corrections to the ground-state dominance (Lifshitz et al. 1978). Therefore, the density profile of chain ends provides a sensitive, quantitative test for the accuracy of the SCF theory. Chain ends are enriched at the center of the interface and there is a concomitant depletion zone further away from the interface. On increasing the chain length, we observe that the depletion zone moves further away from the center of the interface, thus indicating that the length scale of the profile is set by the molecular chain extension R_e. The depletion effect is the more pronounced the stiffer (right side of the profile) or longer the chains are. The SCF theory provides a quantitatively accurate description of these subtle effects.

This detailed comparison of interface properties between MC simulations and SCF calculations demonstrates that the SCF theory is a very versatile theory that is able to quantitatively predict a large variety of structural and thermodynamic properties of interfaces in polymer blends. For such a quantitative comparison, it is necessary to encode local correlations, which are neglected in the SCF theory, by an effective Flory–Huggins parameter $\tilde{\chi}$, and by utilizing the detailed chain conformational statistics on all length scales. The effective Flory–Huggins parameter can consistently be measured in the simulations by comparing the predictions of the mean-field theory to the simulations away from the critical point. The chain conformations can be described by the Gaussian chain model with an end-to-end distance R_e extracted from the simulations, or explicit chain conformations can be utilized in the SCF calculations via a partial enumeration scheme.

The quantitative comparison between MC simulations and SCF theory also reveals the effect of long-range fluctuations. In the vicinity of the critical point, long-range composition fluctuations alter the qualitative behavior; and at interfaces, fluctuations of the local interface position (capillary waves) lead to a broadening of the profiles. While those effects can be quite substantial in simulations or experiments for finite chain length, their importance will be much reduced if we increase the invariant polymerization index $\mathcal{N}$. The relative shift of the critical temperature away from the mean-field value and the relative increase of the interface width due to capillary waves formally decrease like $1/\sqrt{\mathcal{N}}$.

3.3
Liquid–Vapor Interfaces

In this section, we discuss the properties of liquid–vapor interfaces. The phenomenology of liquid–liquid phase separation in dense binary polymer blends and liquid–vapor coexistence in polymers (or incompressible polymer–solvent mixtures) is similar. In fact, the Flory–Huggins theory (cf. Eq. 3.44) has been employed to describe incompressible polymer–solvent mixtures by treating the solvent as a short chain. This reproduces many qualitative features of the liquid–vapor (or polymer–solvent) separation, namely the observation that the critical density decreases with chain length, $\phi_c \sim 1/\sqrt{N}$, and that the critical temperature approaches a chain-length-independent limit, the Θ temperature, with a $1/\sqrt{N}$ correction. We shall point out, however, two important differences.

(1) Instead of an incompressible polymer–solvent mixture, one often considers a pure polymer system where different monomeric units attract each other. These interactions are effective attractions that result from integrating out the degrees of freedom of the solvent, and mirror the fact that monomeric units attract each other more strongly than monomeric units attract solvent particles. While such a compressible, one-component system yields qualitatively similar phase diagrams, there is an important difference. In the Flory–Huggins description, the third- and higher-order virial coefficients exclusively stem from the entropy of mixing (and therefore these are not independent coefficients); while in the compressible, one-component system, higher-order virial coefficients are generated by the interaction of the multiple, monomeric units. This is a much more realistic description, which we shall adopt in the following.

(2) In contrast to a binary polymer blend, where the interactions on the scale of a monomeric unit are typically small, $\sim k_B T/N$, the interactions at the onset of liquid–vapor phase separation in a polymer solution are on the order of $k_B T$. Thus, the characteristic "intrinsic" width of interfaces in polymer–solvent systems is on the order of the monomeric unit itself, and the details of the molecular architecture and the local fluid packing mat-

ter. Polymer–solvent mixtures cannot be described by a few, coarse-grained parameters (like the Flory–Huggins parameter), and generally one cannot expect the behavior of liquid–vapor interfaces to exhibit the same degree of universality as the properties of interfaces in binary polymer blends.

Therefore, we do not aim to define effective parameters by comparing the prediction of the SCF theory to simulations or experiments; rather, we pursue the more ambitious goal of describing the local structure. The problem of the literal SCF theory is that, if we start from the partition function of the microscopic model and perform the saddle-point approximation, we will lose important information about short-range correlations and local structure. To avoid this loss while retaining the general form of the theoretical approach, we can replace the bare interactions between segments by an excess interaction free-energy functional $\Delta F_{\text{ex}}[\phi]$, such that the minimization of the free energy with respect to the density includes correlation effects (Frisch and Lebowitz 1965; Evans 1982). This scheme heavily builds on techniques of the theory of liquids. It can be shown that there exists such an excess interaction free-energy functional but, in general, its explicit form is unknown. Thus, the goal is to construct a good approximation for the excess interaction free-energy functional.

3.3.1
Formalism

Similar to Section 3.2.3 we start from the canonical partition function of n chains in a volume $\mathcal{V}$ (Müller and MacDowell 2000; Müller et al. 2003b):

$$\mathcal{Z} \propto \frac{1}{n!} \int \prod_{\alpha=1}^{n} \tilde{\mathcal{D}} \boldsymbol{R}_\alpha \, \exp\left[-\frac{\Delta F_{\text{ex}}[\hat{\phi}]}{k_{\text{B}} T} \right] \qquad (3.120)$$

Here ΔF_{ex} denotes the excess interaction free-energy functional, which is chosen such that it provides an accurate description of the local structure *after* the saddle-point approximation is performed. It depends on the microscopic segmental number density

$$\hat{\phi}(\boldsymbol{r}) = \sum_{\alpha=1}^{n} \sum_{i=1}^{N} \delta(\boldsymbol{r} - \boldsymbol{r}_{\alpha,i}) \qquad (3.121)$$

In Eq. (3.120), $\tilde{\mathcal{D}} \boldsymbol{R}_\alpha$ sums over all conformations α of the polymer molecule, with the statistical weight accounting for all interactions that are not intermolecular, such as external fields (e.g. at surfaces) and intramolecular, bonding, interactions. To rewrite this partition function of interacting chains in terms of independent chains in an external, but fluctuating, field, we introduce the collective density $\Phi(\boldsymbol{r})$ and the conjugated field $U(\boldsymbol{r})$. Inserting the Fourier representation of the δ function (Hong and Noolandi 1981)

$$\int \mathcal{D}\Phi\, \delta(\Phi(\boldsymbol{r}) - \hat{\phi}(\boldsymbol{r})) \sim \int \mathcal{D}\Phi \int_{-i\infty}^{i\infty} \mathcal{D}U \exp\left[\int d\boldsymbol{r}\, U(\boldsymbol{r})(\Phi(\boldsymbol{r}) - \hat{\phi}(\boldsymbol{r}))\right]$$
$$\sim 1, \qquad (3.122)$$

into Eq. (3.120), we obtain

$$\mathcal{Z} \propto \frac{1}{n!} \int \mathcal{D}\Phi\, \mathcal{D}U \prod_{\alpha=1}^{n} \tilde{\mathcal{D}}\boldsymbol{R}_\alpha \exp\left(-\frac{\Delta F_{\text{ex}}[\phi]}{k_{\text{B}}T}\right) \exp\left(\int d\boldsymbol{r}\, U(\boldsymbol{r})(\Phi - \hat{\phi})\right)$$

$$\propto \frac{1}{n!} \int \mathcal{D}\Phi\, \mathcal{D}U \exp\left(-\frac{\Delta F_{\text{ex}}[\phi]}{k_{\text{B}}T} + \int d\boldsymbol{r}\, U\Phi\right)$$

$$\times \int \prod_{\alpha=1}^{n} \tilde{\mathcal{D}}\boldsymbol{R}_\alpha \exp\left(-\int d\boldsymbol{r}\, U\hat{\phi}\right) \qquad (3.123)$$

$$\propto \int \mathcal{D}\Phi\, \mathcal{D}U \exp\left(-\frac{\Delta F_{\text{ex}}[\phi]}{k_{\text{B}}T} + \int d\boldsymbol{r}\, U\Phi\right) \frac{(\mathcal{Q}[U])^n}{n!}$$

$$\equiv \int \mathcal{D}\Phi\, \mathcal{D}U \exp\left(-\frac{\mathcal{F}[U, \Phi]}{k_{\text{B}}T}\right) \qquad (3.124)$$

where the single-chain partition function in the external field, $U(\boldsymbol{r})$, is defined by

$$\mathcal{Q}[U] = \int \tilde{\mathcal{D}}\boldsymbol{R} \exp\left[-\sum_{i=1}^{N} U(\boldsymbol{r}_i)\right] \qquad (3.125)$$

and the free-energy functional, $\mathcal{F}[U, \Phi]$, takes the form

$$\frac{\mathcal{F}[U, \Phi]}{k_{\text{B}}T} = n \ln\left(\frac{n}{e\mathcal{Q}[U]}\right) + \frac{\Delta F_{\text{ex}}[\phi]}{k_{\text{B}}T} - \int d\boldsymbol{r}\, U\Phi \qquad (3.126)$$

The functional integral over the imaginary field U and the real segmental density Φ in Eq. (3.124) cannot be exactly evaluated. Thus, we approximate it by its saddle-point value. To this end, we extremize the free-energy functional $\mathcal{F}[U, \Phi]$ with respect to the fields U and Φ. The saddle-point values of the fields are denoted by lower-case letters u and ϕ:

$$\frac{\mathcal{D}\mathcal{F}[U, \Phi]}{\mathcal{D}\Phi(\boldsymbol{r})} = 0 \implies u(\boldsymbol{r}) = \frac{1}{k_{\text{B}}T} \frac{\mathcal{D}\Delta F_{\text{ex}}[\phi]}{\mathcal{D}\phi(\boldsymbol{r})} \qquad (3.127)$$

Note that the saddle-point value of the field is real. The variation with respect to U yields

$$\frac{\mathcal{D}\mathcal{F}[U,\Phi]}{\mathcal{D}U(\boldsymbol{r})} = 0 \implies \phi(\boldsymbol{r}) = -n\frac{\mathcal{D}\ln\mathcal{Q}[u]}{\mathcal{D}u(\boldsymbol{r})} \qquad (3.128)$$

The last expression identifies $\phi(\boldsymbol{r})$ as the Boltzmann average of the single-chain density in the external field u; it is also the estimate of the density distribution of the polymer segments within the saddle-point approximation. Given an expression for the excess interaction free-energy functional ΔF_{ex}, the two Eqs. (3.127) and (3.128) constitute a closed set of equations from which density profiles and thermodynamic properties are obtained. Substituting the saddle-point values, u, and, ϕ, back into the free-energy functional Eq. (3.126), we obtain a mean-field estimate for the free energy F and the chemical potential μ:

$$\frac{F}{k_\text{B}T} = n\ln\left(\frac{n}{e\mathcal{Q}[u]}\right) + \frac{\Delta F_{\text{ex}}[\phi]}{k_\text{B}T} - \int \text{d}\boldsymbol{r}\, u\phi \qquad (3.129)$$

$$\frac{\mu}{k_\text{B}T} \equiv \frac{1}{k_\text{B}T}\frac{\partial F}{\partial n} = \frac{1}{k_\text{B}T}\frac{\mathcal{D}\mathcal{F}[\phi, u[\phi]]}{\mathcal{D}\phi(\boldsymbol{r})} = \ln\left(\frac{n}{\mathcal{Q}[u[\phi]]}\right) \qquad (3.130)$$

From these expressions we calculate the grand-canonical potential, Ω:

$$\frac{\Omega}{k_\text{B}T} = \frac{F - \mu n}{k_\text{B}T} = -n + \frac{\Delta F_{\text{ex}}[\phi]}{k_\text{B}T} - \frac{1}{k_\text{B}T}\int \text{d}\boldsymbol{r}\, \phi(\boldsymbol{r})\frac{\mathcal{D}\Delta F_{\text{ex}}[\phi]}{\mathcal{D}\phi(\boldsymbol{r})} \qquad (3.131)$$

In the Appendix (Section 3.5.2) we demonstrate that identical equations can be derived by density-functional techniques (Chandler et al. 1986; Sen et al. 1994; Woodward 1991; Yethiraj and Woodward 1995; Yethiraj 1998; Frischknecht et al. 2002; Yu and Wu 2002; Patra and Yethiraj 2003; Frischknecht and Curro 2004). The latter derivation shows that the general form of the equations remains valid beyond the saddle-point approximation, and that there exists an excess free-energy functional, ΔF_{ex}, that yields the exact density profile. However, the excess free-energy functional cannot be identified with the bare intermolecular interactions (as one would within the literal mean-field approximation), but it incorporates correlation effects. Unfortunately, the explicit form of ΔF_{ex} is unknown, and one has to resort to an approximation.

3.3.2
Approximating the Excess Interaction Free-Energy Functional ΔF_{ex}

Bead–Spring Model for Liquid–Vapor Interfaces
The form of the excess interaction free-energy functional ΔF_{ex} depends on the details of the model, for it describes correlation effects on short distances.

In the following, we outline a scheme to find an approximation for a coarse-grained bead–spring model. We use an off-lattice model (Müller and MacDowell 2000) that incorporates only the relevant features of polymeric materials: the excluded volume of the segments, chain connectivity, and an attractive interaction between monomers that drives the phase separation. Our model is similar in spirit to the bead–spring model of Kremer and Grest (1990). Monomeric units are modeled as Lennard-Jones particles:

$$V_{\text{LJ}}(r) = 4\epsilon\left[\left(\frac{\sigma}{r}\right)^{12} - \left(\frac{\sigma}{r}\right)^{6}\right] + \frac{127\epsilon}{4096} \quad \text{for} \quad r < 2 \times 2^{1/6}\sigma \quad (3.132)$$

The potential is cut off at twice the minimum distance and shifted so as to produce a continuous potential. The short-range repulsive part models the excluded volume of a monomer, whereas the attractive portion mimics van der Waals attractions and results in liquid–vapor coexistence. In addition to the Lennard-Jones potential, neighboring monomers along a chain interact via a finitely extensible nonlinear elastic (FENE) potential

$$V_{\text{bond}}(r) = -15\epsilon(R_0/\sigma)^2 \ln\left(1 - \frac{r^2}{R_0^2}\right) \quad \text{with} \quad R_0 = 1.5\sigma \quad (3.133)$$

The parameters are chosen such that the Lennard-Jones potential between non-bonded segments prefers a distance $r_{\text{nb}} \approx 1.12\sigma$, while the most favorable distance between bonded neighbors is slightly smaller, $r_{\text{b}} \approx 0.96\sigma$. Molecular-dynamics simulations by Bennemann et al. (1998) have shown that the occurrence of two incompatible length scales leads to a density-driven glass transition around a monomer number density $\phi\sigma^3 \approx 1.0$ rather than the "buildup" of long-range crystalline order (as emerges in the corresponding monomer fluid). Most of the data have been obtained for chain length $N = 10$ and $k_{\text{B}}T/\epsilon = 1.68$. For these parameters, a liquid of monomer number density $\phi\sigma^3 = 0.61$ coexists with a vapor of density $\phi\sigma^3 = 0.0083$. The temperature is low enough such that the vapor pressure over the liquid is nearly zero and the vapor phase has a vanishingly small polymer content. On the other hand, the density of the liquid is low enough to allow for an efficient grand-canonical simulation. In this regime, properties do not strongly depend on chain length.

Weighted Density Approximation for the Excess Free-Energy Function ΔF_{ex}
Since liquid–vapor interfaces are narrow, local correlations are important for making quantitative predictions. On the one hand, the local chain architecture has to be incorporated to describe the conformations of short loops and the orientation on different length scales at the narrow interface. Using explicit chain conformations, extracted from a bulk simulation of the liquid, we can

capture those intramolecular correlation effects on all length scales and for arbitrary chain architectures. In Section 3.2.6 we have demonstrated that this yields an excellent description of the detailed chain conformations at spatial inhomogeneities. We shall note that the chain conformations in a spatially homogeneous phase are always described by the unperturbed distribution $\tilde{\mathcal{D}}\boldsymbol{R}$, and are independent of temperature or density. The unperturbed distribution of single chains is input to the calculation. Thus, this scheme cannot describe the changes of chain conformations between the homogeneous liquid, where the chains adopt Gaussian conformations on large length scales, and the vapor phase, where the single-chain conformations are collapsed. Far below the Θ temperature, however, the density of the vapor phase is vanishingly low and the conformations in the vapor phase are chiefly determined by balancing the attractive and repulsive contributions to the equation of state. Thus, the conformational entropy of chains in the vapor gives only a minor contribution to the free energy. In the vicinity of the Θ temperature and for not too large chain lengths, the chain conformations vary only very gradually as a function of density.

The fluid-like packing of segments in the dense liquid is also important, because it determines the cohesive energy and pressure of the liquid. If the intramolecular interactions (i.e. bonding potentials) do not strongly perturb the fluid-like structure, the packing of the segments at liquid-like densities will resemble the structure of a fluid of unconnected monomers. The fluid structure of monomeric fluids has attracted abiding interest, and there exist a variety of sophisticated and elaborate techniques to describe their properties. These short-range correlations are described by the excess free-energy function ΔF_{ex}, and we shall employ a weighted density approximation (Curtin and Ashcroft 1985; Yethiraj 1998). Studies of monomeric fluids show that it is important to distinguish between the strong short-range repulsions and weak longer-range attractions (van Swol and Henderson 1991; Katsov and Weeks 2001). The former contribution determines the packing structure of the liquid and chiefly depends on density. The latter contribution controls the pressure and thus the phase behavior as a function of temperature, but it has only a minor influence on the grouping of segments into neighboring shells.

Several choices for the excess free-energy density have been explored (Chandler et al. 1986; Yethiraj 1998; Müller and MacDowell 2000; Yu and Wu 2002; Müller et al. 2003b; Patra and Yethiraj 2003). In the following we use the ansatz (van Swol and Henderson 1991; Müller et al. 2003b)

$$\frac{\Delta F_{\text{ex}}[\phi]}{k_\text{B} T} = \int \mathrm{d}\boldsymbol{r}\, \phi(\boldsymbol{r}) \left[f_{\text{hc}}(\bar{\phi}_{\text{hc}}(\boldsymbol{r})) + f_{\text{att}}(\bar{\phi}_{\text{att}}(\boldsymbol{r})) \right] \quad (3.134)$$

where $\bar{\phi}_{\text{hc}}(\boldsymbol{r})$ and $\bar{\phi}_{\text{att}}(\boldsymbol{r})$ are weighted densities, defined by

$$\bar{\phi}_{\text{hc}}(\boldsymbol{r}) = \int \mathrm{d}\boldsymbol{r}'\, w_{\text{hc}}(\boldsymbol{r}-\boldsymbol{r}')\phi(\boldsymbol{r}')$$

$$\bar{\phi}_{\text{att}}(\boldsymbol{r}) = \int \mathrm{d}\boldsymbol{r}'\, w_{\text{att}}(\boldsymbol{r}-\boldsymbol{r}')\phi(\boldsymbol{r}') \qquad (3.135)$$

and $w_{\text{hc}}(r)$ and $w_{\text{att}}(r)$ are weighting functions that characterize the spatial extension of the interactions. They satisfy the normalization condition

$$\int \mathrm{d}\boldsymbol{r}\, w_{\text{hc}}(\boldsymbol{r}) = \int \mathrm{d}\boldsymbol{r}\, w_{\text{att}}(\boldsymbol{r}) = 1. \qquad (3.136)$$

The theory is completely specified by the two thermodynamic functions of state, f_{hc} and f_{att}, and the two weighting functions, $w_{\text{hc}}(r)$ and $w_{\text{att}}(r)$.

Given an explicit expression for the excess free-energy functional, we can use Eq. (3.127) to calculate the external field u that mimics the effect of the surrounding chains:

$$u(\boldsymbol{r}) = f_{\text{hc}}(\bar{\phi}_{\text{hc}}(\boldsymbol{r})) + \int \mathrm{d}\boldsymbol{r}'\, w_{\text{hc}}(\boldsymbol{r}-\boldsymbol{r}')\phi(\boldsymbol{r}')\frac{\mathrm{d}f_{\text{hc}}}{\mathrm{d}\bar{\phi}_{\text{hc}}}(\boldsymbol{r}')$$

$$+ f_{\text{att}}(\bar{\phi}_{\text{att}}(\boldsymbol{r})) + \int \mathrm{d}\boldsymbol{r}'\, w_{\text{att}}(\boldsymbol{r}-\boldsymbol{r}')\phi(\boldsymbol{r}')\frac{\mathrm{d}f_{\text{att}}}{\mathrm{d}\bar{\phi}_{\text{att}}}(\boldsymbol{r}') \qquad (3.137)$$

The thermodynamic function f_{hc} and the weighting function w_{hc} refer to the strong repulsive (hard-core) interactions, while the functions f_{att} and w_{att} are associated with the effect of the longer-range attractions. Approximations for the four functions will be discussed in the following subsection.

Bulk Phase Behavior: Wertheim's Perturbation Theory
As in the study of interfaces in polymer blends, it proves useful first to describe the phase behavior of the homogeneous system before studying interface properties. From Eqs. (3.131) and (3.134), we determine the pressure P:

$$\frac{P\mathcal{V}}{k_{\text{B}}T} = -\frac{\Omega\mathcal{V}}{k_{\text{B}}T} \qquad (3.138)$$

$$= n + \int \mathrm{d}\boldsymbol{r}\,\mathrm{d}\boldsymbol{r}'\,\phi(\boldsymbol{r})\left[w_{\text{hc}}(\boldsymbol{r}-\boldsymbol{r}')\frac{\mathrm{d}f_{\text{hc}}}{\mathrm{d}\phi}(\boldsymbol{r}') + w_{\text{att}}(\boldsymbol{r}-\boldsymbol{r}')\frac{\mathrm{d}f_{\text{att}}}{\mathrm{d}\phi}(\boldsymbol{r}')\right]\phi(\boldsymbol{r}')$$

In the bulk, the density does not depend on the position $\boldsymbol{r}$ in space, and the equation simplifies to

$$\frac{P}{k_{\text{B}}T} = \frac{\phi}{N} + \phi^2 \frac{\mathrm{d}(f_{\text{hc}}+f_{\text{att}})}{\mathrm{d}\phi} \qquad (3.139)$$

The first term corresponds to the ideal contribution, while the second contributions arises from the excess free-energy functional. Equation (3.139) shows that the weighting functions do not influence the bulk thermodynamic properties, which are completely specified by the function $f_{\rm hc}(\phi) + f_{\rm att}(\phi)$.

In principle, one could measure the equation of state for the system of interest and thereby determine the function $f_{\rm hc}(\phi) + f_{\rm att}(\phi)$. Unfortunately, the density profile across a liquid–vapor interface or at a surface adopts values that fall into the miscibility gap, i.e. one needs the thermodynamic function also in a region where the macroscopic bulk system is unstable with respect to phase separation. To describe the hypothetical, homogeneous state inside the miscibility gap, one would have to rely on extrapolating the data from the one-phase region into the miscibility gap.

Alternatively, one can try to predict the properties of the homogeneous system starting from the structure of the liquid of unconnected monomers. This is the idea of Wertheim's theory (Wertheim 1987), which has become a popular scheme to describe the equation of state of polymer liquids (Chapman et al. 1988; Stell and Zhou 1989; Johnson et al. 1993; Vega et al. 2002). The reliability of this approach can be assessed by comparing to simulation data in the one-phase region, and this scheme allows for a systematic extrapolation into the miscibility gap.

In its most basic form, Wertheim's theory simplifies to a thermodynamic perturbation theory (TPT1) with respect to the bonding potential $V_{\rm bond}$. The free-energy difference between that of the polymer liquid, F, and that of the liquid of unconnected monomers, $F_{\rm mono}$, at the same density and temperature can be calculated as follows:

$$\frac{F - F_{\rm mono}}{k_{\rm B}T}$$

$$= -\ln \frac{(1/n!)\int \prod_\alpha d\bm{R}_\alpha e^{-\left[\sum_{(\alpha,i)<(\beta,j)} V_{\rm LJ}(\bm{r}_{\alpha,i}-\bm{r}_{\beta,j})+\sum_{\alpha,i} V_{\rm bond}(\bm{r}_{\alpha,i}-\bm{r}_{\alpha,i+1})\right]/k_{\rm B}T}}{[1/(nN)!]\int \prod_\alpha d\bm{R}_\alpha\, e^{-\left[\sum_{(\alpha,i)<(\beta,j)} V_{\rm LJ}(\bm{r}_{\alpha,i}-\bm{r}_{\beta,j})\right]/k_{\rm B}T}}$$

$$= -\ln\left(\frac{(nN)!}{n!}\right) - \ln\left\langle \prod_{\alpha=1}^{n}\prod_{i=1}^{N-1}\exp\left[-\frac{V_{\rm bond}(\bm{r}_{\alpha,i}-\bm{r}_{\alpha,i+1})}{k_{\rm B}T}\right]\right\rangle_{\rm mono} \quad (3.140)$$

where the last average has to performed in a liquid of unconnected segments. To first order, one can approximate the last term in the form (TPT1):

$$\left\langle \prod_{\alpha=1}^{n}\prod_{i=1}^{N-1}\exp\left[-\frac{V_{\rm bond}(\bm{r}_{\alpha,i}-\bm{r}_{\alpha,i+1})}{k_{\rm B}T}\right]\right\rangle_{\rm mono}$$

3.3 Liquid–Vapor Interfaces

$$\approx \left(\frac{1}{V}\int d\boldsymbol{r}\, g_{\mathrm{mono}}(\boldsymbol{r})\exp\left[-\frac{V_{\mathrm{bond}}(\boldsymbol{r})}{k_{\mathrm{B}}T}\right]\right)^{n(N-1)} \quad (3.141)$$

where $g_{\mathrm{mono}}(r)$ denotes the pair correlation function of the fluid of unconnected segments. This approximation results in the following relation between the free-energy density of the Lennard-Jones monomer fluid and the polymer solution:

$$\frac{F}{Vk_{\mathrm{B}}T} = \frac{\phi}{N}\ln\left(\frac{\phi}{eN}\right) + \left[\frac{F_{\mathrm{mono}}}{Vk_{\mathrm{B}}T} - \phi\ln\left(\frac{\phi}{e}\right)\right]$$

$$-\phi\left(1-\frac{1}{N}\right)\ln\left[\int d\boldsymbol{r}\, g_{\mathrm{mono}}(\boldsymbol{r})\,\mathrm{e}^{-V_{\mathrm{bond}}(\boldsymbol{r})/k_{\mathrm{B}}T}\right] \quad (3.142)$$

The first correction describes the loss of translational entropy by bonding N segments to form a polymer, while the second correction simply describes the free-energy costs of building $n(N-1)$ bonds within first-order thermodynamic perturbation theory (TPT1). Given the structure, g_{mono}, and thermodynamics, F_{mono}, of the fluid of unconnected monomers, we can calculate the excess free energy ΔF_{ex} of the homogeneous polymer system:

$$\frac{\Delta F_{\mathrm{ex}}}{Vk_{\mathrm{B}}T} = \phi[f_{\mathrm{hc}}(\phi) + f_{\mathrm{att}}(\phi)] \quad (3.143)$$

$$= \frac{\Delta F_{\mathrm{mono}}}{Vk_{\mathrm{B}}T} - \phi\left(1-\frac{1}{N}\right)\ln\left[\int d\boldsymbol{r}\, g_{\mathrm{mono}}(\boldsymbol{r})\,\mathrm{e}^{-V_{\mathrm{bond}}(\boldsymbol{r})/k_{\mathrm{B}}T}\right]$$

where $\Delta F_{\mathrm{mono}} = F_{\mathrm{mono}} - k_{\mathrm{B}}TnN\ln(\phi/e)$ denotes the excess free energy of the reference system of unconnected segments. The reduction of the pressure due to bonding is then obtained from Eq. (3.143):

$$\frac{P}{\phi k_{\mathrm{B}}T} = \frac{P_{\mathrm{mono}}}{\phi k_{\mathrm{B}}T} \quad (3.144)$$

$$-\left(1-\frac{1}{N}\right)\left\{1 + \phi\frac{\partial}{\partial\phi}\ln\left[\int d\boldsymbol{r}\, g_{\mathrm{mono}}(\boldsymbol{r})\,\mathrm{e}^{-V_{\mathrm{bond}}(\boldsymbol{r})/k_{\mathrm{B}}T}\right]\right\}$$

From the equation of state, one can calculate the phase behavior. In the limit $N \to \infty$, the critical point occurs at small segmental density, $\phi_{\mathrm{c}} \to 0$, and we can expand the properties of the reference fluid of unconnected monomers in powers of ϕ. The pressure of the monomer fluid has a virial expansion:

$$\frac{P_{\mathrm{mono}}}{\phi k_{\mathrm{B}}T} = 1 + b_2(T)\phi + b_3(T)\phi^2 + \cdots \quad (3.145)$$

where $b_2, b_3, \ldots$ denote the virial coefficients of the monomer fluid. Likewise, we can expand the last term in Eq. (3.144) in the form

$$\frac{\partial}{\partial \phi} \ln \left[\int \mathrm{d}\boldsymbol{r}\, g_{\mathrm{mono}}(\boldsymbol{r})\, e^{-V_{\mathrm{bond}}(\boldsymbol{r})/k_{\mathrm{B}}T} \right] = a_2 + a_3 \phi + \cdots \tag{3.146}$$

In the following we truncate the virial expression after the third-order coefficient. The coefficients of the monomer fluid do not depend on the chain length N. Thus, the entire chain-length dependence arises from the N dependence in Eq. (3.144) (MacDowell et al. 2000). Inserting the truncated virial expression of the monomer fluid into Eq. (3.144), we obtain

$$\frac{P}{(\phi/N)k_{\mathrm{B}}T} = 1 + B_2(T)\frac{\phi}{N} + B_3(T)\left(\frac{\phi}{N}\right)^2 \tag{3.147}$$

with

$$B_2 = N^2 \left[b_2 - \left(1 - \frac{1}{N}\right) a_2 \right] \tag{3.148}$$

$$B_3 = N^3 \left[b_3 - \left(1 - \frac{1}{N}\right) a_3 \right] \tag{3.149}$$

where $B_2, B_3, \ldots$ denote the virial coefficients of the polymer solution. The location of the critical point of the polymer fluid is given by the conditions $\partial P/\partial \phi = 0$ and $\partial^2 P/\partial \phi^2 = 0$. Using the truncated virial expansion, one obtains

$$B_2(T_c)^2 = 3 B_3(T_c) \quad \text{and} \quad \frac{\phi_c}{N} = -\frac{B_2(T_c)}{3 B_3(T_c)} = \frac{1}{\sqrt{3 B_3(T_c)}} \tag{3.150}$$

For long chain lengths, $N \to \infty$, the conditions simplify to

$$b_2(\Theta) = a_2(\Theta) \quad \text{and} \quad \phi_c = \frac{1}{\sqrt{3N[b_3(\Theta) - a_3(\Theta)]}} \tag{3.151}$$

Independent of the specific choice of the reference fluid of unconnected monomers, Wertheim's perturbation theory predicts the critical temperature of the polymer solution to converge towards an N-independent temperature, the Θ temperature, in the limit $N \to \infty$ and the critical density decreases like $1/\sqrt{N}$. Expanding the coefficients in Eq. (3.150) with respect to the distance from the Θ temperature, one shows that $\Theta - T_c \sim 1/\sqrt{N}$ (MacDowell et al. 2000). Gratifyingly, the chain-length dependence with which the critical temperature and critical density approach the limit $N \to \infty$ is in accord with the critical behavior predicted by renormalization group calculations. The Θ point corresponds to a tricritical point, where mean-field theory becomes correct in three spatial dimensions (upper critical dimension) up to (large) logarithmic corrections (Duplantier 1987; Frauenkron and Grassberger 1997).

These considerations show that Wertheim's perturbation theory captures the chain-length dependence of the critical properties in the limit $N \to \infty$. It also correctly describes the limit of dilute polymer solutions (i.e. ideal-gas limit) and the behavior of dense polymer liquids, $\phi \sim \mathcal{O}(1)$. In the latter, the structures of the polymer liquid and of the reference fluid of unconnected monomers strongly resemble each other, and the first-order perturbation theory is rather accurate. At intermediate densities (semidilute solutions), however, the polymers overlap but the segment density is small (semidilute solutions: $\phi R_e^3/N \gg 1$ but $\phi \ll 1$). In this case, the structures of the reference fluid (ideal gas) and of the polymer fluid substantially differ, and the first-order perturbation theory cannot describe the long-range correlations induced by the chain connectivity. In good solvents, one expects a scaling behavior of the form

$$\frac{P}{(\phi/N)k_\mathrm{B}T} = 1 + \mathcal{P}\left(\frac{\phi R_e^3}{N}\right) \quad \text{for} \quad \phi \ll 1 \tag{3.152}$$

where $\mathcal{P}$ is a scaling function (Schäfer 1999). For small arguments, the scaling function is linear in its argument, i.e. $B_2 \sim R_e^3 \sim N^{3\nu}$, where the exponent ν characterizes the chain extension in dilute solution. For chains in a good solvent, $T > \Theta$, the exponent is that of a self-avoiding walk, $\nu = 0.588$; right at the Θ temperature one finds $\nu = 1/2$. For large arguments, $\phi R_e^3/N$ (semidilute solution), the pressure depends only on the segment density but becomes independent of chain length. To cancel the chain-length dependence, the function $\mathcal{P}$ scales like $(\phi R_e^3/N)^{1/(3\nu-1)}$ and for the pressure one finds $P \sim \phi^{3\nu/(3\nu-1)}$. This non-trivial power law stems from the fractal behavior of the large-scale conformations of polymers in solutions.

Generally, one can expect Wertheim's equation of state for polymer solutions to be qualitatively reliable at very small densities (ideal gas) and melt-like densities – the quantitative accuracy depends on the choice of the reference fluid and knowledge about its structure and thermodynamics. However, for semidilute solutions of long macromolecules, stronger deviations are anticipated, and a more phenomenological approach that captures the correct scaling behavior might be more appropriate.

To illustrate the accuracy of Wertheim's perturbation theory, we compare its prediction in Fig. 3.15(a) to MC simulations of short chains, $N = 10$. The structure and thermodynamics of the reference fluid of unconnected monomers were obtained from the Ornstein–Zernicke equation using the reference hypernetted chain closure of Lado et al. (1983). Similar results (not

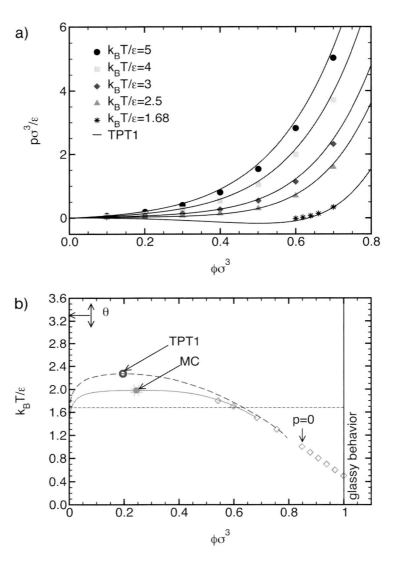

Fig. 3.15 (a) Equation of state and (b) phase diagram of a bead–spring polymer model. The pressure of the bulk system in (a) has been extracted from canonical Monte Carlo simulations using the virial expression (symbols). The lines show the result of Wertheim's perturbation theory. The phase diagram (b) has been obtained from grand-canonical Monte Carlo simulations. The binodal is shown as a full lines, while the corresponding result from Wertheim's perturbation theory is presented by a dashed line. Filled and open circles (arrowed) mark the location of the critical point in the simulations and the TPT1 calculations, respectively. The Θ temperature is indicated by an arrow on the top left-hand side. The diamonds present the results of constant-pressure simulations and indicate the densities that correspond to vanishing pressure (a good approximation for the coexistence curve at low temperatures). (Adapted from MacDowell et al. 2000).

shown) can be obtained utilizing the mean spherical approximation (MacDowell et al. 2000). One observes a good quantitative agreement between the simulation results and the thermodynamic perturbation theory without any adjustable parameter. The last isotherm corresponds to a subcritical temperature, $k_\mathrm{B}T/\epsilon = 1.68$, where we can compare only to the simulation data at large densities outside the miscibility gap. The calculations also provide us with an estimate of the pressure (and excess free energy) of the unstable or metastable states inside the miscibility gap.

The phase diagram of the polymer solution is presented in Fig. 3.15(b). Similar to the behavior of polymer blends, being a mean-field theory, Wertheim's perturbation theory fails to describe long-range density fluctuations. Thus, it overestimates the critical temperature, and the binodal in the vicinity of the critical point is parabolic, while one observes 3D Ising behavior in the simulations. Further away from the critical point, those critical long-range density correlations are less important, and a dense polymer liquid coexists with a dilute polymer vapor. For this practically important regime, Wertheim's theory provides a quantitatively accurate description.

Since, in the SCF calculations, we divide the excess free-energy functional into repulsive and attractive contributions, we also have to specify the system with repulsive hard-core (hc) interactions. For computational convenience, we chose the system of tangentially bonded, hard-sphere chains. The diameter d_hc of the corresponding hard-sphere fluid is obtained from the Lennard-Jones potential via the Barker–Henderson recipe (Barker and Henderson 1967)

$$d_\mathrm{hc} = \int_0^{r_\mathrm{min}} \mathrm{d}r \left[1 - \exp\left(-\frac{V_\mathrm{WCA}(r)}{k_\mathrm{B}T}\right) \right] \quad (3.153)$$

where the integration is extended to the minimum of the Lennard-Jones potential, $r_\mathrm{min} = 2^{1/6}\sigma$, and $V_\mathrm{WCA}(r) = V_\mathrm{LJ}(r) - V_\mathrm{LJ}(r_\mathrm{min})$ for $r < r_\mathrm{min}$. The pressure of the fluid of unconnected hard spheres can accurately be described by the Carnahan–Starling equation of state (Carnahan and Starling 1969):

$$\frac{P}{\phi k_\mathrm{B}T} = \frac{1 + \eta + \eta^2 - \eta^3}{(1-\eta)^3} \quad \text{where} \quad \eta = \frac{\pi \phi d_\mathrm{hc}^3}{6} \quad (3.154)$$

denotes the packing fraction. The excess free energy of the reference system can be obtained from

$$\frac{\Delta F_\mathrm{mono}}{k_\mathrm{B}T} = -\int_\infty^\mathcal{V} \mathrm{d}\mathcal{V}' \left(\frac{P(\phi')}{k_\mathrm{B}T} - \phi' \right)$$

$$= nN \int_0^\phi \mathrm{d}\phi' \frac{1}{\phi'} \left(\frac{P}{\phi' k_\mathrm{B}T} - 1 \right) = \frac{\eta(4-3\eta)}{(1-\eta)^2} \quad (3.155)$$

If the bond length is fixed, the integral in Eq. (3.142) will be proportional to the value of the pair correlation function of the reference system at the bond distance. Since we consider tangentially bonded, hard-sphere chains, the bond length is d_{hc} and the contact value of the pair correlation function in the hard-sphere reference system is related to the pressure via

$$g_{\text{mono}}(d_{\text{hc}}) = \frac{3}{2\pi\phi d_{\text{hc}}^3}\left(\frac{P}{\phi k_B T} - 1\right) = \frac{1 - \eta/2}{(1-\eta)^3} \tag{3.156}$$

Thus, using the Carnahan–Starling equation of state and thermodynamic perturbation theory (TPT1), we obtain for the fluid of tangentially bonded, hard-sphere chains:

$$f_{\text{hc}}(\phi) = \frac{4\eta - 3\eta^2}{(1-\eta)^2} - \left(1 - \frac{1}{N}\right)\ln\left(\frac{1 - \eta/2}{(1-\eta)^3}\right) \tag{3.157}$$

From the thermodynamics of the full system (Eq. 3.143) and that of the tangentially bonded, hard-sphere chains (Eq. 3.157), we calculate the thermodynamic function for the attractive interactions:

$$f_{\text{att}}(\phi) = \frac{\Delta F_{\text{ex}}}{nNk_B T} - f_{\text{hc}}(\phi) \tag{3.158}$$

The two thermodynamic functions f_{hc} and f_{att}, and their sum $\Delta F_{\text{ex}}/(nNk_B T)$, are shown in Fig. 3.16(a). Over the pertinent range of densities, the attractive contribution resembles a linear function of density. This behavior is expected from a van der Waals theory, and shows that the decomposition into repulsive and attractive contributions does not lead to any artifacts.

Short-Range Repulsion versus Longer-Range Attraction
To complete the description of the weighted density functional, we have to specify the weighting functions, w_{hc} and w_{att}. Some guidance for the choice of weighting functions can be obtained from the exact relation between the second functional derivative of F_{ex} with respect to $\phi(r)$ and the direct correlation function, $c(r)$ (Frisch and Lebowitz 1965; Hansen and McDonald 1986; Evans 1982):

$$\frac{1}{k_B T}\frac{\mathcal{D}^2 F_{\text{ex}}}{\mathcal{D}\phi(r)\mathcal{D}\phi(r')}\bigg|_{\phi(r)=\phi(r')=\phi} = -c(r - r') \tag{3.159}$$

where $c(r - r')$ is the direct correlation function of the homogeneous fluid. An approximation for the weighting function can be obtained by forcing the

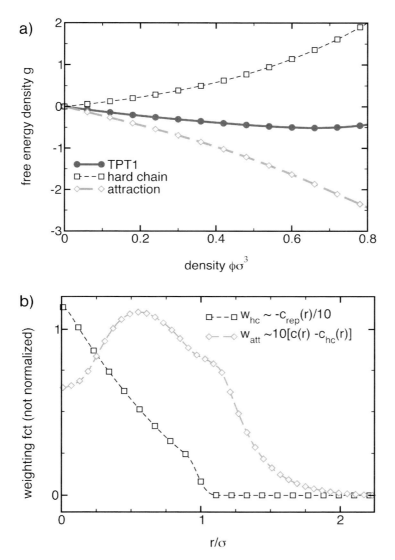

Fig. 3.16 (a) Free-energy densities at $k_B T/\epsilon = 1.68$ as a function of the monomer number density $\phi\sigma^3$. Circles and squares present the results of TPT1 calculations for the system with attractive interactions and the system of tangentially bonded, hard-sphere chains, respectively. The difference f_{att} is represented by the (bottom) curve with diamonds. (b) Unnormalized weighting functions at $k_B T/\epsilon = 1.68$. (Adapted from Müller et al. 2003b)

free-energy functional to satisfy this exact relation. The explicit expression for the excess free-energy functional yields[5]

$$\frac{\mathcal{D}^2 F_{\text{ex}}/k_B T}{\mathcal{D}\phi(\boldsymbol{r})\,\mathcal{D}\phi(\boldsymbol{r}')} = 2\frac{\mathrm{d} f_{\text{hc}}}{\mathrm{d}\bar{\phi}_{\text{hc}}} w_{\text{hc}}(\boldsymbol{r}-\boldsymbol{r}')$$

$$+ \int \mathrm{d}\boldsymbol{r}''\, w_{\text{hc}}(\boldsymbol{r}-\boldsymbol{r}'')\phi\frac{\mathrm{d}^2 f_{\text{hc}}}{\mathrm{d}\bar{\phi}_{\text{hc}}^2} w_{\text{hc}}(\boldsymbol{r}''-\boldsymbol{r}')$$

$$+ 2\frac{\mathrm{d} f_{\text{att}}}{\mathrm{d}\bar{\phi}_{\text{att}}} w_{\text{att}}(\boldsymbol{r}-\boldsymbol{r}')$$

$$+ \int \mathrm{d}\boldsymbol{r}''\, w_{\text{att}}(\boldsymbol{r}-\boldsymbol{r}'')\phi\frac{\mathrm{d}^2 f_{\text{att}}}{\mathrm{d}\bar{\phi}_{\text{att}}^2} w_{\text{att}}(\boldsymbol{r}''-\boldsymbol{r}') \qquad (3.160)$$

This relation shows that the weighting functions have the same spatial range as the direct correlation functions. Ignoring any density dependence of the weighting function and keeping only the leading terms proportional to $\mathrm{d} f_{\text{hc}}/\mathrm{d}\bar{\phi}_{\text{hc}}$ and $\mathrm{d} f_{\text{att}}/\mathrm{d}\bar{\phi}_{\text{att}}$, we obtain

$$w_{\text{hc}}(\boldsymbol{r}) \approx -\frac{c_{\text{hc}}(\boldsymbol{r})}{2\,\mathrm{d} f_{\text{hc}}/\mathrm{d}\phi} \sim c_{\text{hc}}(\boldsymbol{r}) \qquad (3.161)$$

$$w_{\text{att}}(\boldsymbol{r}) \approx -\frac{c(\boldsymbol{r}) - c_{\text{hc}}(\boldsymbol{r})}{2\,\mathrm{d} f_{\text{att}}/\mathrm{d}\phi} \sim c(\boldsymbol{r}) - c_{\text{hc}}(\boldsymbol{r}) \qquad (3.162)$$

where c and c_{hc} denote the direct correlation functions of the Lennard-Jones fluid and a tangentially bonded, hard-sphere chain fluid, respectively.

In Fig. 3.16(b), the direct correlation functions for the repulsive system, $w_{\text{hc}} \sim c_{\text{hc}}$, and the difference, $w_{\text{att}} \sim c - c_{\text{hc}}$, are presented. The polymer reference interaction site model (P-RISM) theory has been used to calculate the direct correlation functions. The P-RISM theory generalizes to polymers the highly successful Ornstein–Zernicke equation for simple fluids. This theory by Schweizer and Curro (1997) defines the direct correlation function via

$$h(k) = S_0(k) c(k) \big(S_0(k) + \phi h(k)\big) \qquad (3.163)$$

in terms of the single-chain structure factor $S_0(k)$ and the Fourier transform of the total intermolecular pair correlation function,

$$h(\boldsymbol{k}) = \int \mathrm{d}\boldsymbol{r}\, \big[g^{\text{inter}}(\boldsymbol{r}) - 1\big] \exp[-\mathrm{i}\boldsymbol{k}\cdot\boldsymbol{r}] \qquad (3.164)$$

---

5) We do not consider a density dependence of the weighting function.

Expressing the intermolecular correlations via the structure factor $S(k)$ of density fluctuations and the single-chain structure factor $S_0(k)$,

$$S(k) = \frac{1}{nN} \left\langle \left| \sum_{\alpha=1}^{n} \sum_{i=1}^{N} \exp[i\boldsymbol{k} \cdot \boldsymbol{r}_{\alpha,i}] \right|^2 \right\rangle = S_0(k) + \phi h(k) \qquad (3.165)$$

one can rewrite the equation above in the RPA-like form (cf. Eq. 3.60):

$$\frac{1}{S(k)} = \frac{1}{S_0(k)} - \phi c(k) \qquad (3.166)$$

The theory requires, as input, the single-chain structure factor $S_0(k)$ and (similar to the Ornstein–Zernicke equation for monomeric liquids) a closure relation between the direct correlation function c and the total correlation function h. Solving these equations for the Lennard-Jones potential and for the hard-sphere chain fluid, one obtains the direct correlation functions, $c(r)$ and c_{rep}, as shown in Fig. 3.16(b). The direct correlation functions are short-range and mirror the spatial extent of the interactions. We note that, fortunately, the results of the SCF calculations are not very sensitive to the detailed form of the weighting function as long as they have the correct spatial range. A rather crude approximation, e.g. a Heaviside step function, already provides a rather good approximation (Yethiraj 1998).

3.3.3
Comparison to Computer Simulations

Profiles in the Vicinity of a Wall
The difficulty in describing liquid–vapor interfaces stems from the short-range correlations, which are reintroduced into the SCF theory via the excess free-energy functional. Those correlations are visible in the density–density pair correlation function or the experimentally accessible structure factor. They also show up in density profiles in the vicinity of a spatial inhomogeneity. In Fig. 3.17 we present the segmental density profiles close to a surface that attracts the liquid via long-range, van der Waals interactions of the form

$$\Phi_{\text{ext}}(z) = A \left[\left(\frac{\sigma}{z}\right)^9 - \left(\frac{\sigma}{z}\right)^3 \right] \qquad (3.167)$$

where z denotes the distance perpendicular to the surface. Similar to the behavior of unconnected monomers, the density vanishes at the surface and rises to the liquid-like bulk density within a distance of a few segment diameters. If the attractive strength A is weak, the density profile resembles that of

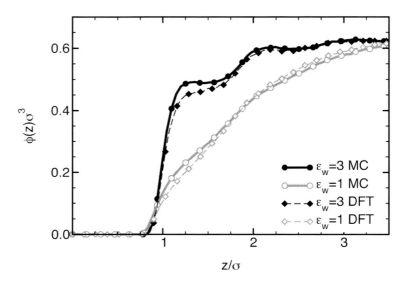

Fig. 3.17 Density profiles in the vicinity of an attractive wall with $A = 3$ and $A = 1$ at $k_\text{B}T = 1.68$. Thick solid lines with circles represent the results of MC simulations of a coarse-grained bead–spring model, while thin dashed lines with diamonds show the results of the SCF calculations. (Adapted from Müller et al. 2003b).

a liquid–vapor interface.[6] The density approaches its bulk value monotonically. If the attraction between the liquid and the surface is larger, the density profile rises more steeply and exhibits oscillations that reflect the packing structure of the liquid. The SCF calculations account for these subtle details of the profiles. Both the period and strength of the density modulation as well as their dependence on the attractive strength are correctly predicted by the SCF theory. Note that one can compare profiles between MC simulations and SCF calculations for large A, because the location where the density reaches half the value of the bulk density occurs very close to the wall, and lateral fluctuations of this position (capillary waves) are suppressed.

In Fig. 3.18 we present the orientation of the molecules on different length scales. The behavior of the end-to-end vector is shown in Fig. 3.18(a). Similar to the behavior at an interface between two immiscible polymers, the end-to-end vector orients parallel to the surface. Further away from the surface, around $z \sim R_\text{e} = 3.66\sigma$, the behavior is reversed, i.e. the perpendicular extension R_z is larger than the lateral one R_{xy}. At larger distances, the distribution of the end-to-end vector becomes isotropic. Note that the lateral extension R_{xy} increases as the center of the polymer approaches the surface. If we used the

---

6) In fact, at $A = 0$ there is a drying transition, i.e. a macroscopically thick vapor layer forms between the substrate and the liquid.

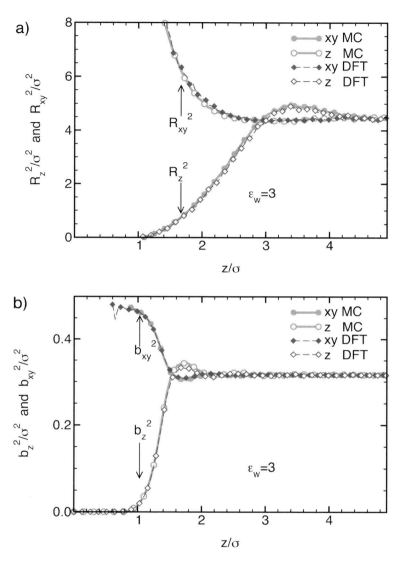

Fig. 3.18 Lateral and perpendicular components of (a) the end-to-end vector and (b) the bond vectors as a function of the distance z from the surface. Thick solid lines with circles represent the results of MC simulations of a coarse-grained bead–spring model, while thin dashed lines with diamonds show the results of the SCF calculations. (Adapted from Müller et al. 2003b).

Gaussian chain model, the lateral and perpendicular components of the end-to-end vector would decouple, and the lateral component would not depend on the distance from the surface. The SCF calculations that use the detailed molecular conformations extracted from an MC simulation of a bulk liquid capture these effects quantitatively.

In Fig. 3.18(b) we show the orientation of the bond vectors. The bond vectors also align parallel to the surface, but the effect extends much less into the liquid than the corresponding effect for the end-to-end vector. In the ultimate vicinity of the surface, the bond vectors align parallel, then the alignment is reversed; and further away, at $z > 2.5\sigma$, the distribution of bond vectors becomes isotropic. The SCF calculations agree quantitatively with the MC simulations in all the details. The good agreement for these conformational properties is chiefly due to the partial enumeration over explicit molecular conformations which incorporates the molecular architecture and intramolecular correlations on all length scales.

Surface and Interface Tension

Similar to the behavior of interfaces in polymer blends, it is difficult to compare profiles across the liquid–vapor interface because of capillary waves. The Fourier spectrum of the local interface position $u(\mathbf{k})$, however, can be utilized to extract the interface tension γ_{LV}. The inset of Fig. 3.19 shows the wavevector dependence of the interface tension

$$\gamma_{\mathrm{LV}}(k) = \frac{k_{\mathrm{B}} T L^2}{k^2 \langle u^2(k) \rangle} \tag{3.168}$$

In the long-wavelength limit, $k \to 0$, the value extracted from the Fourier spectrum agrees quantitatively with the value independently measured in grand-canonical simulations (indicated by the arrow). For larger k-values, the interface tension is reduced, i.e. short-wavelength fluctuations cost less free energy than expected from the capillary-wave Hamiltonian. Formally, this corresponds to a negative bending rigidity of the interface, which has also been predicted by Mecke and Dietrich (1999).

In the main panel of Fig. 3.19 we present the temperature dependence of the interface tension γ_{LV}. At the critical temperature, the interface tension is zero and it increases upon decreasing the temperature. The figure shows the results of grand-canonical MC simulations and two versions of the SCF theory. The first version (designated DFT) divides the excess free-energy functional ΔF_{ex} into repulsive and attractive contributions, each with their own weighting functions (cf. Section 3.3.2), while the second version (DFT(2)) employs a single free-energy density and a single weighting function (Müller and MacDowell 2000). Both versions utilize the same equation of state (i.e. yield the same bulk phase behavior) and identical single-chain conformations.

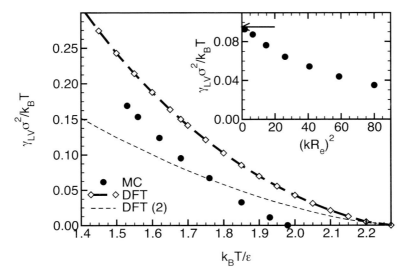

Fig. 3.19 Temperature dependence of the interface tension γ_{LV}. The symbols show the results of MC simulations. The thick dashed line with diamonds presents the result of the SCF calculations using the weighted density functional in Eq. (3.134). The thin dashed line shows the result of SCF calculations using the identical equation of state and chain conformations but a simpler ansatz for the excess free-energy density ΔF_{ex}, which does not distinguish between repulsive and attractive interactions and utilizes only a single weighting function. The inset shows the wavevector dependence of the liquid–vapor interface tension as extracted from the Fourier spectrum of interface fluctuations in an off-lattice bead–spring model. In agreement with theoretical predictions, we find a reduction of the effective interface tension with growing wavevector (negative bending rigidity of the interface). In the limit $k \to 0$, the value extracted from the Fourier spectrum agrees with the value obtained from grand-canonical simulations. (Adapted from Müller and MacDowell 2000; Müller et al. 2003b).

Since the SCF calculations overestimate the critical temperature, the interface tension vanishes at a higher temperature than in the MC simulations and the SCF calculations overestimate the interface tension in the vicinity of the critical point. At lower temperatures, the quantitative agreement between the first version (DFT) and the MC simulations improves, but the DFT calculations always yield a slightly larger value than the MC simulations. This is exactly what one expects from a mean-field calculation that accurately describes the short-range fluid-like correlations but does not incorporate long-range critical density fluctuations. The second version (DFT(2)), however, cannot quantitatively reproduce the temperature dependence of the interface tension: the tension does not increase sufficiently upon decreasing the temperature. This leads to a crossing of the DFT(2) and MC data sets at lower temperatures, $k_B T/\epsilon \approx 1.7$; and upon decreasing the temperature further, the deviation between DFT(2) calculations and MC simulations increases. This highlights the importance of dividing the excess free-energy functional into repulsive and attractive contributions and of using different approximations for each of them.

3.4
Outlook

The SCF theory is a general theoretical framework to investigate the equilibrium properties of systems containing polymers. It is the basis for quantitative, numerical calculations and the starting point for analytical approximations. Due to the molecular extension, one molecule interacts with many neighbors, and long-range fluctuations are strongly suppressed. Therefore, polymer melts are particularly well described by mean-field theories. The saddle-point approximation invoked in the mean-field treatment, however, neglects not only those long-range fluctuations, but also fluctuations on short length scales. The latter are associated with the liquid-like packing of the polymeric fluid, and they remain important independent of chain length. Two different approaches to account for those short-range correlations have been discussed. First, in binary polymer blends, the local fluid-like structure on the scale of a segment and the characteristic length of composition fluctuations R_e are well separated. Thus, the behavior is rather universal, i.e. independent of the details of the fluid-like structure, and short-range correlation effects can be accounted for by identifying the Flory–Huggins parameter via a comparison between the predictions of the SCF theory and experiments or simulations. Second, in polymer solutions, the characteristic length scale of composition fluctuations is comparable to the segment size, and the theory has to incorporate short-range, fluid-like correlations via a density functional for the excess interaction free energy. Those techniques have been developed for and successfully applied to simple fluids, and they can be integrated into the SCF calculations. If one accounts for these short-range correlations, the SCF theory is able to predict quantitatively a large variety of properties on different length scales. These include the orientations of molecules, segmental profiles at surfaces, interface and surface tensions.

We shall emphasize that SCF calculations are orders of magnitude computationally faster than computer simulations of interacting multi-chain system, and thereby allow one to obtain detailed information about complex, macromolecular systems. To illustrate the basic principle, we have restricted ourselves in this chapter to all but the most simple systems, i.e. homopolymer solutions or binary blends of homopolymers, but the theory has been successfully applied to a wide variety of more complex systems: the ordering of diblock copolymers in the bulk (Matsen and Schick 1994; Matsen and Bates 1996) and at surfaces (Matsen 1997; Geisinger et al. 1999; Wang et al. 2002);

polymer brushes (Netz and Schick 1998; Goveas et al. 1997; Matsen 2002; Matsen and Gardiner 2003; Kreer et al. 2004); wetting of polymer solutions and polymer blends (Müller and Binder 1998; Müller et al. 2000b; Müller et al. 2001; Müller et al. 2003b; Müller and MacDowell 2003); and polymer and solvent mixtures (Müller et al. 2002).

The quantitative description of the equilibrium properties is also a good starting point for developing numerical methods to study the kinetics of phase separation of multicomponent polymer systems. This is particularly worthwhile, because equilibrium often cannot be reached on length scales that exceed the molecular extensions (micrometers). In this case, the order is imperfect, and the resulting morphology in a binary polymer blend or the long-range order of microphase-separated domains in copolymer systems is important for many properties of practical interest.

3.5
Appendix

3.5.1
Fluctuations in the Spatially Homogeneous Phase

To assess the accuracy of the saddle-point approximation, one has to quantify the strength of fluctuations of the fields W_+ and W_-. We start from the exact form of the partition function (3.20) in terms of fluctuating, external fields. The saddle-point equations, (Eqs. 3.25 and 3.28), determine the most probable values of the fields, and the strength of their fluctuations can be gauged from the value of the second functional derivatives at the saddle point. Generally, this is a formidable task (Shi et al. 1996), but much insight can already be gained from considering the fluctuations around the homogeneous phase.

To make progress, we have to find an approximation for the single-chain partition functions $\mathcal{Q}_A[iW_+ + W_-]$ and $\mathcal{Q}_B[iW_+ - W_-]$. The random-phase approximation (RPA) (de Gennes 1979) for Gaussian chains provides us with an explicit expression for the partition function of a single Gaussian chain in the field W in the limit of weak fields. Utilizing the Fourier expansion

$$W(\boldsymbol{r}) = \frac{1}{(2\pi)^3} \int d\boldsymbol{k}\, W(\boldsymbol{k}) \exp[i\boldsymbol{k} \cdot \boldsymbol{r}]$$

$$W(\boldsymbol{k}) = \int d\boldsymbol{r}\, W(\boldsymbol{r}) \exp[-i\boldsymbol{k} \cdot \boldsymbol{r}] \qquad (3.169)$$

we can write

$$\frac{1}{N}\sum_{i=1}^{N} W(\boldsymbol{r}_i) = \frac{\rho_0}{N} \int d\boldsymbol{r}\, W(\boldsymbol{r})\hat{\phi}_1(\boldsymbol{r})$$

$$= \frac{\rho_0}{N(2\pi)^3} \int d\boldsymbol{k}\, W(\boldsymbol{k})\hat{\phi}_1(-\boldsymbol{k}) \qquad (3.170)$$

$$\phi_1(\boldsymbol{r}) = \frac{1}{\rho_0}\sum_{i=1}^{N} \delta(\boldsymbol{r}-\boldsymbol{r}_i)$$

and obtain for the single-chain partition function of an A polymer in the field W:

$$\mathcal{Q}_A[W] = \int \tilde{\mathcal{D}}_A \boldsymbol{R}\, \exp\left[-\frac{\rho_0}{N(2\pi)^3}\int d\boldsymbol{k}\, W(\boldsymbol{k})\hat{\phi}_1(-\boldsymbol{k})\right] \qquad (3.171)$$

Next, we expand the exponential in powers of the field W:

$$\mathcal{Q}_A[W] = \mathcal{V}\mathcal{Z}_0 \Bigg[1 - \frac{\rho_0}{N(2\pi)^3}\int d\boldsymbol{k}\, W(\boldsymbol{k})\langle\hat{\phi}_1(-\boldsymbol{k})\rangle_0 \qquad (3.172)$$

$$+ \frac{1}{2}\left(\frac{\rho_0}{N(2\pi)^3}\right)^2 \int d\boldsymbol{k}\, d\boldsymbol{k}'\, W(\boldsymbol{k})W(-\boldsymbol{k})\langle\hat{\phi}_1(-\boldsymbol{k})\hat{\phi}_1(\boldsymbol{k}')\rangle_0 + \cdots \Bigg]$$

where $\langle\cdots\rangle_0$ denotes the average over the single-chain conformations in the absence of an external field. The second term vanished because $\langle\phi_1(\boldsymbol{k})\rangle_0 = 0$ (for $k \neq 0$). Using the Gaussian chain model, we obtain

$$\langle\phi_1(\boldsymbol{k})\phi_1(\boldsymbol{k}')\rangle_0 = \frac{(2\pi)^3}{\mathcal{V}}\left(\frac{N}{\rho_0}\right)^2 g(k)\delta(\boldsymbol{k}-\boldsymbol{k}')$$

where g denotes the Debye function (cf. Eq. 3.62). Inserting this expression into Eq. (3.172), we arrive at the random-phase approximation for the single-chain partition function:

$$\mathcal{Q}[W] = \mathcal{V}\mathcal{Z}_0 \exp\left[\frac{1}{2(2\pi)^3 \mathcal{V}}\int d\boldsymbol{k}\, g(k)W(\boldsymbol{k})W(-\boldsymbol{k})\right] + \mathcal{O}(W^4) \qquad (3.173)$$

Using this result of the RPA in Eq. (3.20), we obtain up to second order in the fields:

$$\frac{\mathcal{F}_{\text{RPA}}[W_+, W_-]}{k_B T (\mathcal{V}/R_e^3)\sqrt{\mathcal{N}}} = \bar{\phi}_A \ln\left(\frac{\bar{\phi}_A \rho_0}{eN\mathcal{Z}_0}\right) + \bar{\phi}_B \ln\left(\frac{\bar{\phi}_B \rho_0}{eN\mathcal{Z}_0}\right)$$

$$- \frac{\bar{\phi}_A}{2(2\pi)^3\mathcal{V}} \int d\boldsymbol{k}\, g(k)\, [iW_+(\boldsymbol{k}) + W_-(\boldsymbol{k})]^2$$

$$- \frac{\bar{\phi}_B}{2(2\pi)^3\mathcal{V}} \int d\boldsymbol{k}\, g(k)\, [iW_+(\boldsymbol{k}) - W_-(\boldsymbol{k})]^2$$

$$+ \frac{1}{2(2\pi)^3 N\mathcal{V}} \int d\boldsymbol{k}\, V_+^{-1}(\boldsymbol{k}) W_+^2(\boldsymbol{k})$$

$$+ \frac{1}{2(2\pi)^3\mathcal{V}} \int d\boldsymbol{k}\, V_-^{-1}(\boldsymbol{k}) W_-^2(\boldsymbol{k}) \qquad (3.174)$$

The last terms are bilinear in the Fourier components of W_+ and W_-, and can be rewritten in the form:

$$\frac{\mathcal{F}_{\text{RPA}}[W_+, W_-]}{k_B T (\mathcal{V}/R_e^3)\sqrt{\mathcal{N}}} = \bar{\phi}_A \ln\left(\frac{\bar{\phi}_A \rho_0}{eN\mathcal{Z}_0}\right) + \bar{\phi}_B \ln\left(\frac{\bar{\phi}_B \rho_0}{eN\mathcal{Z}_0}\right)$$

$$+ \frac{1}{2(2\pi)^3\mathcal{V}} \int d\boldsymbol{k} \left[\frac{[W_+(\boldsymbol{k}) - i\alpha(\boldsymbol{k})W_-(\boldsymbol{k})]^2}{\Delta_+(\boldsymbol{k})} + \frac{[W_-(\boldsymbol{k})]^2}{\Delta_-(\boldsymbol{k})}\right]$$

$$(3.175)$$

Using the identity $V_+^{-1}(\boldsymbol{k}) = 1/V_+(\boldsymbol{k})$ (cf. Eq. 3.16), we obtain for the coefficients:

$$\frac{1}{\Delta_+(\boldsymbol{k})} = \frac{1}{NV_+(\boldsymbol{k})} + g(k) \qquad (3.176)$$

$$\frac{1}{\Delta_-(\boldsymbol{k})} = \frac{1}{V_-(\boldsymbol{k})} - g(k) + \frac{(\bar{\phi}_A - \bar{\phi}_B)^2 g(k)}{1 + 1/[NV_+(\boldsymbol{k})g(k)]} \qquad (3.177)$$

$$\alpha(\boldsymbol{k}) = \frac{(\bar{\phi}_A - \bar{\phi}_B)}{1 + 1/[NV_+(\boldsymbol{k})g(k)]} \qquad (3.178)$$

Here Δ_+ and Δ_- characterize the mean square fluctuations of the fields W_+ and W_-, respectively:

$$\frac{1}{\mathcal{V}^2} \left\langle [W_+(\boldsymbol{k}) - i\alpha(k)W_-(\boldsymbol{k})]^2 \right\rangle = \frac{\Delta_+}{(\mathcal{V}/R_e^3)\sqrt{\mathcal{N}}}$$

$$\frac{1}{\mathcal{V}^2} \left\langle W_-^2(\boldsymbol{k}) \right\rangle = \frac{\Delta_-}{(\mathcal{V}/R_e^3)\sqrt{\mathcal{N}}} \qquad (3.179)$$

The behavior of the fluctuations of the fields W_+ and W_- are very different. First, we note that Δ_+ is always positive, i.e. fluctuations of the field W_+ do not diverge. For wavevectors that characterize the molecular extension, $kR_e \sim \mathcal{O}(1)$, the dominant contribution arises from the second term in Eq. (3.176) and Δ_+ is of order unity. Due to the factor $\sqrt{\mathcal{N}}$ in Eq. (3.179), the long-wavelength fluctuations of $W_+(k)$ are strongly suppressed for large invariant polymerization index $\mathcal{N}$. Second, for short wavelength, $ka \sim \mathcal{O}(1)$, the Debye function is small, $g(k) = 12/(kR_e)^2 \sim 1/N$, and both terms in Eq. (3.176) become comparable. To make this behavior more explicit, we use the simple local version of the potential, $V_+(\boldsymbol{r} - \boldsymbol{r}') = (\rho_0/k_B T \kappa)\delta(\boldsymbol{r} - \boldsymbol{r}')$ or $V_+(k) = \rho_0/k_B T \kappa$, and obtain for $kR_e \gg 1$:

$$\frac{1}{\Delta_+} = \frac{1}{N}\left(\frac{k_B T \kappa}{\rho_0} + \frac{12}{(ka)^2}\right) \tag{3.180}$$

i.e. fluctuations of the field W_+ on the segmental scale are not small and the fluctuations of the field per segment, W_+/N, are independent of chain length. Thus, the mean-field approximation for W_+ is accurate on long length scales, but it fails on short length scales. Since, within mean-field approximation, one cannot accurately describe short-range fluctuations of W_+, one often utilizes an incompressibility constraint or a local potential V_+, which is technically more convenient. The failure of the saddle-point approximation for W_+ for $ka \sim 1$ parallels the neglect of local density fluctuations (packing effects) in the SCF theory.

In contrast to Δ_+, the coefficient Δ_- can change its sign. For long-wavelength fluctuations, we can neglect the terms proportional to V_+ and replace V_- by a local interaction, $V_-(k) = \chi N/2$. Then, we obtain for $qR_e \sim \mathcal{O}(1)$:

$$\frac{1}{\Delta_-} = \frac{2}{\chi N} - g(k)[1 - (\bar{\phi}_A - \bar{\phi}_B)^2]$$

$$= \frac{2\bar{\phi}_A \bar{\phi}_B g(k)}{\chi N}\left[\frac{1}{\bar{\phi}_A g(k)} + \frac{1}{\bar{\phi}_B g(k)} - 2\chi N\right] \tag{3.181}$$

The term in the last bracket can become zero. This corresponds to a divergence of the fluctuations of the field W_-. The instability occurs first at $k = 0$. It signals the instability of the homogeneous state against long-wavelength concentration functions – the spinodal. In the vicinity of the spinodal and, a fortiori, in the vicinity of the critical point, the saddle-point approximation for the field W_- breaks down at long wavelengths. Note, however, that the breakdown of the mean-field approximation due to the divergence of long-wavelength fluctuations of W_- is very different from the breakdown of the saddle-point approximation for W_+ (and also W_-) at short length scales. The strength of the long-wavelength fluctuations is still controlled by $\mathcal{N}$, i.e. at a given state point, specified by the average composition $\bar{\phi}_A$ and incompatibility χN, the long-wavelength fluctuations are only of the order $\langle W_-^2 \rangle \sim 1/\sqrt{\mathcal{N}}$ and decrease upon increasing the invariant polymerization index. The concomitant failure of the SCF theory at the spinodal is described by the Ginzburg criterion (see Eq. 3.67 and the discussion by Wang (2002)). The larger the invariant polymerization index $\mathcal{N}$, the closer one can approach the spinodal before the SCF theory fails in its ultimate vicinity due to long-range composition fluctuations. These considerations motivate the division of fluctuation effects into short-range $[kR_e \gg ka \sim \mathcal{O}(1)]$ and long-range $[kR_e \sim \mathcal{O}(1)]$. The saddle-point approximation is inaccurate at short length scales; on the length scale of a monomeric unit, fluctuations are important and independent of the invariant polymerization index $\mathcal{N}$. The saddle-point approximation is accurate on large length scales, except for the ultimate vicinity of the spinodal.

Using Eq. (3.36) and the long-wavelength expression for $V_-^{-1}(\boldsymbol{r}) = (2/\chi N)\delta(\boldsymbol{r})$, we write:

$$\frac{S(\boldsymbol{k})}{N} = \frac{\rho_0}{4N\mathcal{V}} \int d\boldsymbol{r}\, d\boldsymbol{r}' \, \langle \phi(\boldsymbol{r})\phi(\boldsymbol{r}') \rangle \exp[i\boldsymbol{k} \cdot (\boldsymbol{r} - \boldsymbol{r}')]$$

$$= \frac{\rho_0}{N\mathcal{V}} \frac{1}{(\chi N)^2} \int d\boldsymbol{r}\, d\boldsymbol{r}' \, \langle W_-(\boldsymbol{r})W_-(\boldsymbol{r}') \rangle \exp[i\boldsymbol{k} \cdot (\boldsymbol{r} - \boldsymbol{r}')] - \frac{1}{2\chi}$$

$$= \frac{\rho_0}{N\mathcal{V}} \frac{1}{(\chi N)^2} \langle W_-(\boldsymbol{k})W_-(-\boldsymbol{k}) \rangle - \frac{1}{2\chi}$$

$$= \frac{\Delta_-}{(\chi N)^2} - \frac{1}{2\chi N} \tag{3.182}$$

The RPA for the collective structure factor of composition fluctuations is then obtained from Eq. (3.181):

$$\frac{N}{S(\boldsymbol{k})} = \frac{1}{\bar{\phi}_A g(k)} + \frac{1}{\bar{\phi}_B g(k)} - 2\chi N \tag{3.183}$$

3.5.2
Alternative Derivation of the SCF Equations for Liquid–Vapor Systems

Similar to Section 3.3.1 we start from the grand-canonical partition function of chains with chemical potential μ, in a volume $\mathcal{V}$:

$$\mathcal{Z} \propto \sum_{n=0}^{\infty} \frac{e^{\mu n/k_B T}}{n!} \int \prod_{\alpha=1}^{n} \tilde{\mathcal{D}} \boldsymbol{R}_\alpha \exp\left[-\frac{\Delta F_{\text{ex}}[\phi]}{k_B T}\right] \tag{3.184}$$

Here ΔF_{ex} denotes the excess interaction free-energy functional, which is chosen such that it provides an accurate description of the local structure *after* the saddle-point approximation is performed. It depends on the segmental number density $\phi(\boldsymbol{r})$. In the above, $\hat{\mathcal{D}}\boldsymbol{R}_\alpha$ sums over all conformations α of the polymer molecule with statistical weight according to all interactions that are not intermolecular, i.e. intramolecular interactions U_{intra} (e.g. bonding) and external fields Φ_{ext} (e.g. at surfaces). We define the microscopic molecular density, which describes the joint distribution of all N segments $\boldsymbol{R} = (\boldsymbol{r}_1, \ldots, \boldsymbol{r}_N)$ of a single molecule, via

$$\hat{\pi}(\boldsymbol{R}) = \hat{\pi}(\boldsymbol{r}_1, \ldots, \boldsymbol{r}_N) = \sum_{\alpha=1}^{n} \prod_{i=1}^{N} \delta(\boldsymbol{r}_i - \boldsymbol{r}_{\alpha,i}) \qquad (3.185)$$

The segmental density can be expressed in terms of the molecular density as

$$\phi(\boldsymbol{r}) = \int d\boldsymbol{R} \sum_{i=1}^{N} \delta(\boldsymbol{r} - \boldsymbol{r}_i) \hat{\pi}(\boldsymbol{R}) = \sum_{\alpha=1}^{n} \sum_{i=1}^{N} \delta(\boldsymbol{r} - \boldsymbol{r}_{\alpha,i}) \qquad (3.186)$$

To rewrite the partition function of interacting chains in terms of independent chains in an external, but fluctuating, field, we introduce the density distribution $\tilde{\mathcal{P}}(\boldsymbol{R})$ of a single molecule, and the conjugated field $\tilde{U}(\boldsymbol{R})$. Inserting the Fourier representation of the δ function (Hong and Noolandi 1981),

$$\int \mathcal{D}\tilde{\mathcal{P}} \, \delta(\tilde{\mathcal{P}}(\boldsymbol{R}) - \hat{\pi}(\boldsymbol{R})) \sim \int \mathcal{D}\tilde{\mathcal{P}} \, \mathcal{D}\tilde{U} \, \exp\left[\int d\boldsymbol{R} \, \tilde{U}(\boldsymbol{R})(\tilde{\mathcal{P}}(\boldsymbol{R}) - \hat{\pi}(\boldsymbol{R}))\right]$$
$$\sim 1 \qquad (3.187)$$

into Eq. (3.184), we obtain

$$\mathcal{Z} \propto \sum_{n=0}^{\infty} \frac{e^{\mu n / k_B T}}{n!} \int \mathcal{D}\tilde{\mathcal{P}} \, \mathcal{D}\tilde{U} \prod_{\alpha=1}^{n} \tilde{\mathcal{D}}\boldsymbol{R}_\alpha \, e^{-\Delta F_{\text{ex}}[\phi]/k_B T} \, e^{\int d\boldsymbol{R}\, \tilde{U}(\boldsymbol{R})(\tilde{\mathcal{P}} - \hat{\pi})}$$

$$\propto \sum_{n=0}^{\infty} \frac{e^{\mu n / k_B T}}{n!} \int \mathcal{D}\tilde{\mathcal{P}} \, \mathcal{D}\tilde{U} \, e^{-\Delta F_{\text{ex}}[\phi]/k_B T + \int d\boldsymbol{R}\, \tilde{U}\tilde{\mathcal{P}}}$$

$$\times \int \prod_{\alpha=1}^{n} \tilde{\mathcal{D}}\boldsymbol{R}_\alpha \, e^{-\int d\boldsymbol{R}\, \tilde{U}\hat{\pi}} \qquad (3.188)$$

$$\propto \int \mathcal{D}\tilde{\mathcal{P}} \, \mathcal{D}\tilde{U} \, e^{-\Delta F_{\text{ex}}[\phi]/k_B T + \int d\boldsymbol{R}\, \tilde{U}\tilde{\mathcal{P}}} \sum_{n=0}^{\infty} \frac{(e^{\mu/k_B T})^n}{n!} \left(\mathcal{Q}[\tilde{U}]\right)^n$$

$$\propto \int \mathcal{D}\tilde{\mathcal{P}} \, \mathcal{D}\tilde{U} \, e^{-\Delta F_{\text{ex}}[\phi]/k_{\text{B}}T + \int d\boldsymbol{r} \, \tilde{U}\tilde{\mathcal{P}} + \exp(\mu/k_{\text{B}}T)\mathcal{Q}[\tilde{U}]}$$

$$\equiv \int \mathcal{D}\tilde{\mathcal{P}} \, \mathcal{D}\tilde{U} \, e^{-\mathcal{G}[\tilde{U},\tilde{\mathcal{P}}]/k_{\text{B}}T} \tag{3.189}$$

where the partition function of a single polymer in the external field $U(\boldsymbol{R})$ is defined by

$$\mathcal{Q}[\tilde{U}] = \int \tilde{\mathcal{D}}\boldsymbol{R} \, \exp[-\tilde{U}(\boldsymbol{R})]$$

$$= \int d\boldsymbol{R} \, e^{-(U_{\text{intra}} + \Phi_{\text{ext}})/k_{\text{B}}T} \exp[-\tilde{U}(\boldsymbol{R})] \tag{3.190}$$

and the grand free-energy functional, $\mathcal{G}[\tilde{U}, \tilde{\mathcal{P}}]$, takes the form

$$\frac{\mathcal{G}[\tilde{U}, \tilde{\mathcal{P}}]}{k_{\text{B}}T} = \frac{\Delta F_{\text{ex}}[\phi]}{k_{\text{B}}T} - \int d\boldsymbol{R} \, \tilde{U}\tilde{\mathcal{P}} - e^{\mu/k_{\text{B}}T} \mathcal{Q}[\tilde{U}] \tag{3.191}$$

The functional integral over the field $\tilde{U}$ and the density distribution $\tilde{\mathcal{P}}$ in Eq. (3.189) cannot be evaluated exactly, and thus we approximate it by its saddle-point value. To this end, we extremize the grand free-energy functional, $\mathcal{G}[\tilde{U}, \tilde{\mathcal{P}}]$, with respect to the fields $\tilde{U}$ and $\tilde{\mathcal{P}}$. The saddle-point values of the fields are denoted by lower-case letters, $\tilde{u}$ and $\tilde{p}$. In the first step, we perform the saddle-point approximation for $\tilde{U}$:

$$\frac{\mathcal{D}\mathcal{G}[\tilde{U}, \tilde{\mathcal{P}}]}{\mathcal{D}\tilde{U}(\boldsymbol{R})} = 0 \implies$$

$$\tilde{\mathcal{P}}(\boldsymbol{R}) = -e^{\mu/k_{\text{B}}T} \frac{\mathcal{D}\mathcal{Q}[\tilde{u}]}{\mathcal{D}\tilde{u}(\boldsymbol{R})}$$

$$= e^{\mu/k_{\text{B}}T} \, e^{-(U_{\text{intra}} + \Phi_{\text{ext}})/k_{\text{B}}T} \exp[-\tilde{u}(\boldsymbol{R})] \tag{3.192}$$

and this yields for the saddle-point value of the field $\tilde{U}$:

$$\tilde{u}(\boldsymbol{R}) = \frac{\mu - U_{\text{intra}}(\boldsymbol{R}) - \Phi_{\text{ext}}(\boldsymbol{R})}{k_{\text{B}}T} - \ln \tilde{\mathcal{P}}(\boldsymbol{R}) \tag{3.193}$$

Inserting this result into the expression for the single-chain partition function (Eq. 3.190), we obtain

$$\mathcal{Q}[\tilde{u}] = e^{-\mu/k_{\text{B}}T} \int d\boldsymbol{R} \, \tilde{\mathcal{P}}(\boldsymbol{R}) \tag{3.194}$$

and the grand-canonical free-energy functional, $\Omega[\tilde{\mathcal{P}}] = \mathcal{G}[\tilde{u}[\mathcal{P}], \tilde{\mathcal{P}}]$, is given by

$$\Omega[\tilde{\mathcal{P}}] = k_B T \int d\boldsymbol{R}\, \tilde{\mathcal{P}}(\boldsymbol{R})[\ln \tilde{\mathcal{P}}(\boldsymbol{R}) - 1] + \Delta F_{\text{ex}}[\phi]$$

$$+ \int d\boldsymbol{R}\, \tilde{\mathcal{P}}(\boldsymbol{R})[\tilde{U}_{\text{intra}} + \Phi_{\text{ext}} - \mu] \qquad (3.195)$$

Equation (3.195) is the starting point of density-functional calculations for polymeric systems. Density-functional theory asserts that there exists an excess interaction free-energy functional, $\Delta F_{\text{ex}}[\phi]$, such the extremal value of Eq. (3.195) describes the exact density profile. Thus, although we have used a mean-field approximation to arrive at Eq. (3.195), the general structure of the equation is valid beyond the mean-field approximation. However, $\Delta F_{\text{ex}}[\phi]$ does not describe the bare interactions (as it would in a literal mean-field approximation) but an effective interaction that accounts for local correlations.

In the second step, we determine the extremum of this grand-canonical free-energy functional, $\Omega[\tilde{\mathcal{P}}]$, with respect to $\tilde{\mathcal{P}}(\boldsymbol{R})$.

Using

$$\frac{\mathcal{D}\Delta F_{\text{ex}}[\phi]}{\mathcal{D}\tilde{\mathcal{P}}(\boldsymbol{R})} = \int d\boldsymbol{r}\, \frac{\mathcal{D}\Delta F_{\text{ex}}[\phi]}{\mathcal{D}\phi(\boldsymbol{r})} \frac{\mathcal{D}\phi(\boldsymbol{r})}{\mathcal{D}\tilde{\mathcal{P}}(\boldsymbol{R})}$$

$$= \int d\boldsymbol{r}\, \frac{\mathcal{D}\Delta F_{\text{ex}}[\phi]}{\mathcal{D}\phi(\boldsymbol{r})} \sum_{i=1}^{N} \delta(\boldsymbol{r} - \boldsymbol{r}_i) = \sum_{i=1}^{N} \frac{\mathcal{D}\Delta F_{\text{ex}}[\phi]}{\mathcal{D}\phi(\boldsymbol{r}_i)} \qquad (3.196)$$

we obtain for the molecular density $\tilde{p}(\boldsymbol{R})$, which makes the grand-canonical free-energy functional (3.195) an extremum:

$$\frac{\mathcal{D}\Omega}{\mathcal{D}\tilde{\mathcal{P}}(\boldsymbol{R})} = 0 \implies$$

$$\tilde{p}(\boldsymbol{R}) \propto e^{\mu/k_B T} \exp\left[-\frac{U_{\text{intra}}(\boldsymbol{R}) + \Phi_{\text{ext}}(\boldsymbol{R})}{k_B T} - \frac{1}{k_B T}\sum_{i=1}^{N} \frac{\mathcal{D}\Delta F_{\text{ex}}[\phi]}{\mathcal{D}\phi(\boldsymbol{r}_i)}\right]$$

$$(3.197)$$

Inserting this expression for the saddle-point value of $\tilde{\mathcal{P}}$ into Eq. (3.193), we can simplify the expression for the external field $u(\boldsymbol{R})$ to

$$\tilde{u}(\boldsymbol{R}) = \frac{1}{k_B T} \sum_{i=1}^{N} \frac{\mathcal{D}\Delta F_{\text{ex}}[\phi]}{\mathcal{D}\phi(\boldsymbol{r}_i)} = \sum_{i=1}^{N} u(\boldsymbol{r}_i) \qquad (3.198)$$

i.e. the external field that mimics the interaction with the surrounding molecules is the sum of the interactions of the segments in the external field $u(\boldsymbol{r}) = (1/k_\mathrm{B}T)[\mathcal{D}\Delta F_\mathrm{ex}/\mathcal{D}\phi(\boldsymbol{r})]$ (cf. Eq. 3.127).

Using

$$\int \mathrm{d}\boldsymbol{R}\, \tilde{p}(\boldsymbol{R})\tilde{u}(\boldsymbol{R}) = \int \mathrm{d}\boldsymbol{R}\, \tilde{p}(\boldsymbol{R}) \int \mathrm{d}\boldsymbol{r} \sum_{i=1}^{N} \delta(\boldsymbol{r}-\boldsymbol{r}_i) u(\boldsymbol{r})$$

$$= \int \mathrm{d}\boldsymbol{r}\, u(\boldsymbol{r}) \int \mathrm{d}\boldsymbol{R} \sum_{i=1}^{N} \delta(\boldsymbol{r}-\boldsymbol{r}_i) p(\boldsymbol{R}) \quad (3.199)$$

$$= \int \mathrm{d}\boldsymbol{r}\, u(\boldsymbol{r}) \phi(\boldsymbol{r})$$

and

$$\mathcal{Q}[u] = \mathrm{e}^{-\mu/k_\mathrm{B}T} \langle n \rangle \quad (3.200)$$

(cf. Eq. 3.194), we obtain for the grand-canonical free energy:

$$\frac{\Omega^*}{k_\mathrm{B}T} \equiv \frac{\Omega[\tilde{p}]}{k_\mathrm{B}T} \equiv \frac{\mathcal{G}[\tilde{u},\tilde{p}]}{k_\mathrm{B}T} = -\langle n \rangle + \frac{\Delta F_\mathrm{ex}[\phi]}{k_\mathrm{B}T} - \int \mathrm{d}\boldsymbol{r}\, \phi(\boldsymbol{r}) u(\boldsymbol{r}) \quad (3.201)$$

Within the mean-field approximation, the saddle-point value $\tilde{p}(\boldsymbol{R})$ provides an estimate for the thermal average $\langle \hat{\pi}(\boldsymbol{R}) \rangle$ of the molecular density. To show this, we go back to the partition function (3.188):

$$\langle \hat{\pi}(\boldsymbol{R}') \rangle = \frac{1}{\mathcal{Z}} \sum_{n=0}^{\infty} \frac{\mathrm{e}^{\mu n/k_\mathrm{B}T}}{n!} \int \mathcal{D}\tilde{p}\, \mathcal{D}\tilde{U}\, \mathrm{e}^{-\Delta F_\mathrm{ex}[\phi]/k_\mathrm{B}T + \int \mathrm{d}\boldsymbol{R}\, \tilde{U}\tilde{p}}$$

$$\times \int \prod_{\alpha=1}^{n} \tilde{\mathcal{D}}\boldsymbol{R}_\alpha\, \hat{\pi}(\boldsymbol{R}')\, \mathrm{e}^{-\int \mathrm{d}\boldsymbol{R}\, \tilde{U}\hat{\pi}}$$

$$= \frac{1}{\mathcal{Z}} \sum_{n=0}^{\infty} \frac{\mathrm{e}^{\mu n/k_\mathrm{B}T}}{n!} \int \mathcal{D}\tilde{p}\, \mathcal{D}\tilde{U}\, \mathrm{e}^{-\Delta F_\mathrm{ex}[\phi]/k_\mathrm{B}T + \int \mathrm{d}\boldsymbol{R}\, \tilde{U}\tilde{p}}$$

$$\times \left[-\frac{\mathcal{D}}{\mathcal{D}\tilde{U}(\boldsymbol{R}')} \int \prod_{\alpha=1}^{n} \tilde{\mathcal{D}}\boldsymbol{R}_\alpha\, \mathrm{e}^{-\int \mathrm{d}\boldsymbol{R}\, \tilde{U}\hat{\pi}} \right]$$

$$= \frac{1}{\mathcal{Z}} \sum_{n=0}^{\infty} \frac{\mathrm{e}^{\mu n/k_\mathrm{B}T}}{n!} \int \mathcal{D}\tilde{p}\, \mathcal{D}\tilde{U} \left[\frac{\mathcal{D}}{\mathcal{D}\tilde{U}(\boldsymbol{R}')}\, \mathrm{e}^{-\Delta F_\mathrm{ex}[\phi]/k_\mathrm{B}T + \int \mathrm{d}\boldsymbol{R}\, \tilde{U}\tilde{p}} \right]$$

$$\times \int \prod_{\alpha=1}^{n} \tilde{\mathcal{D}} \boldsymbol{R}_\alpha \, e^{-\int \mathrm{d}\boldsymbol{R}\,\tilde{U}\hat{\pi}}$$

$$= \frac{1}{\mathcal{Z}} \sum_{n=0}^{\infty} \frac{e^{\mu n/k_B T}}{n!} \int \mathcal{D}\tilde{\mathcal{P}}\,\mathcal{D}\tilde{U}\,\tilde{\mathcal{P}}(\boldsymbol{R}')\,e^{-\Delta F_{\mathrm{ex}}[\phi]/k_B T + \int \mathrm{d}\boldsymbol{R}\,\tilde{U}\tilde{\mathcal{P}}} \left(\mathcal{Q}[\tilde{U}]\right)^n$$

Performing the saddle-point integration, we simplify this expression to $\langle \hat{\pi}(\boldsymbol{R}') \rangle = \tilde{p}(\boldsymbol{R}')$. Integrating this expression over the coordinates of all segments, one obtains

$$\int \mathrm{d}\boldsymbol{R}\,\tilde{p}(\boldsymbol{R}) = \left\langle \int \mathrm{d}\boldsymbol{R}\,\hat{\pi}(\boldsymbol{R}) \right\rangle = \langle n \rangle$$

From Eq. (3.200) and the expression for the grand-canonical free energy, we calculate the Helmholtz free energy F:

$$\frac{F}{k_B T} = \frac{\Omega^*}{k_B T} + \frac{\mu \langle n \rangle}{k_B T}$$

$$= \langle n \rangle \ln\left(\frac{\langle n \rangle}{e\mathcal{Q}}\right) + \frac{\Delta F_{\mathrm{ex}}[\phi]}{k_B T} - \int \mathrm{d}\boldsymbol{r}\,\phi(\boldsymbol{r})u(\boldsymbol{r}) \qquad (3.202)$$

with

$$\mathcal{Q} = \int \tilde{\mathcal{D}}\boldsymbol{R}\,\exp\left[-\sum_{i=1}^{N} u(\boldsymbol{r}_i)\right] \qquad (3.203)$$

References

Abrams, C. F., Site, L. D., and Kremer, K., 2003, *Phys. Rev. E* **67**, 021807.
Aksimentiev, A. and Holyst, R., 1998, *Macromol. Theory Simul.* **7**, 447.
Barker, A. and Henderson, D., 1967, *J. Chem. Phys.* **47**, 4714.
Baschnagel, J. and Binder, K., 1995, *Macromolecules* **28**, 6808.
Baschnagel, J., Binder, K., Doruker, P., Gusev, A. A., Hahn, O., Kremer, K., Mattice, W. L., Müller-Plathe, F., Murat, M., Paul, W., Santos, S., Suter, U. W., and Tries, V., 2000, *Adv. Chem. Phys.* **152**, 41.
Bates, F. S., Wignall, G. D., and Koehler, W. C., 1985, *Phys. Rev. Lett.* **55**, 2425.
Bates, F. S., Rosedale, J. H., Stepanek, P., Lodge, T. P., Wiltzius, P., Fredrickson, G. H., and Helm, P. P., 1990, *Phys. Rev. Lett.* **65**, 1893.
Bates, F. S., Schultz, M., Rosedale, J., and Almdal, K., 1992, *Macromolecules* **25**, 5547.

Bedeaux, D. and Weeks, J. D., 1985, *J. Chem. Phys.* **82**, 972.
Bennemann, C., Paul, W., Binder, K., and Dünweg, B., 1998, *Phys. Rev. E* **57**, 843.
Binder, K., 1984, *Phys. Rev. A* **29**, 341.
Binder, K., 1995, *Monte Carlo and Molecular Dynamics Simulations in Polymer Science.* Oxford University Press, New York.
Binder, K., Baschnagel, J., and Paul, W., 2003, *Prog. Polym. Sci.* **28**, 115.
Bitsanis, I. and Hadziiouanou, G., 1990, *J. Chem. Phys.* **92**, 3824.
Buff, F. P., Lovett, R. A., and Stillinger, F. H., 1965, *Phys. Rev. Lett.* **15**, 621.
Buta, D. and Freed, K. F., 2002, *J. Chem. Phys.* **116**, 10959.
Cahn, R. W., Haasen, P., and Kramer, E. J., 1993, *Materials Science and Technology, A Comprehensive Treatment.* VCH Verlag, Weinheim.
Carmesin, I. and Kremer, K., 1988, *Macromolecules* **21**, 2819.
Carnahan, N. F. and Starling, K. E., 1969, *J. Chem. Phys.* **51**, 635.
Cavallo, A., Müller, M., and Binder, K., 2003, *Europhys. Lett.* **61**, 214.
Chandler, D., McCoy, J. D., and Singer, S. J., 1986, *J. Chem. Phys.* **85**, 5971.
Chapman, W. G., Jackson, G., and Gubbins, K. S., 1988, *Mol. Phys.* **65**, 1057.
Chung, G. C., Kornfield, J. A., and Smith, S. D., 1994, *Macromolecules* **27**, 964.
Creton, C., Kramer, E. J., and Hadziioannou, G., 1991, *Macromolecules* **24**, 1846.
Crist, B., 1998, *Macromolecules* **31**, 5853.
Curtin, A. and Ashcroft, N. W., 1985, *Phys. Rev. A* **32**, 2909.
de Gennes, P.-G., 1972, *Phys. Lett.* **38A**, 339.
de Gennes, P.-G., 1977, *J. Phys. (Paris) Lett.* **38**, L441.
de Gennes, P.-G., 1979, *Scaling Concepts in Polymer Physics.* Cornell University Press, Ithaca, NY.
des Cloizeaux, J. and Jannink, G., 1990, *Polymers in Solution: Their Modeling and Structure.* Oxford Science Publications, Oxford.
Deutsch, H.-P. and Binder, K., 1992a, *Macromolecules* **25**, 6214.
Deutsch, H.-P. and Binder, K., 1992b, *Europhys. Lett.* **18**, 667.
Doi, M. and Edwards, S. F., 1986, *The Theory of Polymer Dynamics.* Clarendon Press, Oxford.
Düchs, D., Ganesan, V., Fredrickson, G. H., and Schmid, F., 2003, *Macromolecules* **36**, 9237.
Dudowicz, J., Freed, K. F., and Douglas, J. F., 2002, *J. Chem. Phys.* **116**, 9983.
Duplantier, B., 1987, *J. Chem. Phys.* **86**, 4233.
Escobedo, F. A. and dePablo, J. J., 1996, *Mol. Phys.* **87**, 347.
Escobedo, F. A. and dePablo, J. J., 1999, *Macromolecules* **32**, 900.
Evans, R., 1982, Density functionals in the theory of non-uniform fluids, in *Fundamentals of Inhomogeneous Fluids*, ed. D. Henderson. Marcel Dekker, New York.

Flory, P. J., 1941, *J. Chem. Phys.* **9**, 660.
Flory, P. J., 1969, *Statistical Mechanics of Chain Molecules*. Wiley-Interscience, New York.
Frauenkron, H. and Grassberger, P., 1997, *J. Chem. Phys.* **107**, 9599.
Freed, K. F., 1987, *Renormalization Group Theory of Macromolecules*. Wiley-Interscience, New York.
Freed, K. F. and Dudowicz, J., 1998, *Macromolecules* **31**, 6681.
Frisch, H. L. and Lebowitz, J. L., 1965, *Equilibrium Theory of Classical Fluids*. Benjamin, New York.
Frischknecht, A. L. and Curro, J. G., 2004, *J. Chem. Phys.* **121**, 2788.
Frischknecht, A. L., Weinhold, J. D., Salinger, A. G., Curro, J. G., Frink, L. J. D., and McCoy, J. D., 2002, *J. Chem. Phys.* **117**, 10385.
Garbassi, F., Morra, M., and Occhiello, E., 2000, *Polymer Surface: From Physics to Technology*. John Wiley, Chichester.
Gehlsen, M. D. and Bates, F. S., 1994, *Macromolecules* **27**, 3611.
Gehlsen, M. D., Rosedale, J. H., Bates, F. S., Wignall, G. D., and Almdal, K., 1992, *Phys. Rev. Lett.* **68**, 2452.
Geisinger, T., Müller, M., and Binder, K., 1999, *J. Chem. Phys.* **111**, 5241.
Ginzburg, V. L., 1960, *Sov. Phys. Solid State* **1**, 1824.
Goveas, J. L., Milner, S. T., and Russel, W. B., 1997, *Macromolecules* **30**, 5541.
Grayce, C. J., Yethiraj, A., and Schweizer, K. S., 1994, *J. Chem. Phys.* **100**, 6857.
Grest, G. S., Lacasse, M.-D., Kremer, K., and Gupta, A. M., 1996, *J. Chem. Phys.* **105**, 10583.
Hansen, J.-P. and McDonald, I. R., 1986, *Theory of Simple Liquids*. Academic Press, San Diego, CA.
Heine, D., Wu, D., Curro, J., and Grest, G., 2003, *J. Chem. Phys.* **118**, 914.
Helfand, E., 1975, *J. Chem. Phys.* **62**, 999.
Helfand, E. and Sapse, A. M., 1975, *J. Chem. Phys.* **62**, 1329.
Helfand, E. and Tagami, Y., 1972, *J. Chem. Phys.* **56**, 3592.
Helfrich, W., 1973, *Z. Naturforsch.* **28c**, 693.
Hong, K. M. and Noolandi, J., 1981, *Macromolecules* **14**, 727.
Huggins, M. L., 1941, *J. Chem. Phys.* **9**, 440.
Joanny, J. F., 1978, *J. Phys. A* **11**, L117.
Johnson, J. K., Zollweg, J. A., and Gubbins, K. E., 1993, *Mol. Phys.* **3**, 591.
Kamath, S., Colby, R. H., and Kumar, S. K., 2003, *Macromolecules* **36**, 8567.
Katsov, K. and Weeks, J. D., 2001, *J. Phys. Chem. B* **10**, 6738.
Kerle, T., Klein, J., and Binder, K., 1996, *Phys. Rev. Lett.* **77**, 1318.
Kratky, O. and Porod, G., 1949, *Rec. Trav. Chim.* **68**, 1106.
Kreer, T., Metzger, S., Müller, M., Binder, K., and Baschnagel, J., 2004, *J. Chem. Phys.* **120**, 4012.
Kremer, K. and Grest, G., 1990, *J. Chem. Phys.* **92**, 5057.

Kremer, K. and Müller-Plathe, F., 2000, *MRS Bull.* **26**, 169.
Kumar, S. K., 1994, *Macromolecules* **27**, 260.
Kumar, S. K., Colby, R. H., Anastasiadis, S. H., and Fytas, G., 1996, *J. Chem. Phys.* **105**, 3777.
Lacasse, M. D., Grest, G. S., and Levine, A. J., 1998, *Phys. Rev. Lett.* **80**, 309.
Lado, F., Foiles, S. M., and Ashcroft, N. W., 1983, *Phys. Rev. A* **45**, 2374.
Leibler, L., 1982, *Macromolecules* **15**, 1283.
Lifshitz, I. M., Grosberg, A. Y., and Khokhlov, A. R., 1978, *Rev. Mod. Phys.* **50**, 683.
Lodge, T. P. and McLeish, T. C. B., 2000, *Macromolecules* **33**, 5278.
Louis, A. A., 2002, *J. Phys.: Condens. Matter* **14**, 9187.
MacDowell, L. G., Müller, M., Vega, C., and Binder, K., 2000, *J. Chem. Phys.* **113**, 419.
Mackie, A. D., Panagiotopoulos, A. Z., and Kumar, S. K., 1995, *J. Chem. Phys.* **102**, 1014.
Matsen, M. W., 1997, *J. Chem. Phys.* **106**, 7781.
Matsen, M. W., 2002, *J. Chem. Phys.* **117**, 2351.
Matsen, M. W. and Bates, F. S., 1996, *Macromolecules* **29**, 1091.
Matsen, M. W. and Gardiner, J. M., 2003, *J. Chem. Phys.* **118**, 3775.
Matsen, M. W. and Schick, M., 1994, *Phys. Rev. Lett.* **72**, 2660.
Mattice, W. L. and Suter, U. W., 1994, *Conformational Theory of Large Molecules: the Rotational Isomeric State Model in Macromolecular Systems*. Wiley-Interscience, New York.
Maurits, N. M. and Fraaije, J. G. E. M., 1997, *J. Chem. Phys.* **107**, 5879.
Mecke, K. R. and Dietrich, S., 1999, *Phys. Rev. E* **59**, 6766.
Milner, S. T., 1997, *MRS Bull.* **22**, 38.
Morse, D. and Fredrickson, G., 1995, *Phys. Rev. Lett.* **73**, 3235.
Müller, M., 1995, *Macromolecules* **28**, 6556.
Müller, M., 1998, *Macromolecules* **31**, 9044.
Müller, M., 1999, *Macromol. Theory Simul.* **8**, 343.
Müller, M., 2001, Mesoscopic and continuum models, in *Encyclopedia of Physical Chemistry and Chemical Physics*, eds. J. H. Moore and N. D. Spencer. IOP Publishing, Bristol.
Müller, M. and Binder, K., 1995, *Macromolecules* **28**, 1825.
Müller, M. and Binder, K., 1998, *Macromolecules* **31**, 8323.
Müller, M. and MacDowell, L., 2000, *Macromolecules* **33**, 3902.
Müller, M. and MacDowell, L. G., 2003, *J. Phys.: Condens. Matter* **15**, 609.
Müller, M. and Schick, M., 1996a, *J. Chem. Phys.* **105**, 8885.
Müller, M. and Schick, M., 1996b, *Macromolecules* **29**, 8900.
Müller, M. and Schick, M., 1998, *Phys. Rev. E* **57**, 6973.
Müller, M. and Schmid, F., 2005, *Adv. Polym. Sci.* **185**, 1.
Müller, M. and Werner, A., 1997, *J. Chem. Phys.* **107**, 10764.

Müller, M., Binder, K., and Oed, W., 1995, *J. Chem. Soc., Faraday Trans.* **91**, 2369.
Müller, M., Albano, E. V., and Binder, K., 2000a, *Phys. Rev. E* **62**, 5281.
Müller, M., Binder, K., and Albano, E. V., 2000b, *Eur. Phys. Lett.* **50**, 724.
Müller, M., Binder, K., and Albano, E. V., 2001, *Int. J. Mod. Phys. B* **15**, 1867.
Müller, M., MacDowell, L. G., Virnau, P., and Binder, K., 2002, *J. Chem. Phys.* **117**, 5480.
Müller, M., Katsov, K., and Schick, M., 2003a, *J. Polym. Sci.: Part B: Polym. Phys.* **41**, 1441.
Müller, M., MacDowell, L. G., and Yethiraj, A., 2003b, *J. Chem. Phys.* **118**, 2929.
Müller-Plathe, F., 2002, *Chem. Phys. Chem.* **3**, 754.
Netz, R. R. and Schick, M., 1998, *Macromolecules* **31**, 5105.
Ngai, K.-L. and Plazek, D. J., 1995, *Rubber Chem. Technol.* **68**, 376.
Nielsen, S. O., Lopez, C. F., Srinivas, G., and Klein, M. L., 2004, *J. Phys.: Condens. Matter* **16**, 481.
Panagiotopoulos, A. Z., Wong, V., and Floriano, M. A., 1998, *Macromolecules* **32**, 912.
Patra, C. N. and Yethiraj, A., 2003, *J. Chem. Phys.* **118**, 4702.
Paul, D. R. and Newman, S., 1978, *Polymer Blends*. Academic Press, San Diego, CA.
Paul, W., Binder, K., Heermann, D. W., and Kremer, K., 1991a, *J. Chem. Phys.* **95**, 7726.
Paul, W., Binder, K., Heermann, D. W., and Kremer, K., 1991b, *J. Phys. II (France)* **1**, 37.
Peterson, K. A., Stein, A. D., and Fayer, D. M., 1990, *Macromolecules* **23**, 111.
Reister, E., Müller, M., and Binder, K., 2001, *Phys. Rev. E* **61**, 041804.
Saito, N., Takahashi, K., and Yunoki, Y., 1967, *J. Phys. Soc. Jpn.* **22**, 219.
Sariban, A. and Binder, K., 1988, *Macromolecules* **21**, 711.
Sariban, A. and Binder, K., 1991, *Macromolecules* **24**, 578.
Schäfer, L., 1999, *Excluded Volume Effects in Polymer Solutions*. Springer, Berlin.
Schmid, F. and Müller, M., 1995, *Macromolecules* **28**, 8639.
Schwahn, D., Meier, G., Mortensen, K., and Janssen, S., 1994, *J. Phys. II (France)* **4**, 837.
Schweizer, K. and Curro, J., 1997, *Adv. Chem. Phys.* **98**, 1.
Semenov, A. N., 1993, *Macromolecules* **26**, 6617.
Semenov, A. N., 1994, *Macromolecules* **27**, 2732.
Semenov, A. N., 1996, *J. Phys. II (France)* **6**, 1759.
Semenov, A. N. and Johner, A., 2003, *Eur. Phys. J. E* **12**, 469.
Sen, S., Cohen, J. M., McCoy, J. D., and Curro, J. G., 1994, *J. Chem. Phys.* **101**, 9010.

Sferrazza, M., Xiao, C., Jones, R. A. L., Bucknall, D. G., and J. Penfold, J. W., 1997, *Phys. Rev. Lett.* **78**, 3693.
Shelly, J. C., Shelly, M. Y., Reeder, R. C., Bandyopadhyay, S., and Klein, M., 2001, *J. Phys. Chem. B* **105**, 4464.
Shi, A. C., Noolandi, J., and Desai, R. C., 1996, *Macromolecules* **29**, 6487.
Shull, K. R., Mayes, A., and Russell, T. P., 1993, *Macromolecules* **26**, 3939.
Singh, C., Schweizer, K. S., and Yethiraj, A., 1995, *J. Chem. Phys.* **102**, 2187.
Site, L. D., Abrams, C. F., Alavi, A., and Kremer, K., 2002, *Phys. Rev. Lett.* **89**, 156103.
Smit, B., Karaborni, S., and Siepmann, J. I., 1995, *J. Chem. Phys.* **102**, 2126.
Stell, G. and Zhou, Y., 1989, *J. Chem. Phys.* **91**, 3618.
Szleifer, I. and Carignano, M. A., 1996, *Adv. Chem. Phys.* **94**, 742.
Szleifer, I., Ben-Shaul, A., and Gelbhart, W. M., 1986, *J. Chem. Phys.* **85**, 5345.
Taylor, G. I., 1932, *Proc. R. Soc. London A* **138**, 41.
Tries, V., Paul, W., Baschnagel, J., and Binder, K., 1997, *J. Chem. Phys.* **106**, 738.
van Konynenburg, P. H. and Scott, R. L., 1980, *Phil. Trans. A* **298**, 495.
van Swol, F. and Henderson, J. R., 1991, *Phys. Rev. A* **43**, 2932.
Vega, C., McBride, C., and MacDowell, L. G., 2002, *Phys. Chem. Chem. Phys.* **4**, 853.
Wang, Q., Nealey, P. F., and de Pablo, J. J., 2002, *Macromolecules* **35**, 9563.
Wang, Z. G., 2002, *J. Chem. Phys.* **117**, 481.
Weeks, J. D., 1977, *J. Chem. Phys.* **67**, 3106.
Weinhold, J. D., Curro, J. G., Habenschuss, A., and Londono, J. D., 1999, *Macromolecules* **32**, 7276.
Werner, A., Schmid, F., Müller, M., and Binder, K., 1997, *J. Chem. Phys.* **107**, 8175.
Werner, A., Schmid, F., Müller, M., and Binder, K., 1999, *Phys. Rev. E* **59**, 728.
Wertheim, M. S., 1987, *J. Chem. Phys.* **87**, 7323.
Wilding, N. B., Müller, M., and Binder, K., 1996, *J. Chem. Phys.* **105**, 802.
Wittmer, J. P., Meyer, H., Baschnagel, J., Johner, A., Obukhov, S., Mattoni, L., Müller, M., and Semenov, A., 2004, *Phys. Rev. Lett.* **93**, 147801.
Woodward, C. E., 1991, *J. Chem. Phys.* **94**, 3183.
Yethiraj, A., 1994, *J. Chem. Phys.* **101**, 2489.
Yethiraj, A., 1998, *J. Chem. Phys.* **109**, 3269.
Yethiraj, A., 2002, *Adv. Chem. Phys.* **121**, 89.
Yethiraj, A. and Woodward, C. E., 1995, *J. Chem. Phys.* **102**, 5499.
Yu, Y.-X. and Wu, J., 2002, *J. Chem. Phys.* **117**, 2368.

Index

a

additives   12
amphipathic   6
amphiphile   6

b

bead–spring model   191, 222, 249, 256
bicontinuous structure   7
bilayer membrane   7
binodal   205, 211, 217, 229, 257
– point   140
block copolymer microstructure   90, 158
bond fluctuation model   188, 191, 192, 201, 209, 215, 230
bond length   38
boundary conditions   125
Bragg condition   143
Brazovskii fluctuations   170

c

capillary waves   208, 214, 230, 231, 236, 245, 262, 264
chain entropy   9
chemical potential   148, 202
classical path   102, 105
classical phases   158
coarse-grained parameters   184, 213, 246
colloidal suspensions   4
complex phases   158

complexity   14
composites   14
composition fluctuations   169, 198, 204, 205, 211, 214, 217, 223, 229, 236, 244, 266, 271
conformational asymmetry   162
constraint release   77
contour length   56
– fluctuations   72
copolymer   87
course-grained segment   93
course-grained trajectory   97
critical point   140, 184, 187, 205, 210, 214, 216, 224, 236, 244, 253, 257, 265, 270

d

Debye function   44, 47, 48, 50, 189, 206, 268
deformability   1
degree of polymerization   89, 96
depletion interaction   13
diffusion coefficient   68
diffusion equation   100
dimensionless segment concentration   97
Dirac δ function   173
dispersion colloids   4
DNA   5
double-tangent construction   139

Soft Matter, Vol. 1: Polymer Melts and Mixtures. Edited by G. Gompper and M. Schick
Copyright © 2006 WILEY-VCH Verlag GmbH & Co. KGaA, Weinheim
ISBN: 3-527-30500-9

e

effective interaction   12
elastic free energy   101
end-segment distribution   116
end-to-end distance   38, 56, 58
end-to-end vector   92, 183, 187, 206, 209, 226, 229, 240, 263
Euler–Lagrange equation   175
exclusion zone   166

f

fd virus   5, 14
Flory–Huggins free energy   138, 202, 214, 245
Flory–Huggins parameter   181, 186, 187, 202, 210, 212, 214, 217, 222, 230, 236, 239, 266
Fourier transform   109
freely jointed chain   92
functional   172
– derivative   173
– integration   175

g

Gaussian chain   42
– model   94, 186
ground-state dominance   107

h

hard-core potential   12
homopolymer   87, 182
Hubbard–Stratonovich transformation   170, 193, 247
hydrophilic   6
hydrophobic   6

i

incompressibility constraint   125, 134
indistinguishability   135
interaction χ parameter   134
interfacial tension   145, 148

interfacial width   145, 147, 148, 158, 163
internal energy   134
intrinsic profile   214, 235
invariant polymerization index   96, 187, 204, 207, 212, 217, 220, 225, 245, 270

k

Kuhn length   42

l

Landau–Ginzburg free energy   108, 113, 169
Lennard-Jones potential   133
lever rule   139
lipid   6
– bilayer   7
living polymerization   23
living polymers   14
local reptation   55
loss modulus   37

m

macrophase separation   138
melt viscosity   68
membrane lipid   6
method of steepest descent   123
micelles   8
microemulsion   8
microphase separation   152
molecular orientation   185, 242, 249, 262, 264
monomer   87

n

nucleation and growth   140

o

orthonormal basis functions   106

p

P-RISM theory   212, 224, 229, 260
packing effects   189, 193, 212, 222, 227, 245, 250, 262, 266
pair correlation function   28
phosphatidyl choline   6
plateau modulus   64
polymer blend   180, 182, 186, 187, 190, 198, 206, 211, 226, 230, 244, 266
polymer brush   90, 114
polymer interface   90
primitive path   56

r

radius of gyration   41
random-phase approximation   111, 267
red blood cell   3
reptation   55
rheology   66
Rouse dynamics   45, 55, 56, 58, 60, 62, 63, 66, 67

s

saddle point   123
scattering function   141, 155
self-consistent field condition   137, 149, 198, 247
self-correlation function   28
shear modulus   1
sifting property   174
sodium dodecyl sulfonate   6
soft matter   1, 11
– composites   14
spectral method   106, 127, 143, 156
spinodal   204, 205, 270
– decomposition   140
– point   140
sponge phase   8
spring–bead model   43, 45
statistical segment length   94, 183, 188, 210, 222, 239
Stirling approximation   136
storage modulus   37
surfactant   6
suspensions   4

t

time–temperature superposition   38
tobacco mosaic virus   5
transfer matrix   129
tube concept   55, 61–63
tube survival probability   63

u

unit-cell approximation   162
universality   13

w

weak and strong segregation limits   203, 207, 209, 214, 235
worm-like chain model   167, 239